White Rose Maths

White Rose Maths Edition

Power Maths

Year 5A
A Guide to Teaching for Mastery

Series Editor: Tony Staneff
Lead author: Josh Lury

Pearson

Contents

Introduction to the author team

Power Maths arises from the work of maths mastery experts who are committed to proving that, given the right mastery mindset and approach, **everyone can do maths**. Based on robust research and best practice from around the world, *Power Maths* was developed in partnership with a group of UK teachers to make sure that it not only meets our children's wide-ranging needs but also aligns with the National Curriculum in England.

Power Maths – White Rose Maths edition

This edition of *Power Maths* has been developed and updated by:

Tony Staneff, Series Editor and Author

Vice Principal at Trinity Academy, Halifax, Tony also leads a team of mastery experts who help schools across the UK to develop teaching for mastery via nationally recognised CPD courses, problem-solving and reasoning resources, schemes of work, assessment materials and other tools.

Josh Lury, Lead Author

Josh is a specialist maths teacher, author and maths consultant with a passion for innovative and effective maths education.

The first edition of *Power Maths* was developed by a team of experienced authors, including:

- **Tony Staneff and Josh Lury**
- **Trinity Academy Halifax** (Michael Gosling CEO, Emily Fox, Kate Henshall, Rebecca Holland, Stephanie Kirk, Stephen Monaghan and Rachel Webster)
- **David Board, Belle Cottingham, Jonathan East, Tim Handley, Derek Huby, Neil Jarrett, Stephen Monaghan, Beth Smith, Tim Weal, Paul Wrangles** – skilled maths teachers and mastery experts
- **Cherri Moseley** – a maths author, former teacher and professional development provider
- **Professors Liu Jian and Zhang Dan**, Series Consultants and authors, and their team of mastery expert authors: **Wei Huinv, Huang Lihua, Zhu Dejiang, Zhu Yuhong, Hou Huiying, Yin Lili, Zhang Jing, Zhou Da and Liu Qimeng**

  Used by over 20 million children, Professor Liu Jian's textbook programme is one of the most popular in China. He and his author team are highly experienced in intelligent practice and in embedding key maths concepts using a C-P-A approach.

- **A group of 15 teachers and maths co-ordinators**

  We consulted our teacher group throughout the development of *Power Maths* to ensure we are meeting their real needs in the classroom.

What is *Power Maths*?

Created especially for UK primary schools, and aligned with the new National Curriculum, *Power Maths* is a whole-class, textbook-based mastery resource that empowers every child to understand and succeed. *Power Maths* rejects the notion that some people simply 'can't do' maths. Instead, it develops growth mindsets and encourages hard work, practice and a willingness to see mistakes as learning tools.

Best practice consistently shows that mastery of small, cumulative steps builds a solid foundation of deep mathematical understanding. *Power Maths* combines interactive teaching tools, high-quality textbooks and continuing professional development (CPD) to help you equip children with a deep and long-lasting understanding. Based on extensive evidence, and developed in partnership with practising teachers, *Power Maths* ensures that it meets the needs of children in the UK.

Power Maths and Mastery

Power Maths makes mastery practical and achievable by providing the structures, pathways, content, tools and support you need to make it happen in your classroom.

To develop mastery in maths, children must be enabled to acquire a deep understanding of maths concepts, structures and procedures, step by step. Complex mathematical concepts are built on simpler conceptual components and when children understand every step in the learning sequence, maths becomes transparent and makes logical sense. Interactive lessons establish deep understanding in small steps, as well as effortless fluency in key facts such as tables and number bonds. The whole class works on the same content and no child is left behind.

Power Maths

- ⚡ Builds every concept in small, progressive steps
- ⚡ Is built with interactive, whole-class teaching in mind
- ⚡ Provides the tools you need to develop growth mindsets
- ⚡ Helps you check understanding and ensure that every child is keeping up
- ⚡ Establishes core elements such as intelligent practice and reflection

The *Power Maths* approach

Everyone can!

Founded on the conviction that every child can achieve, *Power Maths* enables children to build number fluency, confidence and understanding, step by step.

Child-centred learning

Children master concepts one step at a time in lessons that embrace a concrete-pictorial-abstract (C-P-A) approach, avoid overload, build on prior learning and help them see patterns and connections. Same-day intervention ensures sustained progress.

Continuing professional development

Embedded teacher support and development offer every teacher the opportunity to continually improve their subject knowledge and manage whole-class teaching for mastery.

Whole-class teaching

An interactive, whole-class teaching model encourages thinking and precise mathematical language and allows children to deepen their understanding as far as they can.

What's different in the new edition?

If you have previously used the first editions of *Power Maths*, you might be interested to know how this edition is different. All of the improvements described below are based on feedback from *Power Maths* customers.

Changes to units and the progression

⚡ The order of units has been slightly adjusted, creating closer alignment between adjacent year groups, which will be useful for mixed age teaching.

⚡ The flow of lessons has been improved within units to optimise the pace of the progression and build in more recap where needed. For key topics, the sequence of lessons gives more opportunities to build up a solid base of understanding. Other units have fewer lessons than before, where appropriate, making it possible to fit in all the content.

⚡ Overall, the lessons put more focus on the most essential content for that year, with less time given to non-statutory content.

⚡ The progression of lessons matches the steps in the new White Rose Maths schemes of learning.

Lesson resources

⚡ There is a Quick recap for each lesson in the Teacher Guide, which offers an alternative lesson starter to the Power Up for cases where you feel it would be more beneficial to surface prerequisite learning than general number fluency.

⚡ In the **Discover** and **Share** sections there is now more of a progression from 1 a) to 1 b). Whereas before, 1 b) was mainly designed as a separate question, now 1 a) leads directly into 1 b). This means that there is an improved whole-class flow, and also an opportunity to focus on the logic and skills in more detail. As a teacher, you will be using 1 a) to lead the class into the thinking, then 1 b) to mould that thinking into the core new learning of the lesson.

⚡ In the **Share** section, for KS1 in particular, the number of different models and representations has been reduced, to support the clarity of thinking prompted by the flow from 1 a) into 1 b).

⚡ More fluency questions have been built into the guided and independent practice.

⚡ Pupil pages are as easy as possible for children to access independently. The pages are less full where this supports greater focus on key ideas and instructions. Also, more freedom is offered around answer format, with fewer boxes scaffolding children's responses; squared paper backgrounds are used in the Practice Books where appropriate. Artwork has also been revisited to ensure the highest standards of accessibility.

New components

480 Individual Practice Games are available in *ActiveLearn* for practising key facts and skills in Years 1 to 6. These are designed in an arcade style, to feel like fun games that children would choose to play outside school. They can be accessed via the Pupil World for homework or additional practice in school – and children can earn rewards. There are Support, Core and Extend levels to allocate, with Activity Reporting available for the teacher. There is a Quick Guide on *ActiveLearn* and you can use the Help area for support in setting up child accounts.

There is also a new set of lesson video resources on the Professional Development tile, designed for in-school training in 10- to 20-minute bursts. For each part of the *Power Maths* lesson sequence, there is a slide deck with embedded video, which will facilitate discussions about how you can take your *Power Maths* teaching to the next level.

Your *Power Maths* resources

Pupil Textbooks

Discover, Share and Think together sections promote discussion and introduce mathematical ideas logically, so that children understand more easily.

Using a Concrete-Pictorial-Abstract approach, clear mathematical models help children to make connections and grasp concepts.

Appealing scenarios stimulate curiosity, helping children to identify the maths problem and discover patterns and relationships for themselves.

Friendly, supportive characters help children develop a growth mindset by prompting them to think, reason and reflect.

To help you teach for mastery, *Power Maths* comprises a variety of high-quality resources.

The coherent *Power Maths* lesson structure carries through into the vibrant, high-quality textbooks. Setting out the core learning objectives for each class, the lesson structure follows a carefully mapped journey through the curriculum and supports children on their journey to deeper understanding.

Pupil Practice Books

The Practice Books offer just the right amount of intelligent practice for children to complete independently in the final section of each lesson.

Practice questions are finely tuned to move children forward in their thinking and to reveal misconceptions.

The practice questions are for everyone – each question varies one small element to move children on in their thinking.

Calculations are connected so that children think about the underlying concept.

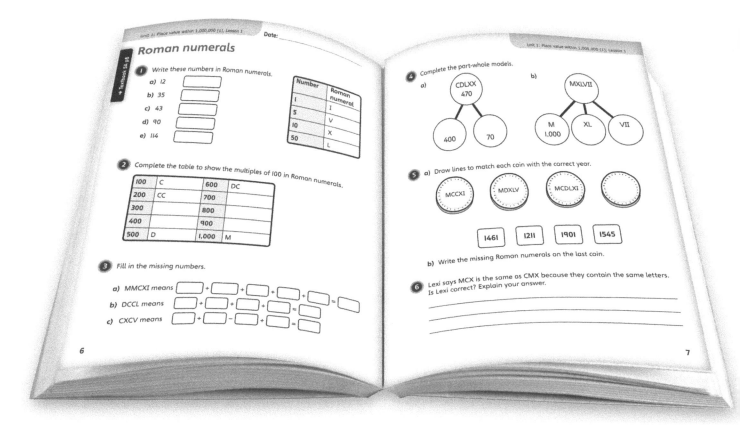

The *Power Maths* characters support and encourage children to think and work in different ways.

CHALLENGE questions allow children to delve deeper into a concept.

Think differently questions encourage children to use reasoning as well as their mathematical knowledge to reach a solution.

Reflect questions reveal the depth of each child's understanding before they move on.

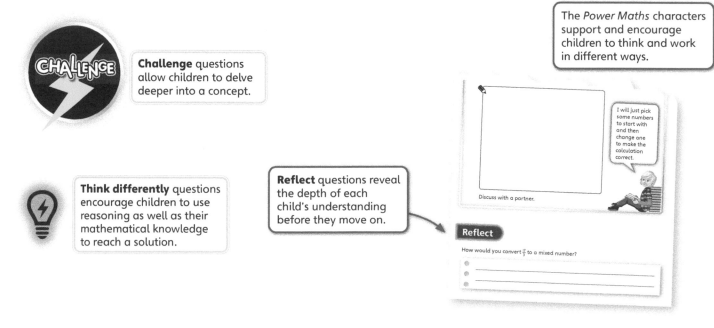

Online subscription

The online subscription will give you access to additional resources and answers from the Textbook and Practice Book.

eTextbooks

Digital versions of *Power Maths* Textbooks allow class groups to share and discuss questions, solutions and strategies. They allow you to project key structures and representations at the front of the class, to ensure all children are focusing on the same concept.

Teaching tools

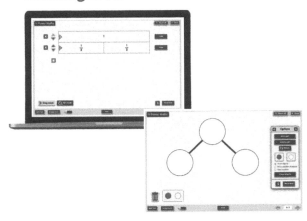

Here you will find interactive versions of key *Power Maths* structures and representations.

Power Ups

Use this series of daily activities to promote and check number fluency.

Online versions of Teacher Guide pages

PDF pages give support at both unit and lesson levels. You will also find help with key strategies and templates for tracking progress.

Unit videos

Watch the professional development videos at the start of each unit to help you teach with confidence. The videos explore common misconceptions in the unit, and include intervention suggestions as well as suggestions on what to look out for when assessing mastery in your students.

End of unit Strengthen and Deepen materials

The Strengthen activity at the end of every unit addresses a key misconception and can be used to support children who need it. The Deepen activities are designed to be low ceiling/high threshold and will challenge those children who can understand more deeply. These resources will help you ensure that every child understands and will help you keep the class moving forward together. These printable activities provide an optional resource bank for use after the assessment stage.

Individual Practice Games

These enjoyable games can be used at home or at school to embed key number skills (see page 6).

Professional Development videos and slides

These slides and videos of *Power Maths* lessons can be used for ongoing training in short bursts or to support new staff.

The *Power Maths* teaching model

At the heart of *Power Maths* is a clearly structured teaching and learning process that helps you make certain that every child masters each maths concept securely and deeply. For each year group, the curriculum is broken down into core concepts, taught in units. A unit divides into smaller learning steps – lessons. Step by step, strong foundations of cumulative knowledge and understanding are built.

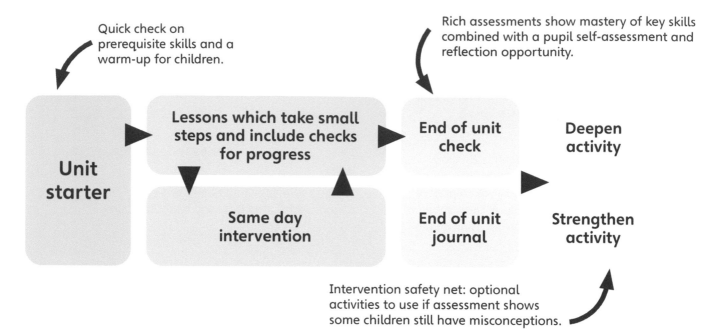

Quick check on prerequisite skills and a warm-up for children.

Rich assessments show mastery of key skills combined with a pupil self-assessment and reflection opportunity.

Unit starter → Lessons which take small steps and include checks for progress → Same day intervention → End of unit check → End of unit journal → Deepen activity → Strengthen activity

Intervention safety net: optional activities to use if assessment shows some children still have misconceptions.

Unit starter

Each unit begins with a unit starter, which introduces the learning context along with key mathematical vocabulary and structures and representations.

- The Textbooks include a check on readiness and a warm-up task for children to complete.
- Your Teacher Guide gives support right from the start on important structures and representations, mathematical language, common misconceptions and intervention strategies.
- Unit-specific videos develop your subject knowledge and insights so you feel confident and fully equipped to teach each new unit. These are available via the online subscription.

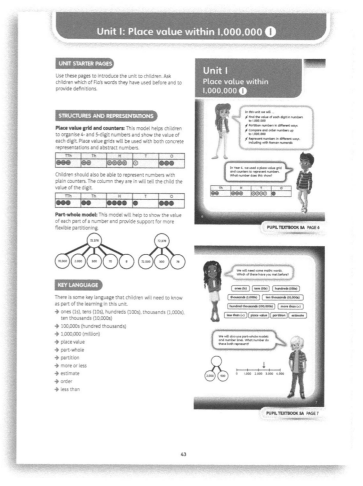

Lesson

Once a unit has been introduced, it is time to start teaching the series of lessons.

- Each lesson is scaffolded with Textbook and Practice Book activities and begins with a Power Up activity (available via online subscription) or the Quick recap activity in the Teacher Guide (see page 15).

- *Power Maths* identifies lesson by lesson what concepts are to be taught.

- Your Teacher Guide offers lots of support for you to get the most from every child in every lesson. As well as highlighting key points, tricky areas and how to handle them, you will also find question prompts to check on understanding and clarification on why particular activities and questions are used.

Same-day intervention

Same-day interventions are vital in order to keep the class progressing together. This can be during the lesson as well as afterwards (see page 28). Therefore, *Power Maths* provides plenty of support throughout the journey.

- Intervention is focused on keeping up now, not catching up later, so interventions should happen as soon as they are needed.

- Practice section questions are designed to bring misconceptions to the surface, allowing you to identify these easily as you circulate during independent practice time.

- Child-friendly assessment questions in the Teacher Guide help you identify easily which children need to strengthen their understanding.

End of unit check and journal

For each unit, the End of unit check in the Textbook lets you see which children have mastered the key concepts, which children have not and where their misconceptions lie. The Practice Books also include an End of unit journal in which children can reflect on what they have learned. Each unit also offers Strengthen and Deepen activities, available via the online subscription.

The Teacher Guide offers different ways of managing the End of unit assessments as well as giving support with handling misconceptions.

The End of unit check presents multiple-choice questions. Children think about their answer, decide on a solution and explain their choice.

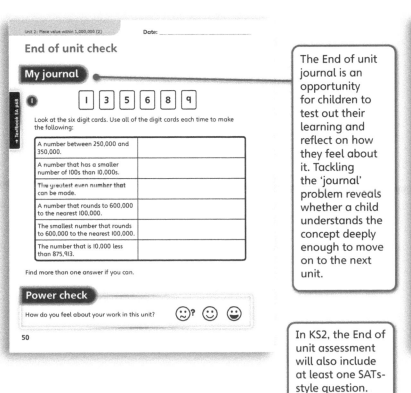

The End of unit journal is an opportunity for children to test out their learning and reflect on how they feel about it. Tackling the 'journal' problem reveals whether a child understands the concept deeply enough to move on to the next unit.

In KS2, the End of unit assessment will also include at least one SATs-style question.

The *Power Maths* lesson sequence

At the heart of *Power Maths* is a unique lesson sequence designed to empower children to understand core concepts and grow in confidence. Embracing the National Centre for Excellence in the Teaching of Mathematics' (NCETM's) definition of mastery, the sequence guides and shapes every *Power Maths* lesson you teach.

Flexibility is built into the *Power Maths* programme so there is no one-to-one mapping of lessons and concepts and you can pace your teaching according to your class. While some children will need to spend longer on a particular concept (through interventions or additional lessons), others will reach deeper levels of understanding. However, it is important that the class moves forward together through the termly schedules.

Power Up 🕐 5 minutes

Each lesson begins with a Power Up activity (available via the online subscription) which supports fluency in key number facts.

The whole-class approach depends on fluency, so the Power Up is a powerful and essential activity.

The Quick recap is an alternative starter, for when you think some or all children would benefit more from revisiting pre-requisite work (see page 15).

TOP TIP
If the class is struggling with the task, revisit it later and check understanding.

Power Ups reinforce the two key things that are essential for success: times-tables and number bonds.

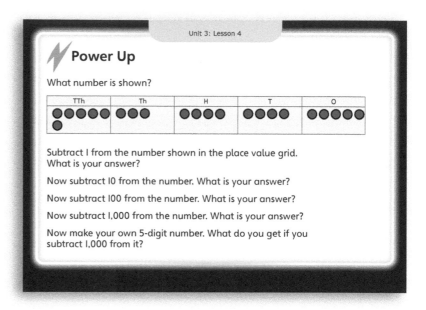

Unit 3: Lesson 4

⚡ Power Up

What number is shown?

TTh	Th	H	T	O
●●●●● ●●● ●		●●●●	●●●●	●●●●●

Subtract I from the number shown in the place value grid. What is your answer?

Now subtract 10 from the number. What is your answer?

Now subtract 100 from the number. What is your answer?

Now subtract 1,000 from the number. What is your answer?

Now make your own 5-digit number. What do you get if you subtract 1,000 from it?

Discover 🕐 10 minutes

A practical, real-life problem arouses curiosity. Children find the maths through story telling.

A real-life scenario is provided for the **Discover** section but feel free to build upon these with your own examples that are more relevant to your class, or get creative with the context.

TOP TIP
Discover works best when run at tables, in pairs with concrete objects.

Question ❶ a) tackles the key concept and question ❶ b) digs a little deeper. Children have time to explore, play and discuss possible strategies.

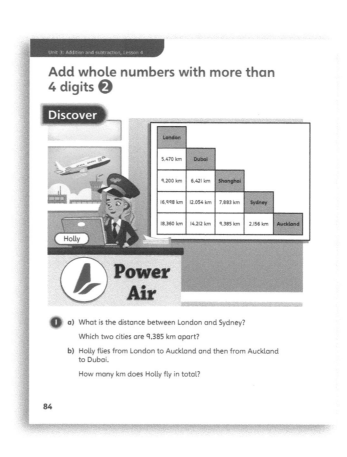

Unit 3: Addition and subtraction, Lesson 4

Add whole numbers with more than 4 digits ❷

Discover

London				
5,470 km	Dubai			
9,200 km	6,421 km	Shanghai		
16,998 km	12,054 km	7,883 km	Sydney	
18,360 km	14,212 km	9,385 km	2,156 km	Auckland

Holly

Power Air

❶ a) What is the distance between London and Sydney?
   Which two cities are 9,385 km apart?

b) Holly flies from London to Auckland and then from Auckland to Dubai.
   How many km does Holly fly in total?

84

Share 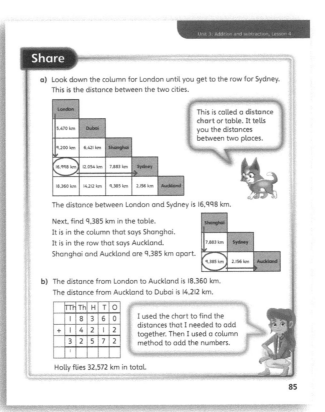 10 minutes

Teacher-led, this interactive section follows the **Discover** activity and highlights the variety of methods that can be used to solve a single problem.

TOP TIP

Pairs sharing a textbook is a great format for **Share**!

Your Teacher Guide gives target questions for children. The online toolkit provides interactive structures and representations to link concrete and pictorial to abstract concepts.

Bring children to the front to share and celebrate their solutions and strategies.

Think together

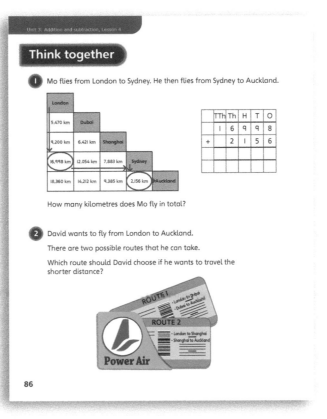 10 minutes

Children work in groups on the carpet or at tables, using their textbooks or eBooks.

TOP TIP

Make sure children have mini whiteboards or pads to write on if they are not at their tables.

Using the Teacher Guide, model question ① for your class.

Question ② is less structured. Children will need to think together in their groups, then discuss their methods and solutions as a class.

In question ③ children try working out the answer independently. The openness of the **Challenge** question helps to check depth of understanding.

Practice ⏱ 15 minutes

↑ Textbook 5A p88

Date: _____

Subtract whole numbers with more than 4 digits ❶

1 Solve the following calculations.

a) 24,592 – 3,470 = ☐

TTh	Th	H	T	O
⊙⊙	⊙⊙⊙⊙	⊙⊙⊙⊙⊙	⊙⊙⊙⊙⊙ ⊙⊙⊙⊙	⊙⊙

	TTh	Th	H	T	O
	2	4	5	9	2
–		3	4	7	0

b) 51,340 – 30,720 =

TTh	Th	H	T	O
⊙⊙⊙⊙⊙⊙		⊙⊙⊙	⊙⊙⊙⊙	

	TTh	Th	H	T	O
	5	1	3	4	0
–	3	0	7	2	0

c)

	TTh	Th	H	T	O
		4	3	6	5
–		2	4	2	3

e) 15,712 – 6,000 = ☐

f) 26,318 – 11,148 = ☐

d)

	TTh	Th	H	T	O
		7	6	1	8	5
–			5	2	2	4

64

Reflect ⏱ 5 minutes

4 Fill in the missing digits. ⚡

a)

	TTh	Th	H	T	O
	2	6	1	8	
–		4		2	
	2		4	5	0

b)

	TTh	Th	H	T	O
			9	9		3
–	1		6	2	7
	3	5		5	

5 Three chests each contain some treasure.   CHALLENGE
• The first chest contains 18,455 coins.
• The second chest has 4,200 fewer coins than the first chest.
• The third chest has 5,120 fewer coins than the second chest.

How many coins are in each treasure chest?

Reflect

Explain how to work out 32,728 – 14,605.

● _____
● _____
● _____
● _____

66

Sidebar notes (left column):

Using their Practice Books, children work independently while you circulate and check on progress.

Questions follow small steps of progression to deepen learning.

TOP TIP
Some children could work separately with a teacher or assistant.

'Spot the mistake' questions are great for checking misconceptions.

The **Reflect** section is your opportunity to check how deeply children understand the target concept.

Sidebar notes (right column):

Are some children struggling? If so, work with them as a group, using mathematical structures and representations to support understanding as necessary.

There are no set routines: for real understanding, children need to think about the problem in different ways.

The Practice Books use various approaches to check that children have fully understood each concept.

Looking like they understand is not enough! It is essential that children can show they have grasped the concept.

Using the *Power Maths* Teacher Guide

Think of your Teacher Guides as *Power Maths* handbooks that will guide, support and inspire your day-to-day teaching. Clear and concise, and illustrated with helpful examples, your Teacher Guides will help you make the best possible use of every individual lesson. They also provide wrap-around professional development, enhancing your own subject knowledge and helping you to grow in confidence about moving your children forward together.

There is a Teacher Guide per year group for every term, with unit and lesson level guidance and support.

Never feel stuck! You will find ideas for introducing every unit and lesson and questions to encourage teacher reflection before and after each lesson.

Tips and advice on key elements such as C-P-A approaches, misconceptions, language, modelling growth mindsets and same day intervention.

Annotations for every Textbook and Practice Book page, providing prompts for key questions to ask to expose understanding and explanations as to why key questions have been chosen.

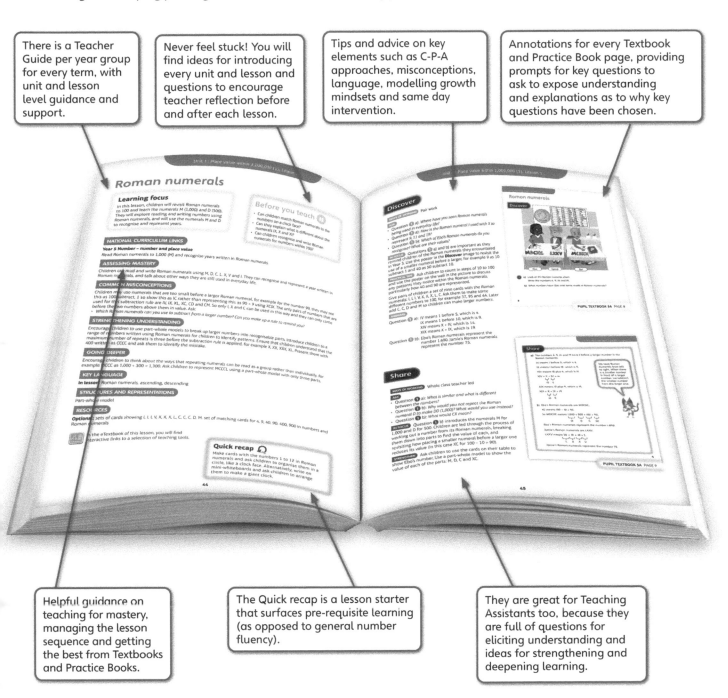

Helpful guidance on teaching for mastery, managing the lesson sequence and getting the best from Textbooks and Practice Books.

The Quick recap is a lesson starter that surfaces pre-requisite learning (as opposed to general number fluency).

They are great for Teaching Assistants too, because they are full of questions for eliciting understanding and ideas for strengthening and deepening learning.

At the end of each unit, your Teacher Guide helps you identify who has fully grasped the concept, who has not and how to move every child forward. This is covered later in the Assessment strategies section.

Power Maths Year 5, yearly overview

Textbook	Strand	Unit		Number of lessons
Textbook A / Practice Workbook A (Term 1)	Number – number and place value	1	Place value within 1,000,000 (1)	8
	Number – number and place value	2	Place value within 1,000,000 (2)	6
	Number – addition and subtraction	3	Addition and subtraction	12
	Number – multiplication and division	4	Multiplication and division (1)	10
	Number – fractions (including decimals and percentages)	5	Fractions (1)	8
	Number – fractions (including decimals and percentages)	6	Fractions (2)	11
Textbook B / Practice Workbook B (Term 2)	Number – multiplication and division	7	Multiplication and division (2)	10
	Number – fractions (including decimals and percentages)	8	Fractions (3)	7
	Number – fractions (including decimals and percentages)	9	Decimals and percentages	15
	Measurement	10	Measure – perimeter and area	8
	Statistics	11	Graphs and tables	6
Textbook C / Practice Workbook C (Term 3)	Geometry – properties of shapes	12	Geometry – properties of shapes	12
	Geometry – position and direction	13	Geometry – position and direction	6
	Number – fractions (including decimals and percentages)	14	Decimals	15
	Number – number and place value	15	Negative numbers	4
	Measurement	16	Measure – converting units	10
	Measurement	17	Measure – volume and capacity	3

Power Maths Year 5, Textbook 5A (Term 1) overview

Strand	Unit		Lesson number	Lesson title	NC Objective 1	NC Objective 2
Number – number and place value	Unit 1	Place value within 1,000,000 (1)	1	Roman numerals	Read Roman numerals to 1000 (M) and recognise years written in Roman numerals.	
Number – number and place value	Unit 1	Place value within 1,000,000 (1)	2	Numbers to 10,000	Read, write, order and compare numbers to at least 1 000 000 and determine the value of each digit	
Number – number and place value	Unit 1	Place value within 1,000,000 (1)	3	Numbers to 100,000	Read, write, order and compare numbers to at least 1 000 000 and determine the value of each digit	
Number – number and place value	Unit 1	Place value within 1,000,000 (1)	4	Numbers to 1,000,000	Read, write, order and compare numbers to at least 1 000 000 and determine the value of each digit	
Number – number and place value	Unit 1	Place value within 1,000,000 (1)	5	Read and write 5- and 6-digit numbers	Read, write, order and compare numbers to at least 1 000 000 and determine the value of each digit	
Number – number and place value	Unit 1	Place value within 1,000,000 (1)	6	Powers of 10	Count forwards or backwards in steps of powers of 10 for any given number up to 1 000 000	
Number – number and place value	Unit 1	Place value within 1,000,000 (1)	7	10/100/1,000/ 10,000/100,000 more or less	Count forwards or backwards in steps of powers of 10 for any given number up to 1 000 000	
Number – number and place value	Unit 1	Place value within 1,000,000 (1)	8	Partition numbers to 1,000,000	Read, write, order and compare numbers to at least 1 000 000 and determine the value of each digit	
Number – number and place value	Unit 2	Place value within 1,000,000 (2)	1	Number line to 1,000,000	Read, write, order and compare numbers to at least 1,000,000 and determine the value of each digit	
Number – number and place value	Unit 2	Place value within 1,000,000 (2)	2	Compare and order numbers to 100,000	Read, write, order and compare numbers to at least 1,000,000 and determine the value of each digit	
Number – number and place value	Unit 2	Place value within 1,000,000 (2)	3	Compare and order numbers to 1,000,000	Read, write, order and compare numbers to at least 1,000,000 and determine the value of each digit	
Number – number and place value	Unit 2	Place value within 1,000,000 (2)	4	Round numbers to the nearest 100,000	Round any number up to 1,000,000 to the nearest 10, 100, 1,000, 10,000 and 100,000	
Number – number and place value	Unit 2	Place value within 1,000,000 (2)	5	Round numbers to the nearest 10,000	Round any number up to 1,000,000 to the nearest 10, 100, 1,000, 10,000 and 100,000	
Number – number and place value	Unit 2	Place value within 1,000,000 (2)	6	Round numbers to the nearest 10, 100 and 1,000	Round any number up to 1,000,000 to the nearest 10, 100, 1,000, 10,000 and 100,000	
Number – addition and subtraction	Unit 3	Addition and subtraction	1	Mental strategies (addition)	Add and subtract numbers mentally with increasingly large numbers	
Number – addition and subtraction	Unit 3	Addition and subtraction	2	Mental strategies (subtraction)	Add and subtract numbers mentally with increasingly large numbers	
Number – addition and subtraction	Unit 3	Addition and subtraction	3	Add whole numbers with more than 4 digits (1)	Add and subtract whole numbers with more than 4 digits, including using formal written methods (columnar addition and subtraction)	
Number – addition and subtraction	Unit 3	Addition and subtraction	4	Add whole numbers with more than 4 digits (2)	Add and subtract whole numbers with more than 4 digits, including using formal written methods (columnar addition and subtraction)	

Strand	Unit		Lesson number	Lesson title	NC Objective 1	NC Objective 2
Number – addition and subtraction	Unit 3	Addition and subtraction	5	Subtract whole numbers with more than 4 digits (1)	Add and subtract whole numbers with more than 4 digits, including using formal written methods (columnar addition and subtraction)	
Number – addition and subtraction	Unit 3	Addition and subtraction	6	Subtract whole numbers with more than 4 digits (2)	Add and subtract whole numbers with more than 4 digits, including using formal written methods (columnar addition and subtraction)	
Number – addition and subtraction	Unit 3	Addition and subtraction	7	Round to check answers	Use rounding to check answers to calculations and determine, in the context of a problem, levels of accuracy	
Number – addition and subtraction	Unit 3	Addition and subtraction	8	Inverse operations (addition and subtraction)	Estimate and use inverse operations to check answers to a calculation	
Number – addition and subtraction	Unit 3	Addition and subtraction	9	Multi-step addition and subtraction problems (1)	Solve addition and subtraction multi-step problems in contexts, deciding which operations and methods to use and why	
Number – addition and subtraction	Unit 3	Addition and subtraction	10	Multi-step addition and subtraction problems (2)	Solve addition and subtraction multi-step problems in contexts, deciding which operations and methods to use and why	
Number – addition and subtraction	Unit 3	Addition and subtraction	11	Solve missing number problems	Solve addition and subtraction multi-step problems in contexts, deciding which operations and methods to use and why	
Number – addition and subtraction	Unit 3	Addition and subtraction	12	Solve comparison problems	Solve addition and subtraction multi-step problems in contexts, deciding which operations and methods to use and why	
Number – multiplication and division	Unit 4	Multiplication and division (1)	1	Multiples	Identify multiples and factors, including finding all factor pairs of a number, and common factors of two numbers	
Number – multiplication and division	Unit 4	Multiplication and division (1)	2	Common multiples	Identify multiples and factors, including finding all factor pairs of a number, and common factors of two numbers	
Number – multiplication and division	Unit 4	Multiplication and division (1)	3	Factors	Identify multiples and factors, including finding all factor pairs of a number, and common factors of two numbers	
Number – multiplication and division	Unit 4	Multiplication and division (1)	4	Common factors	Identify multiples and factors, including finding all factor pairs of a number, and common factors of two numbers	
Number – multiplication and division	Unit 4	Multiplication and division (1)	5	Prime numbers	Know and use the vocabulary of prime numbers, prime factors and composite (non-prime) numbers	
Number – multiplication and division	Unit 4	Multiplication and division (1)	6	Square numbers	Recognise and use square numbers and cube numbers, and the notation for squared (2) and cubed (3)	
Number – multiplication and division	Unit 4	Multiplication and division (1)	7	Cube numbers	Recognise and use square numbers and cube numbers, and the notation for squared (2) and cubed (3)	
Number – multiplication and division	Unit 4	Multiplication and division (1)	8	Multiply by 10, 100 and 1,000	Multiply and divide whole numbers and those involving decimals by 10, 100 and 1000	

Strand	Unit		Lesson number	Lesson title	NC Objective 1	NC Objective 2
Number – multiplication and division	Unit 4	Multiplication and division (1)	9	Divide by 10, 100 and 1,000	Multiply and divide whole numbers and those involving decimals by 10, 100 and 1000	
Number – multiplication and division	Unit 4	Multiplication and division (1)	10	Multiples of 10, 100 and 1,000	Multiply and divide whole numbers and those involving decimals by 10, 100 and 1000	
Number – fractions (including decimals and percentages)	Unit 5	Fractions (1)	1	Equivalent fractions 1	Identify, name and write equivalent fractions of a given fraction, represented visually, including tenths and hundredths	
Number – fractions (including decimals and percentages)	Unit 5	Fractions (1)	2	Equivalent fractions 2 – unit and non-unit fractions	Identify, name and write equivalent fractions of a given fraction, represented visually, including tenths and hundredths	
Number – fractions (including decimals and percentages)	Unit 5	Fractions (1)	3	Equivalent fractions 3 – families of equivalent fractions	Identify, name and write equivalent fractions of a given fraction, represented visually, including tenths and hundredths	
Number – fractions (including decimals and percentages)	Unit 5	Fractions (1)	4	Improper fractions to mixed numbers	Recognise mixed numbers and improper fractions and convert from one form to the other and write mathematical statements > 1 as a mixed number [for example, $\frac{2}{5} + \frac{4}{5} = \frac{6}{5} = 1\frac{1}{5}$]	
Number – fractions (including decimals and percentages)	Unit 5	Fractions (1)	5	Mixed numbers to improper fractions	Recognise mixed numbers and improper fractions and convert from one form to the other and write mathematical statements > 1 as a mixed number [for example, $\frac{2}{5} + \frac{4}{5} = \frac{6}{5} = 1\frac{1}{5}$]	
Number – fractions (including decimals and percentages)	Unit 5	Fractions (1)	6	Compare fractions less than 1	Compare and order fractions whose denominators are all multiples of the same number	
Number – fractions (including decimals and percentages)	Unit 5	Fractions (1)	7	Order fractions less than 1	Compare and order fractions whose denominators are all multiples of the same number	
Number – fractions (including decimals and percentages)	Unit 5	Fractions (1)	8	Compare and order fractions greater than 1	Compare and order fractions whose denominators are all multiples of the same number	
Number – fractions (including decimals and percentages)	Unit 6	Fractions (2)	1	Add and subtract fractions	Add and subtract fractions with the same denominator and denominators that are multiples of the same number	
Number – fractions (including decimals and percentages)	Unit 6	Fractions (2)	2	Add fractions within 1	Add and subtract fractions with the same denominator and denominators that are multiples of the same number	
Number – fractions (including decimals and percentages)	Unit 6	Fractions (2)	3	Add fractions with total greater than 1	Add and subtract fractions with the same denominator and denominators that are multiples of the same number	Recognise mixed numbers and improper fractions and convert from one form to the other and write mathematical statements > 1 as a mixed number [for example, $\frac{2}{5} + \frac{4}{5} = \frac{6}{5} = 1\frac{1}{5}$]

Strand	Unit		Lesson number	Lesson title	NC Objective 1	NC Objective 2
Number – fractions (including decimals and percentages)	Unit 6	Fractions (2)	4	Add to a mixed number	Add and subtract fractions with the same denominator and denominators that are multiples of the same number	Recognise mixed numbers and improper fractions and convert from one form to the other and write mathematical statements > 1 as a mixed number [for example, $\frac{2}{5} + \frac{4}{5} = \frac{6}{5} = 1\frac{1}{5}$]
Number – fractions (including decimals and percentages)	Unit 6	Fractions (2)	5	Add two mixed numbers	Add and subtract fractions with the same denominator and denominators that are multiples of the same number	Recognise mixed numbers and improper fractions and convert from one form to the other and write mathematical statements > 1 as a mixed number [for example, $\frac{2}{5} + \frac{4}{5} = \frac{6}{5} = 1\frac{1}{5}$]
Number – fractions (including decimals and percentages)	Unit 6	Fractions (2)	6	Subtract fractions within 1	Add and subtract fractions with the same denominator and denominators that are multiples of the same number	Recognise mixed numbers and improper fractions and convert from one form to the other and write mathematical statements > 1 as a mixed number [for example, $\frac{2}{5} + \frac{4}{5} = \frac{6}{5} = 1\frac{1}{5}$]
Number – fractions (including decimals and percentages)	Unit 6	Fractions (2)	7	Subtract from a mixed number	Add and subtract fractions with the same denominator and denominators that are multiples of the same number	Recognise mixed numbers and improper fractions and convert from one form to the other and write mathematical statements > 1 as a mixed number [for example, $\frac{2}{5} + \frac{4}{5} = \frac{6}{5} = 1\frac{1}{5}$]
Number – fractions (including decimals and percentages)	Unit 6	Fractions (2)	8	Subtract from a mixed number – breaking the whole	Add and subtract fractions with the same denominator and denominators that are multiples of the same number	Recognise mixed numbers and improper fractions and convert from one form to the other and write mathematical statements > 1 as a mixed number [for example, $\frac{2}{5} + \frac{4}{5} = \frac{6}{5} = 1\frac{1}{5}$]
Number – fractions (including decimals and percentages)	Unit 6	Fractions (2)	9	Subtract two mixed numbers	Add and subtract fractions with the same denominator and denominators that are multiples of the same number	Recognise mixed numbers and improper fractions and convert from one form to the other and write mathematical statements > 1 as a mixed number [for example, $\frac{2}{5} + \frac{4}{5} = \frac{6}{5} = 1\frac{1}{5}$]
Number – fractions (including decimals and percentages)	Unit 6	Fractions (2)	10	Solve fraction problems	Add and subtract fractions with the same denominator and denominators that are multiples of the same number	
Number – fractions (including decimals and percentages)	Unit 6	Fractions (2)	11	Solve multi-step fraction problems	Add and subtract fractions with the same denominator and denominators that are multiples of the same number	

Mindset: an introduction

Global research and best practice deliver the same message: learning is greatly affected by what learners perceive they can or cannot do. What is more, it is also shaped by what their parents, carers and teachers perceive they can do. Mindset – the thinking that determines our beliefs and behaviours – therefore has a fundamental impact on teaching and learning.

Everyone can!

Power Maths and mastery methods focus on the distinction between 'fixed' and 'growth' mindsets (Dweck, 2007).[1] Those with a fixed mindset believe that their basic qualities (for example, intelligence, talent and ability to learn) are pre-wired or fixed: 'If you have a talent for maths, you will succeed at it. If not, too bad!' By contrast, those with a growth mindset believe that hard work, effort and commitment drive success and that 'smart' is not something you are or are not, but something you become. In short, everyone can do maths!

Key mindset strategies

A growth mindset needs to be actively nurtured and developed. *Power Maths* offers some key strategies for fostering healthy growth mindsets in your classroom.

It is okay to get it wrong

Mistakes are valuable opportunities to re-think and understand more deeply. Learning is richer when children and teachers alike focus on spotting and sharing mistakes as well as solutions.

Praise hard work

Praise is a great motivator, and by focusing on praising effort and learning rather than success, children will be more willing to try harder, take risks and persist for longer.

Mind your language!

The language we use around learners has a profound effect on their mindsets. Make a habit of using growth phrases, such as, 'Everyone can!', 'Mistakes can help you learn' and 'Just try for a little longer'. The king of them all is one little word, 'yet'... I can't solve this...yet!' Encourage parents and carers to use the right language too.

Build in opportunities for success

The step-by-small-step approach enables children to enjoy the experience of success. In addition, avoid ability grouping and encourage every child to answer questions and explain or demonstrate their methods to others.

[1]Dweck, C (2007) *The New Psychology of Success*, Ballantine Books: New York

The *Power Maths* characters

The *Power Maths* characters model the traits of growth mindset learners and encourage resilience by prompting and questioning children as they work. Appearing frequently in the Textbooks and Practice Books, they are your allies in teaching and discussion, helping to model methods, alternatives and misconceptions, and to pose questions. They encourage and support your children, too: they are all hardworking, enthusiastic and unafraid of making and talking about mistakes.

Meet the team!

Creative Flo is open-minded and sometimes indecisive. She likes to think differently and come up with a variety of methods or ideas.

Determined Dexter is resolute, resilient and systematic. He concentrates hard, always tries his best and he'll never give up – even though he doesn't always choose the most efficient methods!

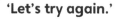

'Let's try again.'

'Mistakes are cool!'

'Have I found all of the solutions?'

'Let's try it this way…'

'Can we do it differently?'

'I've got another way of doing this!'

'I'm going to try this!'

'I know how to do that!'

'Want to share my ideas?'

Curious Ash is eager, interested and inquisitive, and he loves solving puzzles and problems. Ash asks lots of questions but sometimes gets distracted.

'What if we tried this…?'

'I wonder…'

'Is there a pattern here?'

Miaow!

Sparks the Cat

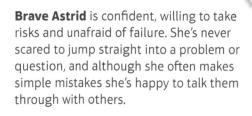

Brave Astrid is confident, willing to take risks and unafraid of failure. She's never scared to jump straight into a problem or question, and although she often makes simple mistakes she's happy to talk them through with others.

Mathematical language

Traditionally, we in the UK have tended to try simplifying mathematical language to make it easier for young children to understand. By contrast, evidence and experience show that by diluting the correct language, we actually mask concepts and meanings for children. We then wonder why they are confused by new and different terminology later down the line! *Power Maths* is not afraid of 'hard' words and avoids placing any barriers between children and their understanding of mathematical concepts. As a result, we need to be deliberate, precise and thorough in building every child's understanding of the language of maths. Throughout the Teacher Guides you will find support and guidance on how to deliver this, as well as individual explanations throughout the pupil Textbooks.

Use the following key strategies to build children's mathematical vocabulary, understanding and confidence.

Precise and consistent

Everyone in the classroom should use the correct mathematical terms in full, every time. For example, refer to 'equal parts', not 'parts'. Used consistently, precise maths language will be a familiar and non-threatening part of children's everyday experience.

Full sentences

Teachers and children alike need to use full sentences to explain or respond. When children use complete sentences, it both reveals their understanding and embeds their knowledge.

Stem sentences

These important sentences help children express mathematical concepts accurately, and are used throughout the *Power Maths* books. Encourage children to repeat them frequently, whether working independently or with others. Examples of stem sentences are:

'4 is a part, 5 is a part, 9 is the whole.'

'There are .... groups. There are .... in each group.'

Key vocabulary

The unit starters highlight essential vocabulary for every lesson. In the pupil books, characters flag new terminology and the Teacher Guide lists important mathematical language for every unit and lesson. New terms are never introduced without a clear explanation.

Mathematical signs

Mathematical signs are used early on so that children quickly become familiar with them and their meaning. Often, the *Power Maths* characters will highlight the connection between language and particular signs.

The role of talk and discussion

When children learn to talk·purposefully together about maths, barriers of fear and anxiety are broken down and they grow in confidence, skills and understanding. Building a healthy culture of 'maths talk' empowers their learning from day one.

Explanation and discussion are integral to the *Power Maths* structure, so by simply following the books your lessons will stimulate structured talk. The following key 'maths talk' strategies will help you strengthen that culture and ensure that every child is included.

Sentences, not words

Encourage children to use full sentences when reasoning, explaining or discussing maths. This helps both speaker and listeners to clarify their own understanding. It also reveals whether or not the speaker truly understands, enabling you to address misconceptions as they arise.

Working together

Working with others in pairs, groups or as a whole class is a great way to support maths talk and discussion. Use different group structures to add variety and challenge. For example, children could take timed turns for talking, work independently alongside a 'discussion buddy', or perhaps play different *Power Maths* character roles within their group.

Think first – then talk

Provide clear opportunities within each lesson for children to think and reflect, so that their talk is purposeful, relevant and focused.

Give every child a voice

Where the 'hands up' model allows only the more confident child to shine, *Power Maths* involves everyone. Make sure that no child dominates and that even the shyest child is encouraged to contribute – and praised when they do.

Assessment strategies

Teaching for mastery demands that you are confident about what each child knows and where their misconceptions lie; therefore, practical and effective assessment is vitally important.

Formative assessment within lessons

The **Think together** section will often reveal any confusions or insecurities; try ironing these out by doing the first **Think together** question as a class. For children who continue to struggle, you or your Teaching Assistant should provide support and enable them to move on.

▶ Performance in practice can be very revealing: check Practice Books and listen out both during and after practice to identify misconceptions.

▶ The **Reflect** section is designed to check on the all-important depth of understanding. Be sure to review how the children performed in this final stage before you teach the next lesson.

End of unit check – Textbook

Each unit concludes with a summative check to help you assess quickly and clearly each child's understanding, fluency, reasoning and problem solving skills. Your Teacher Guide will suggest ideal ways of organising a given activity and offer advice and commentary on what children's responses mean. For example, 'What misconception does this reveal?'; 'How can you reinforce this particular concept?'

Assessment with young children should always be an enjoyable activity, so avoid one-to-one individual assessments, which they may find threatening or scary. If you prefer, the End of unit check can be carried out as a whole-class group using whiteboards and Practice Books.

End of unit check – Practice Book

The Practice Book contains further opportunities for assessment, and can be completed by children independently whilst you are carrying out diagnostic assessment with small groups. Your Teacher Guide will advise you on what to do if children struggle to articulate an explanation – or perhaps encourage you to write down something they have explained well. It will also offer insights into children's answers and their implications for next learning steps. It is split into three main sections, outlined below.

My journal is designed to allow children to show their depth of understanding of the unit. It can also serve as a way of checking that children have grasped key mathematical vocabulary. The question children should answer is first presented in the Textbook in the Think! section. This provides an opportunity for you to discuss the question first as a class to ensure children have understood their task. Children should have some time to think about how they want to answer the question, and you could ask them to talk to a partner about their ideas. Then children should write their answer in their Practice Book, using the word bank provided to help them with vocabulary.

The **Power check** allows pupils to self-assess their level of confidence on the topic by colouring in different smiley faces. You may want to introduce the faces as follows:

I am starting to understand. I need more practice

I don't yet know how to do this

I know how to do this, but I will keep practising

Each unit ends with either a Power play or a Power puzzle. This is an activity, puzzle or game that allows children to use their new knowledge in a fun, informal way.

Progress Tests

There are *Power Maths* Progress Tests for each half term and at the end of the year, including an Arithmetic test and Reasoning test in each case. You can enter results in the online markbook to track and analyse results and see the average for all schools' results. The tests use a 6-step scale to show results against age-related expectation.

How to ask diagnostic questions

The diagnostic questions provided in children's Practice Books are carefully structured to identify both understanding and misconceptions (if children answer in a particular way, you will know why). The simple procedure below may be helpful:

Ask the question, offering the selection of answers provided.

Children take time to think about their response.

Each child selects an answer and shares their reasoning with the group.

Give minimal and neutral feedback (for example, 'That's interesting', or 'Okay').

Ask, 'Why did you choose that answer?', then offer an opportunity to change their mind by providing one correct and one incorrect answer.

Note which children responded and reasoned correctly first time and everyone's final choices.

Reflect that together, we can get the right answer.

Keeping the class together

Traditionally, children who learn quickly have been accelerated through the curriculum. As a consequence, their learning may be superficial and will lack the many benefits of enabling children to learn with and from each other.

By contrast, *Power Maths'* mastery approach values real understanding and richer, deeper learning above speed. It sees all children learning the same concept in small, cumulative steps, each finding and mastering challenge at their own level. Remember that when you teach for mastery, EVERYONE can do maths! Those who grasp a concept easily have time to explore and understand that concept at a deeper level. The whole class therefore moves through the curriculum at broadly the same pace via individual learning journeys.

For some teachers, the idea that a whole class can move forward together is revolutionary and challenging. However, the evidence of global good practice clearly shows that this approach drives engagement, confidence, motivation and success for all learners, and not just the high flyers. The strategies below will help you keep your class together on their maths journey.

Mix it up

Do not stick to set groups at each table. Every child should be working on the same concept, and mixing up the groupings widens children's opportunities for exploring, discussing and sharing their understanding with others.

Recycling questions

Reuse the Textbook and Practice Book questions with concrete materials to allow children to explore concepts and relationships and deepen their understanding. This strategy is especially useful for reinforcing learning in same-day interventions.

Strengthen at every opportunity

The next lesson in a *Power Maths* sequence always revises and builds on the previous step to help embed learning. These activities provide golden opportunities for individual children to strengthen their learning with the support of Teaching Assistants.

Prepare to be surprised!

Children may grasp a concept quickly or more slowly. The 'fast graspers' won't always be the same individuals, nor does the speed at which a child understands a concept predict their success in maths. Are they struggling or just working more slowly?

Same-day intervention

Since maths competence depends on mastering concepts one by one in a logical progression, it is important that no gaps in understanding are ever left unfilled. Same-day interventions – either within or after a lesson – are a crucial safety net for any child who has not fully made the small step covered that day. In other words, intervention is always about keeping up, not catching up, so that every child has the skills and understanding they need to tackle the next lesson. That means presenting the same problems used in the lesson, with a variety of concrete materials to help children model their solutions.

We offer two intervention strategies below, but you should feel free to choose others if they work better for your class.

Within-lesson intervention

The **Think together** activity will reveal those who are struggling, so when it is time for practice, bring these children together to work with you on the first practice questions. Observe these children carefully, ask questions, encourage them to use concrete models and check that they reach and can demonstrate their understanding.

After-lesson intervention

You might like to use the **Think together** questions to recap the lesson with children who are working behind expectations during assembly time. Teaching Assistants could also work with these children at other convenient points in the school day. Some children may benefit from revisiting work from the same topic in the previous year group. Note also the suggestion for recycling questions from the Textbook and Practice Book with concrete materials on page 27.

The role of practice

Practice plays a pivotal role in the *Power Maths* approach. It takes place in class groups, smaller groups, pairs, and independently, so that children always have the opportunities for thinking as well as the models and support they need to practise meaningfully and with understanding.

Intelligent practice

In *Power Maths*, practice never equates to the simple repetition of a process. Instead we embrace the concept of intelligent practice, in which all children become fluent in maths through varied, frequent and thoughtful practice that deepens and embeds conceptual understanding in a logical, planned sequence. To see the difference, take a look at the following examples.

Traditional practice

- Repetition can be rote – no need for a child to think hard about what they are doing

- Praise may be misplaced

- Does this prove understanding?

Intelligent practice

- Varied methods – concrete, pictorial and abstract

- Equation expressed in different ways, requiring thought and understanding

- Constructive feedback

All practice questions are designed to move children on and reveal misconceptions.

Simple, logical steps build onto earlier learning.

C-P-A runs throughout – different ways of modelling and understanding the same concept.

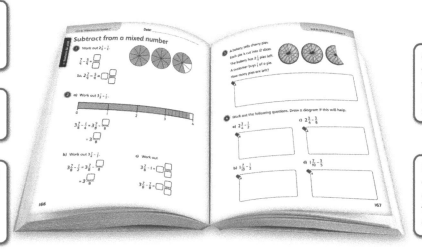

Conceptual variation – children work on different representations of the same maths concept.

Friendly characters offer support and encourage children to try different approaches.

A carefully designed progression

The Practice Books provide just the right amount of intelligent practice for children to complete independently in the final sections of each lesson. It is really important that all children are exposed to the practice questions, and that children are not directed to complete different sections. That is because each question is different and has been designed to challenge children to think about the maths they are doing. The questions become more challenging so children grasping concepts more quickly will start to slow down as they progress. Meanwhile, you have the chance to circulate and spot any misconceptions before they become barriers to further learning.

Homework and the role of parents and carers

While *Power Maths* does not prescribe any particular homework structure, we acknowledge the potential value of practice at home. For example, practising fluency in key facts, such as number bonds and times-tables, is an ideal homework task. You can share the Individual Practice Games for homework (see page 6), or parents and carers could work through uncompleted Practice Book questions with children at either primary stage.

However, it is important to recognise that many parents and carers may themselves lack confidence in maths, and few, if any, will be familiar with mastery methods. A Parents' and Carers' evening that helps them understand the basics of mindsets, mastery and mathematical language is a great way to ensure that children benefit from their homework. It could be a fun opportunity for children to teach their families that everyone can do maths!

Structures and representations

Unlike most other subjects, maths comprises a wide array of abstract concepts – and that is why children and adults so often find it difficult. By taking a concrete-pictorial-abstract (C-P-A) approach, *Power Maths* allows children to tackle concepts in a tangible and more comfortable way.

Non-linear stages

Concrete

Replacing the traditional approach of a teacher working through a problem in front of the class, the concrete stage introduces real objects that children can use to 'do' the maths – any familiar object that a child can manipulate and move to help bring the maths to life. It is important to appreciate, however, that children must always understand the link between models and the objects they represent. For example, children need to first understand that three cakes could be represented by three pretend cakes, and then by three counters or bricks. Frequent practice helps consolidate this essential insight. Although they can be used at any time, good concrete models are an essential first step in understanding.

Pictorial

This stage uses pictorial representations of objects to let children 'see' what particular maths problems look like. It helps them make connections between the concrete and pictorial representations and the abstract maths concept. Children can also create or view a pictorial representation together, enabling discussion and comparisons. The *Power Maths* teaching tools are fantastic for this learning stage, and bar modelling is invaluable for problem solving throughout the primary curriculum.

Abstract

Our ultimate goal is for children to understand abstract mathematical concepts, symbols and notation and of course, some children will reach this stage far more quickly than others. To work with abstract concepts, a child must be comfortable with the meaning of and relationships between concrete, pictorial and abstract models and representations. The C-P-A approach is not linear, and children may need different types of models at different times. However, when a child demonstrates with concrete models and pictorial representations that they have grasped a concept, we can be confident that they are ready to explore or model it with abstract symbols such as numbers and notation.

Use at any time and with any age to support understanding

Variation helps visualisation

Children find it much easier to visualise and grasp concepts if they see them presented in a number of ways, so be prepared to offer and encourage many different representations.

For example, the number six could be represented in various ways:

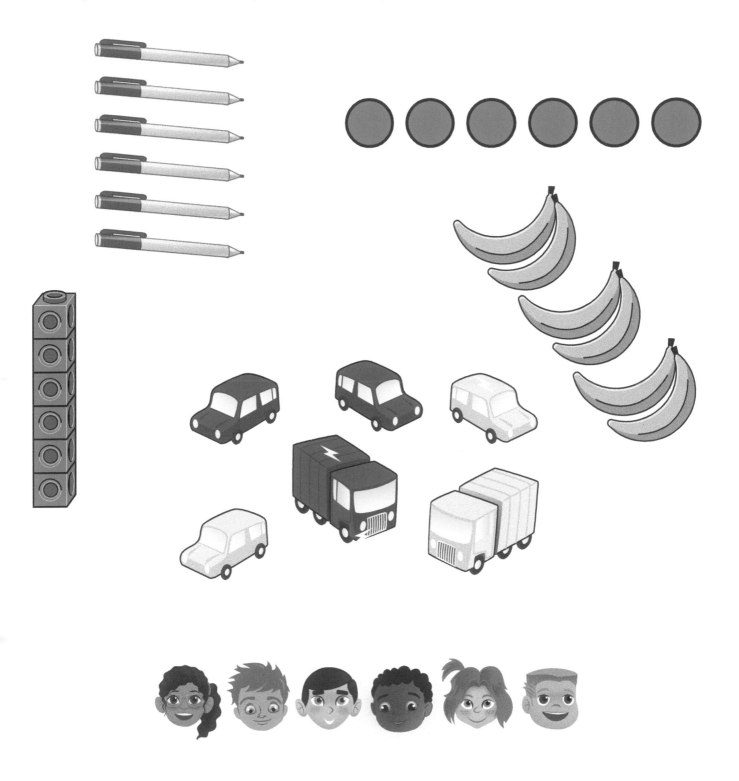

Practical aspects of *Power Maths*

One of the key underlying elements of *Power Maths* is its practical approach, allowing you to make maths real and relevant to your children, no matter their age.

Manipulatives are essential resources for both key stages and *Power Maths* encourages teachers to use these at every opportunity, and to continue the Concrete-Pictorial-Abstract approach right through to Year 6.

The Textbooks and Teacher Guides include lots of opportunities for teaching in a practical way to show children what maths means in real life.

Discover and Share

The **Discover** and **Share** sections of the Textbook give you scope to turn a real-life scenario into a practical and hands-on section of the lesson. Use these sections as inspiration to get active in the classroom. Where appropriate, use the **Discover** contexts as a springboard for your own examples that have particular resonance for your children – and allow them to get their hands dirty trying out the mathematics for themselves.

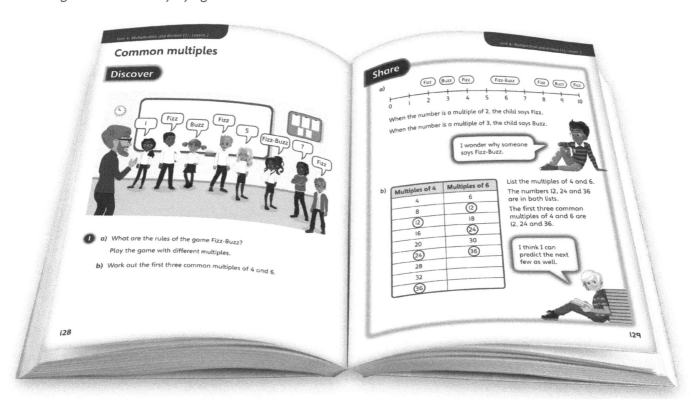

Unit videos

Every term has one unit video which incorporates real-life classroom sequences.

These videos show you how the reasoning behind mathematics can be carried out in a practical manner by showing real children using various concrete and pictorial methods to come to the solution. You can see how using these practical models, such as part-whole and bar models, helps them to find and articulate their answer.

Mastery tips

Mastery Experts give anecdotal advice on where they have used hands-on and real-life elements to inspire their children.

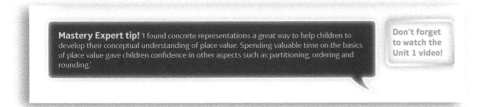

Mastery Expert tip! 'I found concrete representations a great way to help children to develop their conceptual understanding of place value. Spending valuable time on the basics of place value gave children confidence in other aspects such as partitioning, ordering and rounding.'

Don't forget to watch the Unit 1 video!

Concrete-Pictorial-Abstract (C-P-A) approach

Each **Share** section uses various methods to explain an answer, helping children to access abstract concepts by using concrete tools, such as counters. Remember, this isn't a linear process, so even children who appear confident using the more abstract method can deepen their knowledge by exploring the concrete representations. Encourage children to use all three methods to really solidify their understanding of a concept.

Pictorial representation – drawing the problem in a logical way that helps children visualise the maths

Concrete representation – using manipulatives to represent the problem. Encourage children to physically use resources to explore the maths.

Abstract representation – using words and calculations to represent the problem.

Practical tips

Every lesson suggests how to draw out the practical side of the **Discover** context.

You'll find these in the **Discover** section of the Teacher Guide for each lesson.

PRACTICAL TIPS Ask children to count in steps of 10 to 100 and use the poster on the wall in the picture to discuss any patterns they notice within the Roman numerals, particularly how 40 and 90 are represented.

Resources

Every lesson lists the practical resources you will need or might want to use. There is also a summary of all of the resources used throughout the term on page 39 to help you be prepared.

RESOURCES

Mandatory: counters, dice

Optional: base 10 equipment

Working with children below age-related expectation

This section offers advice on using *Power Maths* with children who are significantly behind age-related expectation. Teacher judgement will be crucial in terms of where and why children are struggling, and in choosing the right approach. The suggestions can of course be adapted for children with special educational needs, depending on the specific details of those needs.

General approaches to support children who are struggling

Keeping the pace manageable

Remember, you have more teaching days than *Power Maths* lessons so you can cover a lesson over more than one day, and revisit key learning, to ensure all children are ready to move on. You can use the + and – buttons to adjust the time for each unit in the online planning. The NCETM's Ready-to-Progress criteria can be used to help determine what should be highest priority.

Same-day intervention

You could go over the Textbook pages or revisit the previous year's work if necessary (see Addressing gaps). Remember that same-day intervention can be within the lesson, as well as afterwards (see page 28). As children start their independent practice, you can work with those who found the first part of the lesson difficult, checking understanding using manipulatives.

Fluency sessions

Fit in as much practice as you can for number bonds and times-tables, etc., at other times of the day. If you can, plan a short 'maths meeting' for this in the afternoon. You might choose to use a Power Up you haven't used already.

Addressing gaps

Use material from the same topic in the previous year to consolidate or address gaps in learning, e.g. Textbook pages and Strengthen activities. The End of unit check will help gauge children's understanding.

Pre-teaching

Find a 5- to 10-minute slot before the lesson to work with the children you feel would benefit. The afternoon before the lesson can work well, because it gives children time to think in between. Recap previous work on the topic (addressing any gaps you're aware of) and do some fluency practice, targeting number facts etc. that will help children access the learning.

Focusing on the key concepts

If children are a long way behind, it can be helpful to take a step back and think about the key concepts for children to engage with, not just the fine detail of the objective for that year group (e.g. addition with a specific number of columns). Bearing that in mind, how could children advance their understanding of the topic?

Providing extra support within the lesson

Support in the Teacher Guide

First of all, use the Strengthen support in the Teacher Guide for guided and independent work in each lesson, and share this with Teaching Assistants, where relevant. As you read through the lesson content and corresponding Teacher Guide pages before the lesson, ask yourself what key idea or nugget of understanding is at the heart of the lesson. If children are struggling, this should help you decide what's essential for all children before they move on.

Annotating pages

You can annotate questions to provide extra scaffolding or hints if you need to, but aim to build up children's ability to access questions independently wherever you can. Children tend to get used to the style of the *Power Maths* questions over time.

Quick recap as lesson starter

The Quick recap for each lesson in the Teacher Guide is an alternative starter activity to the Power Up. You might choose to use this with some or all children if you feel they will need support accessing the main lesson.

Consolidation questions

If you think some children would benefit from additional questions at the same level before moving on, write one or two similar questions on the board. (This shouldn't be at the expense of reasoning and problem-solving opportunities: take longer over the lesson if you need to.)

Hard copy Textbooks

The Textbooks help children focus in more easily on the mathematical representations, read the text more comfortably, and revisit work from a previous lesson that you are building on, as well as giving children ownership of their learning journey. In main lessons, it can work well to use the e-Textbook for **Discover** and give out the books when discussing the methods in the **Share** section.

Reading support

It's important that all children are exposed to problem solving and reasoning questions, which often involve reading. For whole-class work you can read questions together. For independent practice you could consider annotating pages to help children see what the question is asking, and stem sentences to help structure their answer. A general focus on specific mathematical language and vocabulary will help children access the questions. You could consider pairing weaker readers with stronger readers, or read questions as a group if those who need support are on the same table.

Providing extra depth and challenge with *Power Maths*

Just as prescribed in the National Curriculum, the goal of *Power Maths* is never to accelerate through a topic but rather to gain a clear, deep and broad understanding. Here are some suggestions to help ensure all children are appropriately challenged as you work with the resources.

Overall approaches

First of all, remember that the materials are designed to help you keep the class together, allowing all children to master a concept while those who grasp it quickly have time to explore it in more depth. Use the Deepen support in the Teacher Guide (see below) to challenge children who work through the questions quickly. Here are some questions and ideas to encourage breadth and depth during specific parts of the lesson, or at any time (where no part of the lesson sequence is specified):

- **Discover**: 'Can you demonstrate your solution another way?'

- **Share**: Make sure every child is encouraged to give answers and engage with the discussion, not just the most confident.

- **Think together**: 'Can you model your answers using concrete materials? Can you explain your solution to a partner?'

- Practice: Allow all children to work through the full set of questions, so that they benefit from the logical sequence.

- **Reflect**: 'Is there another way of working out the answer? And another way?'
  'Have you found all the solutions?'
  'Is that always true?'
  'What's different between this question and that question? And what's the same?'

Note that the **Challenge** questions are designed so that all children can access and attempt them, if they have worked through the steps leading up to them. There may be some children in a given lesson who don't manage to do the **Challenge**, but it is not supposed to be a distinct task for a subset of the class. When you look through the lesson materials before teaching, think about what each question is specifically asking, and compare this with the key learning point for the lesson. This will help you decide which questions you feel it's essential for all children to answer, before moving on. You can at least aim for all children to try the **Challenge**!

Deepen activities and support

The Teacher Guide provides valuable support for each stage of the lesson. This includes Deepen tips for the guided and independent practice sections, which will help you provide extra stretch and challenge within your lesson, without having to organise additional tasks. If you have a Teaching Assistant, they can also make use of this advice. There are also suggestions for the lesson as a whole in the 'Going Deeper' section on the first page of the Teacher Guide section for that lesson. Every class is different, so you can always go a bit further in the direction indicated, if appropriate, and build on the suggestions given.

There is a Deepen activity for each unit. These are designed to follow on from the End of unit check, stretching children who have a firm understanding of the key learning from the unit. Children can work on them independently, which makes it easier for the teacher to facilitate the Strengthen activity for children who need extra support. Deepen activities could also be introduced earlier in the unit if the necessary work has been covered. The Deepen activities are on *ActiveLearn* on the Planning page for each unit, and also on the Resources page).

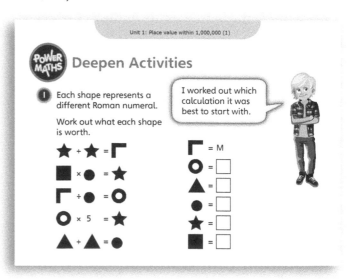

Using the questions flexibly to provide extra challenge

Sometimes you may want to write an extra question on the board or provide this on paper. You can usually do this by tweaking the lesson materials. The questions are designed to form a carefully structured sequence that builds understanding step by step, but, with careful thought about the purpose of each question, you can use the materials flexibly where you need to. Sometimes you might feel that children would benefit from another similar question for consolidation before moving on to the next one, or you might feel that they would benefit from a harder example in the same style. It should be quick and easy to generate 'more of the same' type questions where this is the case.

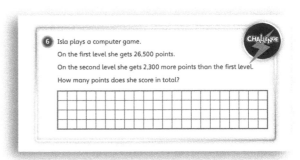

When you see a question like this one (from Unit 3, Lesson 3), it's easy to make harder examples to do afterwards if you need them. Any two numbers will generate a new multi-step problem, and you could make the numbers fiddly rather than round.

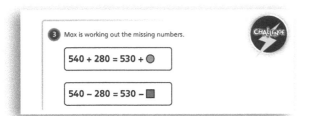

For this example (from Unit 3, Lesson 12), you could ask children to make up their own question(s) for a partner to solve. They could vary the number of digits and the operations. (In fact, for any of these examples you could ask early finishers to create their own question for a partner.)

Here's an example (from Unit 3, Lesson 4) where some of the journeys in the table feature as questions in the lesson, but others don't. Clearly there are any number of extra questions you could ask using the same table (the lesson includes multi-step journeys). Children could calculate journeys for their own itineraries, for example they could look for a multi-step journey that covers between 16,000 and 17,000 km, or they could even look for the shortest itinerary that visits all the destinations.

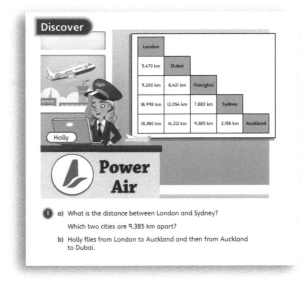

Besides creating additional questions, you should be able to find a question in the lesson that you can adapt into a game or open-ended investigation, if this helps to keep everyone engaged. It could simply be that, instead of answering 5 × 5 etc on the page, they could build a robot with 5 lots of 5 cubes.

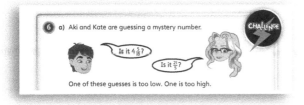

With a question like this (from Unit 5, Lesson 8), children could play a game where they have to guess their partner's mystery number, finding out each time if the guess is too high or too low.

See the bullets above for some general ideas that will help with 'opening out' questions in the books, e.g. 'Can you find all the solutions?' type questions.

Other suggestions

Another way of stretching children is through mixed ability pairs, or via other opportunities for children to explain their understanding in their own way. This is a good way of encouraging children to go deeper into the learning, rather than, for instance, tackling questions that are computationally more challenging but conceptually equivalent in level.

Using *Power Maths* with mixed age classes

Overall approaches

There are many variables between schools that would make it inadvisable to recommend a one-size-fits-all approach to mixed age teaching with *Power Maths*. These include how year groups are merged, availability of Teaching Assistants, experience and preference of teaching staff, range in pupil attainment across years, classroom space and layout, level of flexibility around timetables, and overall organisational structure (whether the school is part of a trust).

Some schools will find it best to timetable separate maths lessons for the different year groups. Others will aim to teach the class together as much as possible using the mixed age planning support on *ActiveLearn* (see the lesson exemplars for ways of organising lessons with strong/medium/weak correlation between year groups). There will also be ways of adapting these general approaches. For example, offset lessons where Year A start their lesson with the teacher, while Year B work independently on the practice from the previous lesson, and then start the next lesson with the teacher while Year A work independently; or teachers may choose to base their provision around the lesson from one year group and tweak the content up/down for the other group.

Key strategies for mixed age teaching

The mixed age teaching webinar on *ActiveLearn* provides advice on all aspects of mixed age teaching, including more detail on the ideas below.

Developing independence over time
Investing time in building up children's independence will pay off in the medium term.

Clear rationale
If someone asked, 'Why did you teach both Unit 3 and 4 in the same lesson/separate lessons?', what would your answer be?

Designing a lesson
1. Identify the core learning for each group
2. Identify any number skills necessary to access the core
3. Consider the flow of concepts and how one core leads to the other

Challenging all children
The questions are designed to build understanding step by step, but with careful thought about the purpose of each question you can tweak them to increase the challenge.

Multiple years combined
With more than two years together, teachers will inevitably need to use the resources flexibly if delivering a single lesson.

Enjoy the positives!

Comparison deepens understanding and there will be lots of opportunities for children, as well as misconceptions to explore. There is also in-built pre-teaching and the chance to build up a concept from its foundations. For teachers there is double the material to draw on! Mixed age teachers require a strong understanding of the progression of ideas across year groups, which is highly valuable for all teachers. Also, it is necessary to engage deeply with the lesson to see how to use the materials flexibly – this is recommended for all teachers and will help you bring your lesson to life!

List of practical resources

Year 5A Mandatory resources

Resource	Lesson
100 square	**Unit 4** Lesson 1 **Unit 4** Lesson 2
base 10 equipment	**Unit 1** Lesson 3 **Unit 4** Lesson 8 **Unit 4** Lessons 9, 10
counters	**Unit 1** Lessons 2, 4, 6 **Unit 3** Lessons 5, 6, 9* **Unit 4** Lessons 3, 4, 5
dice	**Unit 1** Lessons 2, 5 **Unit 5** Lesson 8
digit cards	**Unit 2** Lesson 5 **Unit 4** Lesson 4
digit cards 0–9	**Unit 3** Lesson 6
multilink cubes	**Unit 4** Lesson 7
number lines	**Unit 3** Lesson 11
paper (squared)	**Unit 4** Lesson 3
paper for fraction strips	**Unit 6** Lessons 10, 11
part-whole models	**Unit 3** Lesson 11
place value counters	**Unit 1** Lessons 3, 7, 8 **Unit 2** Lesson 3 **Unit 3** Lessons 3, 4 **Unit 4** Lessons 8, 9, 10
place value grids	**Unit 1** Lessons 5, 6 **Unit 3** Lessons 6, 9*
whiteboards	**Unit 3** Lesson 6 **Unit 4** Lessons 1, 4

Year 5A Optional resources

Resource	Lesson
2D shapes	**Unit 5** Lessons 1, 2
bar model	**Unit 3** Lesson 12
base 10 equipment	**Unit 1** Lessons 2, 4, 7, 8 **Unit 4** Lesson 7
bead strings	**Unit 2** Lesson 1
calculators	**Unit 3** Lesson 11
calendar	**Unit 4** Lesson 10
card shapes (cut into fractions)	**Unit 5** Lesson 8
chessboard	**Unit 4** Lesson 6
clocks (analogue)	**Unit 6** Lesson 3
clock faces (blank)	**Unit 6** Lesson 3
comparison bar models (blank)	**Unit 3** Lesson 10
concrete objects (e.g. pens and pencils)	**Unit 6** Lesson 1
counters	**Unit 4** Lesson 6
counters (blank)	**Unit 4** Lessons 1, 2
cubes	**Unit 4** Lesson 1
dice	**Unit 2** Lessons 2, 3
digit cards	**Unit 2** Lessons 3, 4, 6
fraction cards	**Unit 5** Lessons 6, 7
fraction circles	**Unit 6** Lesson 3
fraction shapes (circles)	**Unit 6** Lessons 7, 8, 9

Year 5A Optional resources – *continued*

Resource	Lesson
fraction strips	**Unit 6** Lessons 3, 4, 5, 7, 8, 9
fraction strips (blank)	**Unit 5** Lesson 6
fraction wall (printed)	**Unit 5** Lessons 6, 7, 8
lengths of ribbon (with fifths, tenths, quarters and twentieths marked on them)	**Unit 6** Lesson 10
measuring jug	**Unit 3** Lesson 10
modelling clay	**Unit 5** Lesson 2
multilink cubes	**Unit 4** Lessons 3, 6
multiplication squares	**Unit 4** Lesson 10
number cards	**Unit 2** Lesson 1 **Unit 3** Lesson 11 **Unit 5** Lesson 8
number cards (set of matching cards for 4, 9, 40, 90, 400, 900 in numbers and Roman numerals)	**Unit 1** Lesson 1
number cards (sets of cards showing I, I, I, V, X, X, X, L, C, C, C, D, M)	**Unit 1** Lesson 1
number lines	**Unit 3** Lessons 7, 12 **Unit 5** Lesson 6
number lines (blank)	**Unit 5** Lessons 7, 8 **Unit 6** Lessons 2, 3, 4, 5, 8, 9, 11
number lines (drywipe)	**Unit 2** Lesson 1
paper (circles of)	**Unit 5** Lesson 5
paper (for folding)	**Unit 5** Lessons 1, 2, 3, 4, 6, 7, 8 **Unit 6** Lessons 1, 2, 6
paper (sheets of, with fractions shaded)	**Unit 5** Lesson 3
paper (squared)	**Unit 4** Lesson 6
paper strips	**Unit 5** Lesson 5 **Unit 6** Lessons 1, 5
part-whole models	**Unit 3** Lesson 8 **Unit 6** Lessons 8, 11
part-whole models (laminated)	**Unit 1** Lessons 5, 6 **Unit 6** Lessons 4, 5
pencils	**Unit 4** Lesson 6
pictures of pots of paint (printed, with different unit fractions listed on them)	**Unit 5** Lesson 4
pictures of shapes (cut into equal fractions)	**Unit 5** Lesson 5
place value counters	**Unit 1** Lessons 5, 6 **Unit 2** Lessons 2, 5 **Unit 3** Lessons 1, 2, 8
place value grids (printed)	**Unit 1** Lesson 7 **Unit 4** Lessons 8, 9
play money	**Unit 1** Lesson 8
play money (notes)	**Unit 1** Lessons 6, 8
puzzle cube	**Unit 4** Lesson 7
rulers	**Unit 4** Lesson 6
scissors	**Unit 5** Lesson 3 **Unit 6** Lessons 3, 8
shape cards	**Unit 3** Lesson 11
sorting circles	**Unit 4** Lessons 1, 2
stopwatch or egg timer	**Unit 3** Lesson 2
toy car parts	**Unit 4** Lesson 8
water	**Unit 3** Lesson 10
whiteboards	**Unit 1** Lesson 8

Getting started with *Power Maths*

As you prepare to put *Power Maths* into action, you might find the tips and advice below helpful.

STEP 1: Train up!

A practical, up-front full day professional development course will give you and your team a brilliant head-start as you begin your *Power Maths* journey. You will learn more about the ethos, how it works and why.

STEP 2: Check out the progression

Take a look at the yearly and termly overviews. Next take a look at the unit overview for the unit you are about to teach in your Teacher Guide, remembering that you can match your lessons and pacing to match your class.

STEP 3: Explore the context

Take a little time to look at the context for this unit: what are the implications for the unit ahead? (Think about key language, common misunderstandings and intervention strategies, for example.) If you have the online subscription, don't forget to watch the corresponding unit video.

STEP 4: Prepare for your first lesson

Familiarise yourself with the objectives, essential questions to ask and the resources you will need. The Teacher Guide offers tips, ideas and guidance on individual lessons to help you anticipate children's misconceptions and challenge those who are ready to think more deeply.

STEP 5: Teach and reflect

Deliver your lesson — and enjoy!

Afterwards, reflect on how it went… Did you cover all five stages?
Does the lesson need more time? How could you improve it?
What percentage of your class do you think mastered the concept?
How can you help those that didn't?

Unit I
Place value within 1,000,000 ①

Mastery Expert tip! 'I found concrete representations a great way to help children to develop their conceptual understanding of place value. Spending valuable time on the basics of place value gave children confidence in other aspects such as partitioning, ordering and rounding.'

Don't forget to watch the Unit 1 video!

WHY THIS UNIT IS IMPORTANT

This unit builds on Year 4 work on numbers within 10,000 to further develop children's sense of larger numbers, working with numbers up to 1,000,000. It is essential for children to have a secure understanding of place value in order to estimate, make decisions when calculating and work with measurements. Therefore, it is important that by the end of this unit, children know how to partition any number. They will continue to represent numbers in different forms. This secure understanding is essential for work in Unit 2, where they will compare and order numbers as well as find their position on number lines.

WHERE THIS UNIT FITS

→ **Unit 1: Place value within 1,000,000 (1)**

→ Unit 2: Place value within 1,000,000 (2)

This unit builds on children's work from Year 4 on 4-digit numbers. Many of the models and images used previously will be further extended to include 5-digit numbers, so that children can flexibly work with all numbers to 1,000,000. In this unit, we will focus primarily on part-whole models and place value counters and grids, as base 10 equipment is not user-friendly for larger numbers.

This unit provides the foundation for further working with numbers up to 1,000,000 and develops fluency with place value to support calculating during the year.

Before they start this unit, it is expected that children:

* know that 4-digit whole numbers are made up of 1,000s, 100s, 10s and 1s, and can represent 3- and 4-digit numbers in different ways (for example, on part-whole models and place value grids)
* understand how to write a 4-digit number as a partitioned number sentence and can partition numbers flexibly. For example, they should know that 2,350 = 2,000 + 300 + 50 or 2,350 = 1,000 + 1,300 + 50.

ASSESSING MASTERY

Children know the value of each digit in numbers up to 1,000,000 and can represent them in different ways. They should be able to write a number in words and as numerals, and be able to partition numbers and write a number as an addition sentence. They should work confidently with equipment such as place value counters and grids.

COMMON MISCONCEPTIONS	STRENGTHENING UNDERSTANDING	GOING DEEPER
Children may think that, in number sentences, they just take the first digit to form the number, for example: 3,465 = 3,000 + 400 + 60 + 5 However, they may think that 400 + 3,000 + 60 + 5 = 4,365 when it does not.	Build numbers with place value counters, then move them to partition differently. Children should find the value of each part before recombining them to confirm they are still equivalent to the whole.	Explore more flexible partitioning into two, three, four, five or six parts. Children should be prepared to prove that their combined value of parts each time is equivalent to the whole. Ask children to compile a set of 'Helpful tips' that can be used by others to compare numbers. Encourage them to think about what mistakes others may make.
When representing numbers on a part-whole model, children may just put the initial digits in, as opposed to the value/size of the number.	Start with smaller numbers, such as 48, and ask children to make them using base 10 equipment. They should notice that this is 40 (4 tens) and 8 (8 ones). Build up to larger numbers.	

Unit I: Place value within 1,000,000 ❶

UNIT STARTER PAGES

Use these pages to introduce the unit to children. Ask children which of Flo's words they have used before and to provide definitions.

STRUCTURES AND REPRESENTATIONS

Place value grid and counters: This model helps children to organise 4- and 5-digit numbers and show the value of each digit. Place value grids will be used with both concrete representations and abstract numbers.

TTh	Th	H	T	O
⬤⬤⬤	⬤⬤	⬤⬤⬤⬤	⬤	⬤⬤⬤

Children should also be able to represent numbers with plain counters. The column they are in will tell the child the value of the digit.

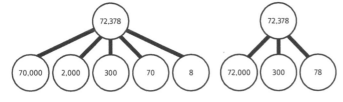

TTh	Th	H	T	O
⬤⬤⬤	⬤⬤	⬤⬤⬤⬤	⬤	⬤⬤⬤

Part-whole model: This model will help to show the value of each part of a number and provide support for more flexible partitioning.

```
        72,378                          72,378
   /  /   |   \  \                    /    |    \
70,000 2,000 300 70 8            72,000  300   78
```

KEY LANGUAGE

There is some key language that children will need to know as part of the learning in this unit.

→ ones (1s), tens (10s), hundreds (100s), thousands (1,000s), ten thousands (10,000s)

→ 100,000s (hundred thousands)

→ 1,000,000 (million)

→ place value

→ part-whole

→ partition

→ more or less

→ estimate

→ order

→ less than

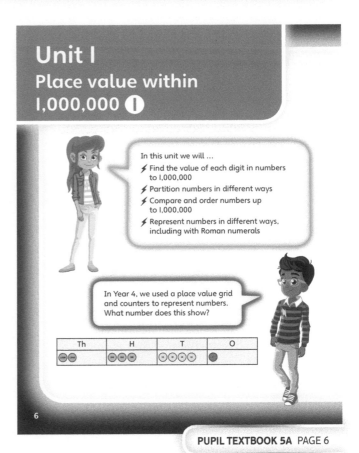

Unit I
Place value within 1,000,000 ❶

In this unit we will …
⚡ Find the value of each digit in numbers to 1,000,000
⚡ Partition numbers in different ways
⚡ Compare and order numbers up to 1,000,000
⚡ Represent numbers in different ways, including with Roman numerals

In Year 4, we used a place value grid and counters to represent numbers. What number does this show?

Th	H	T	O
⬤⬤	⬤⬤⬤	⬤⬤⬤⬤	⬤

6

PUPIL TEXTBOOK 5A PAGE 6

We will need some maths words. Which of these have you met before?

| ones (1s) | tens (10s) | hundreds (100s) |

| thousands (1,000s) | ten thousands (10,000s) |

| hundred thousands (100,000s) | more than (>) |

| less than (<) | place value | partition | estimate |

We will also use part-whole models and number lines. What number do these both represent?

```
     ( )
    /   \
(2,000) (500)
```

```
0   1,000  2,000  3,000  4,000
```

7

PUPIL TEXTBOOK 5A PAGE 7

Roman numerals

Learning focus

In this lesson, children will revisit Roman numerals to 100 and learn the numerals M (1,000) and D (500). They will explore reading and writing numbers using Roman numerals, and will use the numerals M and D to recognise and represent years.

Before you teach

- Can children match Roman numerals to the numbers on a clock face?
- Can they explain what is different about the numerals IX, X and XI?
- Can children recognise and write Roman numerals for numbers within 100?

NATIONAL CURRICULUM LINKS

Year 5 Number – number and place value

Read Roman numerals to 1,000 (M) and recognise years written in Roman numerals.

ASSESSING MASTERY

Children can read and write Roman numerals using M, D, C, L, X, V and I. They can recognise and represent a year written in Roman numerals, and talk about other ways they are still used in everyday life.

COMMON MISCONCEPTIONS

Children may use numerals that are too small before a larger Roman numeral, for example for the number 99, they may see this as 100 subtract, 1 so show this as IC rather than representing this as 90 + 9 using XCIX. The only pairs of numbers that are used for this subtraction rule are IV, IX, XL, XC, CD and CM. So only I, X and C can be used in this way and they can only come before the two numbers above them in value. Ask:

- *Which Roman numerals can you use to subtract from a larger number? Can you make up a rule to remind you?*

STRENGTHENING UNDERSTANDING

Encourage children to use part-whole models to break up larger numbers into recognisable parts. Introduce children to a range of numbers written using Roman numerals for children to identify patterns. Ensure that children understand that the maximum number of repeats is three before the subtraction rule is applied, for example X, XX, XXX, XL. Present them with 400 written as CCCC and ask them to identify the mistake.

GOING DEEPER

Encourage children to think about the ways that repeating numerals can be read as a group rather than individually, for example MCCC as 1,000 + 300 = 1,300. Ask children to represent MCCCL using a part-whole model with only three parts.

KEY LANGUAGE

In lesson: Roman numerals, ascending, descending

STRUCTURES AND REPRESENTATIONS

Part-whole model

RESOURCES

Optional: sets of cards showing I, I, I, V, X, X, X, L, C, C, C, D, M, set of matching cards for 4, 9, 40, 90, 400, 900 in numbers and Roman numerals

 In the eTextbook of this lesson, you will find interactive links to a selection of teaching tools.

Quick recap 🔁

Make cards with the numbers 1 to 12 in Roman numerals and ask children to organise them in a circle, like a clock face. Alternatively, write on mini-whiteboards and ask children to arrange them to make a giant clock.

Discover

Pair work

ASK

- Question ① a): *Where have you seen Roman numerals being used in everyday life?*
- Question ① a): *How is the Roman numeral I used with X to represent 9, 11 and 19?*
- Question ① b): *Which of Ebo's Roman numerals do you recognise? What are their values?*

IN FOCUS Questions ① a) and b) are important as they remind children of the Roman numerals they encountered in Year 3. Use the poster in the **Discover** image to revisit the use of a smaller numeral before a larger, for example 9 as 10 subtract 1 and 40 as 50 subtract 10.

PRACTICAL TIPS Ask children to count in steps of 10 to 100 and use the poster on the wall in the picture to discuss any patterns they notice within the Roman numerals, particularly how 40 and 90 are represented.

Give pairs of children a set of nine cards with the Roman numerals: I, I, I, V, X, X, X, L, C. Ask them to make some different numbers to 100, for example 37, 95 and 44. Later add C, C, D and M so children can make larger numbers.

ANSWERS

Question ① a): IV means 1 before 5, which is 4.
IX means 1 before 10, which is 9.
XIV means X + IV, which is 14.
XIX means X + IX, which is 19.

Question ① b): Ebo's Roman numerals represent the number 1,690. Jamie's Roman numerals represent the number 75.

Share

Whole class teacher led

ASK

- Question ① a): *What is similar and what is different between the numbers?*
- Question ① b): *Why would you not repeat the Roman numeral D to make DD (1,000)? What would you use instead?*
- Question ① b): *What would CX mean?*

IN FOCUS Question ① b) introduces the numerals M for 1,000 and D for 500. Children are led through the process of working out a number from its Roman numerals, breaking them down into parts to find the value of each, and revisiting how placing a smaller numeral before a larger one reduces its value (in this case XC for 100 − 10 = 90).

STRENGTHEN Ask children to use the cards on their table to show Ebo's number. Use a part-whole model to show the value of each of the parts: M, D, C and XC.

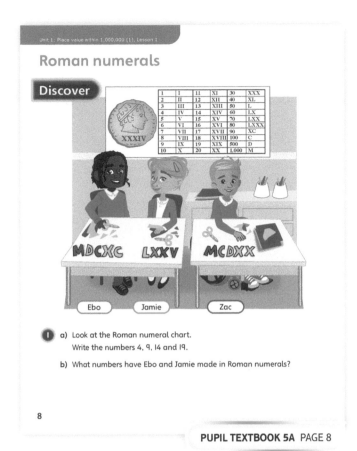

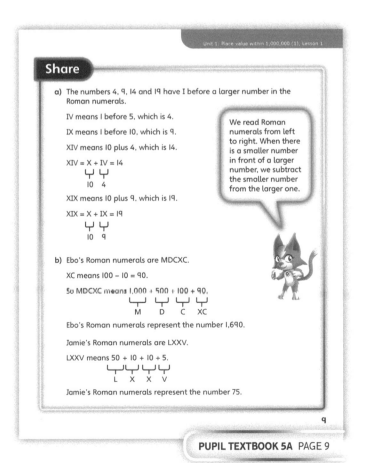

Think together

WAYS OF WORKING Whole class teacher led (I do, We do, You do)

ASK

- Question ❶: *What does C mean? What does X mean? What does it mean when X is after C? What about when it is before?*
- Question ❷: *What is the value of each part? How do you know?*
- Question ❸ a): *I think the baseball poster shows the number 117. What mistake have I made?*

IN FOCUS In question ❶, children focus explicitly on the difference between writing a letter before another or writing it after. In question ❷, a part-whole model is shown to help children recognise the value of the parts that are used to build up numbers in Roman numerals.

STRENGTHEN Ensure that children are secure with the six pairs of Roman numerals used to show the numbers 4, 9, 40, 90, 400, 900 (IV, IX, XL, XC, CD and CM). Play a matching game based on cards, showing numbers and Roman numerals to build up children's familiarity with these forms.

DEEPEN Ask children to explain why some numerals are repeated (such as X) and some are not (such as L). For example, L is not repeated because LL would be 100, which is C.

ASSESSMENT CHECKPOINT Use question ❶ to assess whether children understand the difference between writing Roman numerals before or after other numerals. Use question ❷ to assess whether children can move fluently between numbers and Roman numerals.

ANSWERS

Question ❶: Both numbers feature the same letters but in different places. CX = 100 + 10 = 110. XC = 100 – 10 = 90.

Question ❷: M means 1,000
CD means 500 – 100 = 400
XX means 10 + 10 = 20
1,000 + 400 + 20 = 1,420
Zac's number is 1,420.

Question ❸ a): XCVII means 97.

Question ❸ b): MMIX means 2009.

Question ❸ c): 450 in Roman numerals is CDL.

Question ❸ d): 1791 in Roman numerals is MDCCXCI.

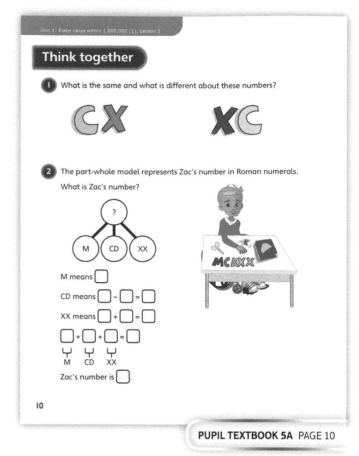

PUPIL TEXTBOOK 5A PAGE 10

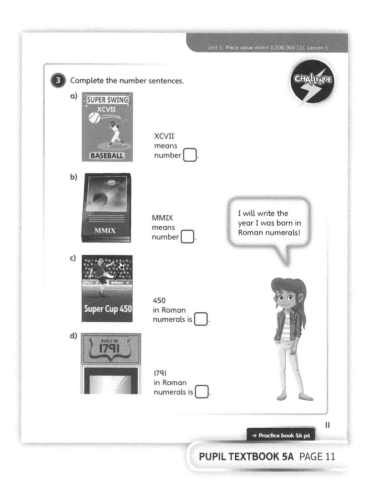

PUPIL TEXTBOOK 5A PAGE 11

Practice

WAYS OF WORKING Independent thinking

IN FOCUS In question ❸, the numbers begin with those that can be simply partitioned using the value of the Roman numerals. Part c) requires children to recognise that two numerals (XC) are grouped together to represent 90. Question ❹ also requires Roman numerals to be grouped together to represent either larger parts (such as LXX) or because a smaller Roman numeral is needed before a larger one to show that the smaller value is subtracted (such as XL).

Questions ❼ and ❽ require children to reason about Roman numerals that can and cannot be before or after other numerals.

STRENGTHEN Encourage children to use the digit cards and part-whole models to help make decisions. They may find it useful to use this strategy to support thinking in question ❼, perhaps by first randomly trying a few digits and then trying to partition the numerals, checking if they make sense or not.

DEEPEN When children have completed all of the questions, ask them to make up a table to show the multiples of 250 from 0 to 2,000 in Roman numerals. Encourage them to look for patterns in their table.

ASSESSMENT CHECKPOINT Use question ❷ to assess whether children can write the multiples of 100 to 1,000 in Roman numerals, so they are able to build larger numbers confidently. Use questions ❹, ❺ and ❻ to check that children can convert between numbers and Roman numerals. Use question ❼ to check that children know the possible order that Roman numerals can be written in and why.

ANSWERS Answers for the **Practice** part of the lesson can be found in the *Power Maths* online subscription.

Reflect

WAYS OF WORKING Independent thinking

IN FOCUS The task revisits the Roman numerals M and D that were introduced in the main lesson, together with L. Children are required to give the value of each and explain how to find the number represented by MDXL.

ASSESSMENT CHECKPOINT Check that children know the value of each Roman numeral and can explain that to find the value of MDXL they need to subtract 10 from 50 because the X comes before the L.

ANSWERS Answers for the **Reflect** part of the lesson can be found in the *Power Maths* online subscription.

After the lesson ⏸

- Can children explain the use of the numeral C in the numbers MC, MCC and MCM?
- Can children write the current year in Roman numerals?
- Can they explain why we do not use the Roman numerals MDCCCC to represent 1,900?

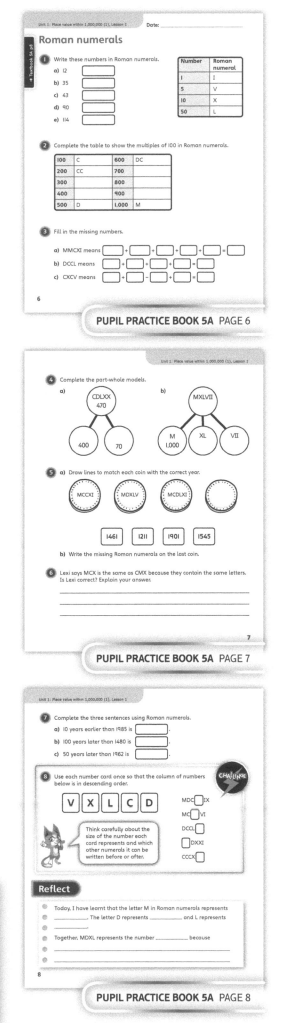

PUPIL PRACTICE BOOK 5A PAGE 6

PUPIL PRACTICE BOOK 5A PAGE 7

PUPIL PRACTICE BOOK 5A PAGE 8

Numbers to 10,000

Learning focus

In this lesson, children will revisit numbers to 10,000 with a focus on place value and count in 1,000s from different numbers. They will build numbers and break them down using what they know about the value of each digit and knowledge of zero as a place holder.

Before you teach

- Can children confidently explain the value of each digit in 3-digit numbers?
- Can they use place value to help them count on and back from any 4-digit number (such as 4,218, 4,219, 4,220)?

NATIONAL CURRICULUM LINKS

Year 5 Number – number and place value

Read, write, order and compare numbers to at least 1,000,000 and determine the value of each digit.

ASSESSING MASTERY

Children can say or write the value of each digit in numbers up to 10,000, representing them in different ways using manipulables, and can partition or build numbers using knowledge of 1,000s, 100s, 10s and 1s (explaining the role of zero as a place holder). Children can explain that when they count on or back in steps of 1,000, only the 1,000s value will change.

COMMON MISCONCEPTIONS

Children often forget to use zero to show when a place has no value; for example, recording 'three thousand and forty-five' as 345, forgetting to use 0 to show there are no 100s. Ask:
- *What is the same about the numbers 345 and 3,045? What is different?*

STRENGTHENING UNDERSTANDING

Give children pairs of 3- and 4-digit numbers to build with place value counters. Each pair should include at least two digits that are the same but have different values, for example 4,521 and 5,362. Ask them to explain what the value of digit 5 is in each number and then repeat this for digit 2.

GOING DEEPER

Ask children to represent 4-digit numbers in different ways; for example in numerals, words, partitioning, manipulables, money, and so on. Ensure that some of the numbers contain one or more zeros, for example 3,501, 6,020 and 4,500.

KEY LANGUAGE

In lesson: place value, digits, thousands (1,000s), hundreds (100s), tens (10s), ones (1s), zero (0), count on, count back

Other language to be used by the teacher: partition

STRUCTURES AND REPRESENTATIONS

Place value grid, part-whole model

RESOURCES

Mandatory: counters, dice

Optional: base 10 equipment

 In the eTextbook of this lesson, you will find interactive links to a selection of teaching tools.

Quick recap 🔎

Make a 3-digit number using counters on a place value grid. Ask children to draw the part-whole model to go alongside the grid and identify the value of each digit. Do the same with two or three numbers, making sure each example has a zero in it.

Discover

Unit 1: Place value within 1,000,000 (1), Lesson 2

Numbers to 10,000

Discover

WAYS OF WORKING Pair work

ASK

- Question **1** a): *What are the scores on the ducks that Lee has already hooked? What do you notice about them?*
- Question **1** a): *How many 1,000s can you see? How many 100s?*
- Question **1** b): *Read out Olivia's total score. How many 100s are in the 100s position?*
- Question **1** b): *What is this question asking you to find? What should you do first?*

IN FOCUS Question **1** b) requires children to first build the number that is shown by the six given ducks, reasoning about the number of 1,000s, 100s, 10s and 1s that can be seen. They need to recognise that the missing scores must change the value of the number to read 3,122.

PRACTICAL TIPS Put some place value counters into a bag to represent the ducks. Play a game with children where they need to take a number of counters from the bag and work out the total. Provide a large place value grid for children to place their counters on.

1 a) What is Lee's score?

   b) Olivia hooks eight ducks.
   Her total score is 3,122.
   What are the missing scores on the two ducks?

12

ANSWERS

Question **1** a): Lee's total score is 2,302.

Question **1** b): The missing scores on the two ducks are 1,000 and 10.

Share

Unit 1: Place value within 1,000,000 (1), Lesson 2

Share

I used a place value grid. There are no 10s, so I will use 0 as a placeholder in the 10s position.

a)
Lee has 2 thousands, 3 hundreds and 2 ones.

Th	H	T	O

Lee's score is 2,302.

b) You can see the following scores on Olivia's ducks.

Th	H	T	O

But Olivia's total score is 3,122.

This is made up of 3 thousands, 1 hundred, 2 tens and 2 ones.

Th	H	T	O

You need 1 more thousand and 1 more ten, so the missing scores on the two ducks are 1,000 and 10.

13

WAYS OF WORKING Whole class teacher led

ASK

- Question **1** a): *Do the place value counters on the grid match Lee's total of 2,302? How do you know?*
- Question **1** a): *The number 2,302 has four digits but it is only shown partitioned into three parts 2,000 + 300 + 2. Explain why.*
- Question **1** b): *What digit changes when you count on 1,000? Why do the other digits not change?*
- Question **1** b): *Why can the missing scores not be 100 and 1?*

IN FOCUS Question **1** a) uses place value counters and a place value grid to represent the scores on the ducks. It focuses on the use of zero as a place holder, ensuring that children recognise that the number being represented is 2,302, not 232. Partitioning of 2,302 using addition is also introduced to show the value of each digit as 2,000 + 300 + 2.

STRENGTHEN Ask children to read the numbers aloud together to identify the value of each digit. Record the numbers in words so that children see how the words reflect the numerals. In question **1** b), children could represent Olivia's total score on another place value grid and compare this with the image in the textbook, explaining what is the same and what is different about them. Look together at counting on 1,000 from 2,112, identifying the digits that change and the ones that stay the same.

Think together

WAYS OF WORKING Whole class teacher led (I do, We do, You do)

ASK

- Question **1** a): *How many 1,000s are there? How many 100s are there? How many 10s are there? What number has Bella made?*
- Question **1** b): *What does it mean when there are no counters in the hundreds column?*
- Question **2**: *Which representations show the value of the ones, tens, hundreds and thousands columns? Which do not? How can you use one representation to complete another?*

IN FOCUS In question **1**, children recap building numbers on a place value grid, immediately being reintroduced to the idea of zero as a place holder. It is important that children understand the value of each digit in the number. In question **2**, children see a variety of representations of the same number and use them to complete missing digits. They should recognise that as all representations show the same number, the number of, for example, ones in each is the same.

STRENGTHEN Provide children with a place value grid and other concrete resources to enable them to build the numbers and support their understanding. Building the numbers in base 10 equipment alongside place value counters can support children in understanding the value of the digits.

DEEPEN In question **3**, ask children for the smallest and greatest number that can be made. Do their answers change if they do not need to use all of the counters?

ASSESSMENT CHECKPOINT Use question **1** to assess children's understanding of the place value of each digit in a 4-digit number. Use question **2** to assess whether children are familiar with the different ways in which numbers can be represented.

ANSWERS

Question **1** a): Bella has made 4,043.

Question **1** b): The value of the 4 is 4 thousands, the value of the 0 is 0 hundreds, the value of the 4 is 4 tens, the value of the 3 is 3 ones.

Question **2**: The number is 5,214.

Question **3** a): 2,143

Question **3** b): 7,210, 4,222 and 5,302.

Question **3** c): A number of possible answers, e.g. 2,311.

Question **3** d): A number of possible answers, e.g. 1,432.

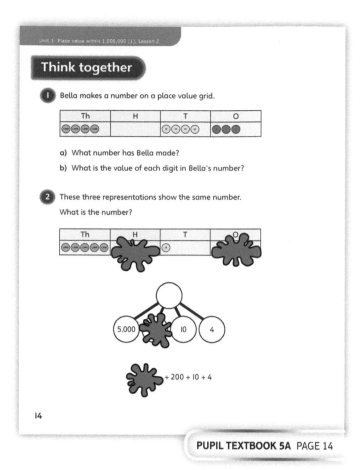

PUPIL TEXTBOOK 5A PAGE 14

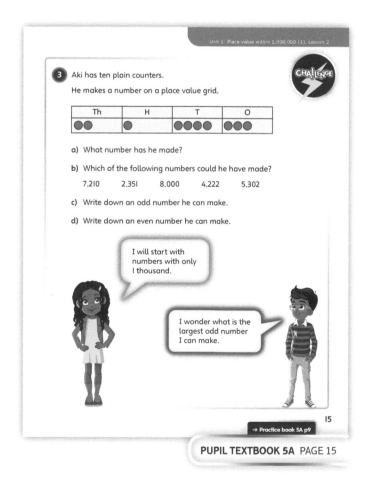

PUPIL TEXTBOOK 5A PAGE 15

Practice

WAYS OF WORKING Independent thinking

IN FOCUS In question ❶, children interpret numbers on place value grids and in a part-whole model. In parts c) and d), it is important they recognise that they need to use zero as a place holder. In question ❷, children are required to draw more place value counters to make a given number. Can they compare each column at a time to see what they need to add? Question ❸ checks understanding of place value of individual digits, whilst question ❹ exposes children to other representations where the parts are not necessarily given in value order.

STRENGTHEN Provide children with place value counters and other concrete resources to build the numbers in the question to support their understanding.

DEEPEN In question ❺, encourage children to explain their reasoning why they have arranged the numbers in that way. Add another digit card and ask them whether this changes any answers or if there is more than one answer for each question.

ASSESSMENT CHECKPOINT Use questions ❶ and ❷ to assess whether children can interpret numbers represented using place value counters and compare these to the numerical form. Use question ❸ to assess children's understanding of place value.

ANSWERS Answers for the **Practice** part of the lesson can be found in the *Power Maths* online subscription.

Reflect

WAYS OF WORKING Pair work

IN FOCUS The **Reflect** part of the lesson requires children to make 4-digit numbers by rolling 4 dice. They could work in pairs to do this. They should be challenged to find all the possible numbers. To make it more challenging, one of the pair shouldn't be able to see what numbers are rolled. Can their partner describe the number accurately enough, using place value language, so that their partner knows what numbers they rolled?

ASSESSMENT CHECKPOINT Children should be able to make and describe 4-digit numbers.

ANSWERS Answers for the **Reflect** part of the lesson can be found in the *Power Maths* online subscription.

After the lesson ❚❚

- Are children confident explaining the value of each digit in numbers up to 10,000?
- Can they represent numbers with manipulables, in words or using partitioning?
- Can they explain why 3,545 will not be in a count of 1,000 starting from 2,544?

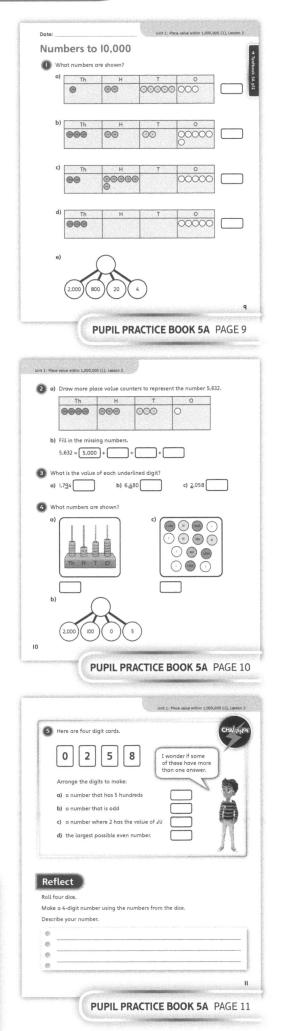

PUPIL PRACTICE BOOK 5A PAGE 9

PUPIL PRACTICE BOOK 5A PAGE 10

PUPIL PRACTICE BOOK 5A PAGE 11

Numbers to 100,000

Learning focus

In this lesson, children will work with numbers to 100,000, focusing on the position and value of each digit. They will represent numbers in different ways and break them down, explaining the value of each part.

Before you teach

- Can children confidently explain the value of each digit in a 4-digit number?
- Can they explain the difference in value between a digit that appears more than once in a number, for example 4,245?

NATIONAL CURRICULUM LINKS

Year 5 Number – number and place value

Read, write, order and compare numbers to at least 1,000,000 and determine the value of each digit.

ASSESSING MASTERY

Children can say or write the value of each digit in numbers up to 100,000 and represent them in different ways. They can partition or build numbers using knowledge of 10,000s, 1,000s, 100s, 10s and 1s, explaining the role of zero as a place holder.

COMMON MISCONCEPTIONS

When a number includes the same digit more than once (such as 45,435), children may not recognise that it is the position of the digit that is important. In this example, count across the columns, establishing that each column is 10 times larger than the previous one. Ask:

- *In the number 45,435, how many times smaller is the value of the digit 4 in the hundreds column than in the ten thousands column?*

STRENGTHENING UNDERSTANDING

Use place value counters on a part-whole model so that children can clearly see the value of each digit. First, ask them to write and represent the whole and then to break it up into parts to identify the value of each digit.

GOING DEEPER

Ask children to make up some clues about a 5-digit number for others to solve. First give some examples. For instance take the number 25,324: *The digit with the smallest value is 4. The digit with the largest value is 2. This digit has a value that is 1,000 times larger than the 10s digit. One of the digits has the value 300. The remaining digit has the value 5,000. What is the number?*

KEY LANGUAGE

In lesson: digit, place value, position, ten thousands (10,000s), thousands (1,000s), hundreds (100s), tens (10s), ones (1s), zero (0), multiple

Other language to be used by the teacher: partition

STRUCTURES AND REPRESENTATIONS

Place value grid, part-whole model, number line

RESOURCES

Mandatory: place value counters, base 10 equipment

 In the eTextbook of this lesson, you will find interactive links to a selection of teaching tools.

Quick recap 🔁

Make a 4-digit number using counters on a place value grid. Ask children to draw the part-whole model to go alongside the grid and identify the value of each digit. Repeat with two or three different numbers, making sure one of them has a zero in it.

Discover

WAYS OF WORKING Pair work

ASK

- Question ① a): *Where can you see how many passengers flew with the airline? How many digits are in the number? What is the value of the digit 2?*
- Question ① b): *What labels can you use to show the value of each position in the number?*
- Question ① b): *How many parts do you think you will need for a part-whole model?*

IN FOCUS The airline figures introduce children to 5-digit numbers, and question ① b) encourages them to use what they have learnt about 4-digit numbers to help partition a 5-digit number and write it in words.

PRACTICAL TIPS Ask children to read out the numbers on the poster that they are confident with. Talk together about the larger numbers, agreeing that the first column on the left is ten times larger than the thousands column to the right and has the value 10,000. Represent the difference using base 10 equipment or place value counters.

ANSWERS

Question ① a): The digit 7 represents 7 ten thousands, or 70,000.

Question ① b):

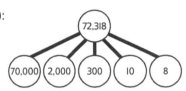

Seventy-two thousand, three hundred and eighteen

Share

WAYS OF WORKING Whole class teacher led

ASK

- Question ① a): *Which part tells you about the value of the digit 7? Which column is it in?*
- Question ① b): *How does using the part-whole model help you to say and write the number?*

IN FOCUS Question ① a) requires children to think about the place value of each digit in the 5-digit number. In question ① b), children are asked to draw a part-whole model for a 5-digit number. They should use the part-whole model to help say and write the number in words.

STRENGTHEN Encourage children to read the numbers in full, for example 72,318 as seventy-two thousand, three hundred and eighteen and not simply as digits 7-2-3-1-8, so that the place value is made explicit.

Numbers to 100,000

Discover

① a) Make the number of passengers on a place value grid. What does the digit 7 represent?

b) Represent the number of passengers on a part-whole model. Now write the number of passengers in words.

16

PUPIL TEXTBOOK 5A PAGE 16

Share

a) 72,318 passengers flew with the airline.

I started by representing the number on a place value grid. Then I could see which column the 7 was in.

TTh	Th	H	T	O

TTh stands for ten thousands.

The digit 7 is in the ten thousands position.

The digit 7 represents 7 ten thousands or 70,000.

b) The airline had 72,318 passengers.

I drew a part-whole model with five parts.

In words, this is seventy-two thousand, three hundred and eighteen.

17

PUPIL TEXTBOOK 5A PAGE 17

Think together

ASK

- Question ❶ a): *How many 10,000s are there? How many 1,000s are there? How many 100s are there? How many 10s are there? How many 1s are there? Does it matter which part you write each number in?*
- Question ❶ b): *What is the same about the part-whole model and the place value grid? What is different? How can you use the part-whole model to tell you what the digit 3 represents?*
- Question ❷: *What does it mean when there are no counters in the tens column?*
- Question ❸ a): *Could you group some of the counters differently but keep them in the same columns? How does this help find the missing numbers?*

IN FOCUS In question ❶, children partition a 5-digit number using a part-whole model and then represent it in a place value grid. They should recognise that whilst they need to include the zeros to show place value in the part-whole model, for example in 200, they do not need to do this in the place value grid. In question ❷, children interpret each digit of a number. They can draw a part-whole model to support them. They also see zero used as a placeholder in the tens column.

STRENGTHEN Provide children with a place value grid and place value counters so they can make the numbers they are interpreting. Encourage them to count out loud in multiples of 100 to identify the value of a given digit.

DEEPEN In question ❸ a), can children explore other ways to partition 43,245 into two parts, three parts, four parts, and so on? In question ❸ b), encourage children to explain how they found the size of the second jump on the number line.

ASSESSMENT CHECKPOINT Use questions ❶ and ❷ to assess whether children can identify the place value of any digit in a 5-digit number. Use question ❸ to assess whether children can partition 5-digit numbers flexibly.

ANSWERS

Question ❶ a):

Question ❶ b):

TTh	Th	H	T	O

Thirteen thousand, two hundred and fifty-eight. The value of the digit 3 is 3,000.

Question ❷ a): 45,206

Question ❷ b): The value of 4 is forty thousands, the value of 5 is five thousands, the value of 2 is two hundreds, the value of 0 is zero tens and the value of 6 is six ones.

Question ❷ c): Forty-five thousand, two hundred and six

Question ❸ a): 40,000 + 3,000 + 200 + 40 + 5
30,000 + 10,000 + 100 + 100 + 40 + 5
43,000 + 200 + 45

Question ❸ b): 23, 407 = 20,000 + 3000 + 400 + 7

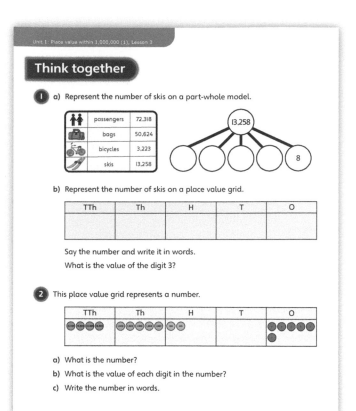

Think together

❶ a) Represent the number of skis on a part-whole model.

b) Represent the number of skis on a place value grid.

TTh	Th	H	T	O

Say the number and write it in words.
What is the value of the digit 3?

❷ This place value grid represents a number.

TTh	Th	H	T	O

a) What is the number?
b) What is the value of each digit in the number?
c) Write the number in words.

18

PUPIL TEXTBOOK 5A PAGE 18

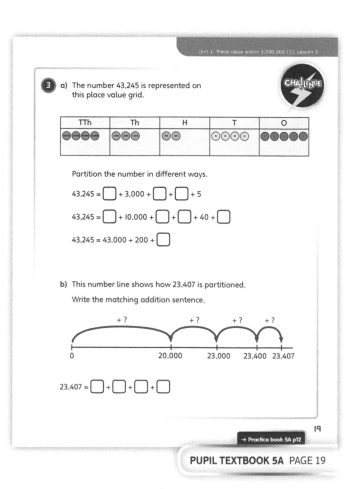

Unit 1: Place value within 1,000,000 (1), Lesson 3

❸ a) The number 43,245 is represented on this place value grid.

CHALLENGE

TTh	Th	H	T	O

Partition the number in different ways.

43,245 = ☐ + 3,000 + ☐ + ☐ + 5

43,245 = ☐ + 10,000 + ☐ + ☐ + 40 + ☐

43,245 = 43,000 + 200 + ☐

b) This number line shows how 23,407 is partitioned.
Write the matching addition sentence.

23,407 = ☐ + ☐ + ☐ + ☐

19

→ Practice book 5A p12

PUPIL TEXTBOOK 5A PAGE 19

Practice

WAYS OF WORKING Independent thinking

IN FOCUS Question ❶ b) emphasises the fact that a digit's value depends on its position by swapping two of the digits and asking children to write the new number. Question ❷ also emphasises this link by requiring children to match each digit in a number with its value.

Question ❻ demands flexible reasoning as children have to find the missing digits that will allow numbers to be made to suit all of the given criteria.

STRENGTHEN Encourage children to represent the numbers in question ❷ on a place value grid or with counters to help them make decisions about the value of the digit 4 in each example.

DEEPEN When children have completed question ❻, ask them to identify two digits that Max could not have used. Then challenge them to change the clues so the solution uses these two digits.

ASSESSMENT CHECKPOINT Use questions ❶ and ❷ to assess whether children can confidently recognise and explain the value of each digit in numbers up to 100,000. Use questions ❸ and ❹ to assess whether children understand how 5-digit numbers can be represented in place value grids and part-whole models. Use question ❺ to assess whether they understand what happens to a number when you add or subtract a multiple of 100, 1,000 or 10,000.

ANSWERS Answers for the **Practice** part of the lesson can be found in the *Power Maths* online subscription.

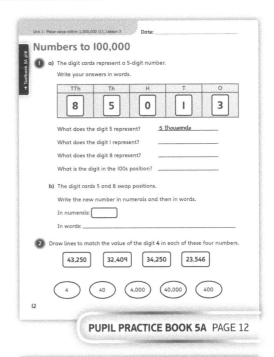

PUPIL PRACTICE BOOK 5A PAGE 12

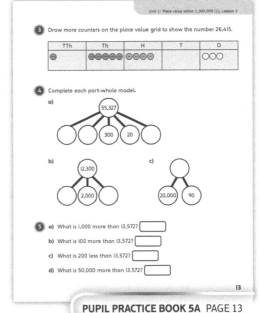

PUPIL PRACTICE BOOK 5A PAGE 13

Reflect

WAYS OF WORKING Independent thinking

IN FOCUS Children are required to use what they know about place value in a 5-digit number. They could use a place value grid or a part-whole model to help them identify the value of each digit.

ASSESSMENT CHECKPOINT Check that children can correctly recognise the place value of each digit.

ANSWERS Answers for the **Reflect** part of the lesson can be found in the *Power Maths* online subscription.

After the lesson ▮▮

- Can children confidently explain the value of each digit in numbers up to 100,000?
- Can they represent numbers with manipulatives, in words and using part-whole models?

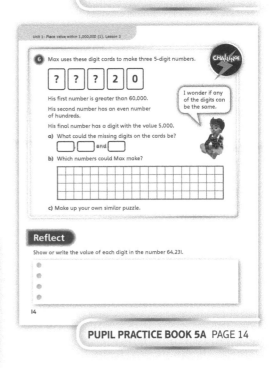

PUPIL PRACTICE BOOK 5A PAGE 14

Numbers to 1,000,000

Learning focus

In this lesson, children will develop their understanding of place value up to the 100,000s, learning to read and write numbers accurately. They will count in steps of 100,000, 10,000, 1,000, 100, 10 and 1.

Before you teach

- How confident are children in their use of a place value grid up to 1,000s?
- What concrete representations will you use to ensure children's deep conceptual understanding of the place value of 6-digit numbers?

NATIONAL CURRICULUM LINKS

Year 5 Number – number and place value

Read, write, order and compare numbers to at least 1,000,000 and determine the value of each digit.

ASSESSING MASTERY

Children can use their knowledge of place value to recognise and name numbers. They are able to count fluently in regular steps of 100,000, 10,000, 1,000, 100, 10 and 1.

COMMON MISCONCEPTIONS

Children may misspell the mathematical vocabulary when trying to write the names of numbers. It may be beneficial to have the vocabulary displayed prominently on the classroom wall for children to refer to. Ask:
- *How do you write 1,000 in words? What about 12,000? Now try writing 400,000 in words.*

STRENGTHENING UNDERSTANDING

Provide opportunities to investigate the place value of numbers up to 100,000s before the lesson. This could be achieved by encouraging children to use base 10 equipment to make numbers, or counters on a place value grid and partition the numbers.

GOING DEEPER

Children could be encouraged to investigate what happens when they add another 100,000 to a number they have created or been given. Ask: *What happens if you add 100,000 to 15,678? Use resources or a picture to show what happens when you add 10,000 to 345,987.*

KEY LANGUAGE

In lesson: digit, ones (1s), tens (10s), hundreds (100s), thousands (1,000s), ten thousands (10,000s), hundred thousands (100,000s), place value

STRUCTURES AND REPRESENTATIONS

Place value grid

RESOURCES

Mandatory: counters

Optional: base 10 equipment

 In the eTextbook of this lesson, you will find interactive links to a selection of teaching tools.

Quick recap

Ask children to try and work out your number. For example, *My number has five digits. It has 3 hundreds, 2 more thousands than hundreds, and the same number of 10s and 1s. All my digits add up to 18. What could my number be?*

Discover

WAYS OF WORKING Pair work

ASK

• Question ① a): *How many sweets does one container hold? How could you represent your thinking? How many sweets would seven containers hold? What about eight? Can you spot any patterns?*
• Question ① b): *What could you use to clearly represent the place value of each digit in the number?*

IN FOCUS Question ① a) gives children an opportunity to begin counting in regular steps of 100,000. Encourage children to predict how the numbers will change with a different number of containers. Question ① b) gives children their first opportunity to investigate the place value of 6-digit numbers.

PRACTICAL TIPS Place value counters could be used to represent the containers of sweets and to help children count up and down in 100,000s, 10,000s and 1,000s.

ANSWERS

Question ① a): Six containers can hold 600,000 sweets.

Question ① b): The board says that 461,905 sweets have been made today.

There are 4 hundred thousands.
There are 6 ten thousands.
There is 1 thousand.
There are 9 hundreds.
There are 0 tens and 5 ones.

Numbers to 1,000,000

Discover

① a) How many sweets can six containers hold?
   b) How many sweets have been made today?
      What does each digit in the number represent?

20

PUPIL TEXTBOOK 5A PAGE 20

Share

WAYS OF WORKING Whole class teacher led

ASK

• Question ① a): *What is each container worth? How does knowing that help you count? What resources did you find useful when representing 100,000? Could they help you to predict how many sweets seven containers would contain, or eight containers?*
• Question ① b): *How is this place value grid different from the ones you have used before?*

IN FOCUS When looking at question ① b), it will be essential to allow children the time to consider how the place value grid is different. Ask children to discuss the headings of the place value grid and, if necessary, remind them of the meanings of the headings they have seen before. Ask children what the new heading could mean and ensure they understand that 4 counters in the hundred thousands column means there are 4 hundred thousands.

Share

Each container holds 100,000 sweets. I counted up in 100,000s.

a) Six containers can hold 600,000 sweets.

0 100,000 200,000 300,000 400,000 500,000 600,000

b) The board says that 461,905 sweets have been made today.

HTh	TTh	Th	H	T	O
4	6	1	9	0	5

HTh in the place value grid means hundred thousands.

There are 4 hundred thousands.
There are 6 ten thousands.
There is 1 thousand.
There are 9 hundreds.
There are 0 tens and 5 ones.

I know that in words this number is four hundred and sixty-one thousand, nine hundred and five.

21

PUPIL TEXTBOOK 5A PAGE 21

Think together

WAYS OF WORKING Whole class teacher led (I do, We do, You do)

ASK

- Question **1**: *How can you represent the containers of sweets? How does your representation help you to identify the number of sweets accurately?*
- Question **2**: *What mathematical vocabulary will you need to use to write the number in words?*
- Question **3**: *What mathematical structure will help you to identify the value of each 5?*

IN FOCUS In question **1**, children will develop their ability to count in steps of 100,000. They should be encouraged to use resources to support their counting. Question **2** will develop children's fluency with the written names of numbers up to the 100,000s. Encouraging children to use place value grids to support them with question **3** will develop their understanding of the place value of 6-digit numbers while helping to develop their fluency with the new place value grid they have met in the lesson.

STRENGTHEN If children need help writing the names of the numbers in question **2**, have the key mathematical vocabulary available to them in written form. This could be as flash cards, a wall display or number lines that show both numerals and written number names.

DEEPEN Use the question after **3** d) to deepen children's understanding of the number concepts they have covered in the lesson so far. Ask: *Is it easier to read or to write a 4-, 5- or 6-digit number? Explain why.* Challenge children to write a 6-digit number that is quicker to write in words than a 4-digit number and ask them how it is possible.

ASSESSMENT CHECKPOINT Can children count in steps of 100,000? Do children use the correct mathematical vocabulary when reading and writing the numbers in question **2**? In question **3**, look out for clear understanding that the 5 in each number is worth a different amount.

ANSWERS

Question **1**: There are 900,000 sweets.

Question **2** a): Seven hundred and twenty-eight thousand, six hundred and eleven

Question **2** b): 370,938

Question **3** a): 50,000

Question **3** b): 5

Question **3** c): 500,000

Question **3** d): 500

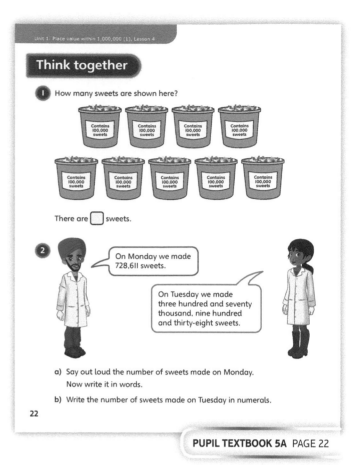

Think together

1 How many sweets are shown here?

There are ☐ sweets.

2 On Monday we made 728,611 sweets.

On Tuesday we made three hundred and seventy thousand, nine hundred and thirty-eight sweets.

a) Say out loud the number of sweets made on Monday. Now write it in words.

b) Write the number of sweets made on Tuesday in numerals.

22

PUPIL TEXTBOOK 5A PAGE 22

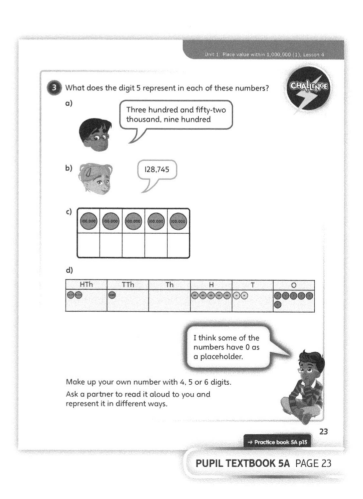

3 What does the digit 5 represent in each of these numbers?

a) Three hundred and fifty-two thousand, nine hundred

b) 128,745

c)

d)

HTh	TTh	Th	H	T	O

I think some of the numbers have 0 as a placeholder.

Make up your own number with 4, 5 or 6 digits. Ask a partner to read it aloud to you and represent it in different ways.

23

→ Practice book 5A p15

PUPIL TEXTBOOK 5A PAGE 23

Practice

WAYS OF WORKING Independent thinking

IN FOCUS Question ② allows children to demonstrate their ability to count in steps of 100,000 up to 1,000,000. Questions ③ and ⑤ provide an opportunity for children to practise reading and writing the key vocabulary from the lesson, using different pictorial and abstract representations.

STRENGTHEN To strengthen understanding of writing the numbers in question ③ in words, it may help to use place value counters on a place value grid to build the numbers first. Ask: *What does each digit represent? How would you write the name of the number that digit is worth?*

DEEPEN Question ⑦ deepens children's understanding of place value by requiring them to use their reasoning ability to demonstrate flexibility with place value. Ask: *Are there only five possible solutions? How can you prove you have found all of the possibilities?*

THINK DIFFERENTLY In question ⑥, children have an opportunity to further develop their understanding of place value. The question requires children to recognise that although the numbers might have some digits in common and in the same order, the value of the 4 in each number may still be different.

ASSESSMENT CHECKPOINT Can children accurately read and write 4-, 5- and 6-digit numbers in words? In question ⑥, can children identify the correct value of the 4 in each number? Do they recognise that the 4 in 240 and 500,240 is worth the same, but the 4 in 314,912 and 77,314 is not?

ANSWERS Answers for the **Practice** part of the lesson can be found in the *Power Maths* online subscription.

Reflect

WAYS OF WORKING Pair work

IN FOCUS This question offers children a final opportunity to practise reading and writing 6-digit numbers.

ASSESSMENT CHECKPOINT Look particularly for children's understanding of the hundred thousands column that has been introduced in this lesson.

ANSWERS Answers for the **Reflect** part of the lesson can be found in the *Power Maths* online subscription.

After the lesson ⏸

- Can children confidently identify the correct place value of the digits in 4-, 5- and 6-digit numbers?
- Are there any written words that children need to become more fluent with?
- How can you support children's future use of the written mathematical vocabulary?

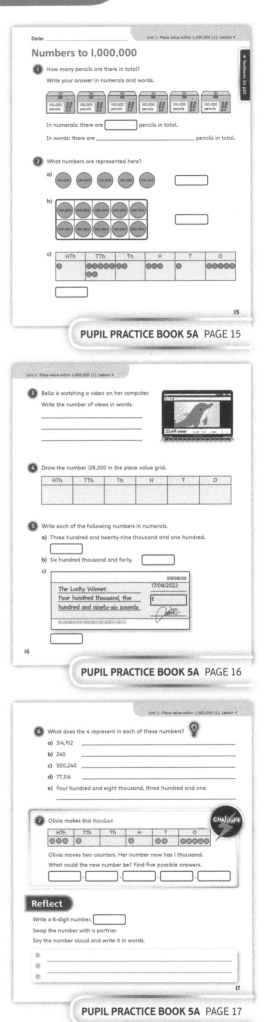

PUPIL PRACTICE BOOK 5A PAGE 15

PUPIL PRACTICE BOOK 5A PAGE 16

PUPIL PRACTICE BOOK 5A PAGE 17

Read and write 5- and 6-digit numbers

Learning focus

In this lesson, children will learn to write 5- and 6-digit numbers using both numerals and words. They will focus on the correct positioning of commas in numbers and understand that this helps them to accurately write numbers in words.

Before you teach

- Can children partition 5- and 6-digit numbers?
- Can children identify the value of a given digit in a 5- and 6-digit number?
- Do children understand what it means when there is a zero in a number?

NATIONAL CURRICULUM LINKS

Year 5 Number – number and place value

Read, write, order and compare numbers to at least 1,000,000 and determine the value of each digit.

ASSESSING MASTERY

Children can write 5- and 6-digit numbers in numerals, placing the comma in the correct place in the numbers. Children can recognise and write numbers written in words.

COMMON MISCONCEPTIONS

Children sometimes think that the comma always comes after the first digit in a number. Ask:
- *How many digits are there in the number? Which digits should have a comma after them?*

They may think that they can write a number in words by simply writing out each digit as a word. Ask:
- *What place value does each digit have? What can you use to help you write the number in words?*

STRENGTHENING UNDERSTANDING

If children are struggling, encourage them to use their knowledge of partitioning from earlier in the unit to help them. By writing numbers in a part-whole model and reading each part separately, they can then use their knowledge of the place value of the digits to support them in writing numbers.

GOING DEEPER

Provide children with numbers represented in different ways and ask them to write the numbers in numerals and words. You could link back to earlier learning on flexible partitioning to challenge them further with this.

KEY LANGUAGE

In lesson: numerals, words, digit

Other language to be used by the teacher: partition

STRUCTURES AND REPRESENTATIONS

Part-whole model, place value grid

RESOURCES

Mandatory: place value grid, dice

Optional: laminated part-whole model, place value counters

 In the eTextbook of this lesson, you will find interactive links to a selection of teaching tools.

Quick recap

Roll a dice four times. Ask children to make a 4-digit number. Ask them to write their number in words. What is the biggest number the class can make? What is the smallest?

Discover

WAYS OF WORKING Pair work

ASK

- Question ① a): *What digit cards has Ambika got? What number has she made? What other digit cards are near her? What numbers could she make? What if she uses all the cards? What numbers can she make?*
- Question ① b): *If a digit is added to the start of a number, what happens to all the other digits? Do they change or stay the same?*

IN FOCUS In question ① a), children practise writing a 4-digit number in words and then read it aloud. In question ① b), they then explore what happens to the value of the digits in the number and the number itself if further digits are added to the start of the number.

PRACTICAL TIPS Provide children with some place value counters and let them arrange them on a place value grid.

ANSWERS

Question ① a): Three thousand, two hundred and forty-five (3,245)

Question ① b): Richard makes thirteen thousand, four hundred and twenty-five (13,425); Ambika makes seven hundred and thirteen thousand, four hundred and twenty-five (713,245).

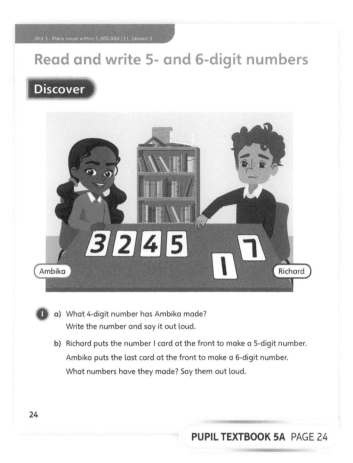

Read and write 5- and 6-digit numbers

Discover

① a) What 4-digit number has Ambika made?
Write the number and say it out loud.

b) Richard puts the number 1 card at the front to make a 5-digit number.
Ambika puts the last card at the front to make a 6-digit number.
What numbers have they made? Say them out loud.

24

PUPIL TEXTBOOK 5A PAGE 24

Share

WAYS OF WORKING Whole class teacher led

ASK

- Question ① a): *What is the value of each digit? How can you see that in the numerals? How can you see it in the words?*
- Question ① b): *Why is the comma always after the 3, even when more digits are added to the start of the number?*

IN FOCUS Here, children see the numbers Ambika and Richard have made in both numerals and words. They should make clear connections between the numeric form and how it is represented in words.

PUPIL TEXTBOOK 5A PAGE 25

Think together

WAYS OF WORKING Whole class teacher led (I do, We do, You do)

ASK

• Question **1**: *Which number did you find hardest to say? Which number did you find easiest to say? Why? What was your strategy?*

• Question **2**: *What happens if you add a digit at the start? What happens if you add a digit at the end? How many different numbers can you make?*

IN FOCUS In question **1**, children practise reading aloud 5- and 6-digit numbers. They should focus on how the positioning of the comma in the numbers helps them when reading numbers aloud. In question **2**, they explore making 5- and 6-digit numbers and what happens to them when other digits are added to the start or the end.

STRENGTHEN Provide children with a part-whole model or a place value grid to support them in identifying, reading and writing the value of each digit.

DEEPEN Ask children to explain why adding a digit at the start of a number does not change the value of the other digits, but adding a digit to the end of the number does. In question **3**, ask them to explain how the second place value grid helps them to read and write numbers.

ASSESSMENT CHECKPOINT Use question **1** to assess whether children can read aloud 5- and 6-digit numbers. Ensure they do not read numbers such as 89,995 as 'eighty-nine, nine, nine, five' but instead as 'eighty-nine thousand, nine hundred and ninety-five'. Use question **2** to assess whether children can confidently read and write 5- and 6-digit numbers and understand the effect that adding in other digits has.

ANSWERS

Question **1**: Children read the numbers: eighty-nine thousand, nine hundred and ninety five; one hundred and twenty thousand, seven hundred and fifty; one hundred and ninety-nine thousand, nine hundred and ninety-nine.

Question **2**: Children's answers may vary, e.g. 3,003, 30,030, 300,300.

Question **3** a): The digits are in the same place, but the column headings are grouped differently.

Question **3** b): Children's answers will vary, e.g. 450,900.

PUPIL TEXTBOOK 5A PAGE 26

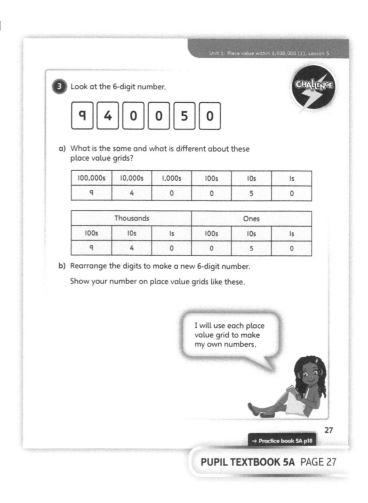

PUPIL TEXTBOOK 5A PAGE 27

Practice

WAYS OF WORKING Independent thinking

IN FOCUS In questions ❶ and ❷, children focus on the correct positioning of the comma within numbers, and should recognise that it comes after the thousands digit each time. In question ❸, children need to match the numbers written in words to those written in numerals and should recognise how the value of each digit can be seen in each representation.

STRENGTHEN Provide children with a part-whole model or a place value grid to support them in identifying, reading or writing the value of each digit.

DEEPEN In question ❹, encourage children to explain why they cannot use a 0 as the first digit in the number.

THINK DIFFERENTLY In question ❺, children need to think strategically to decide which calculation they will work with first.

ASSESSMENT CHECKPOINT Use questions ❶ and ❷ to assess whether children can correctly position a comma in numbers with four or more digits. Use question ❸ to assess whether children can recognise numbers written in words and match them to their numeric form.

ANSWERS Answers for the **Practice** part of the lesson can be found in the *Power Maths* online subscription.

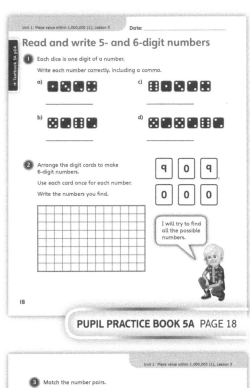

PUPIL PRACTICE BOOK 5A PAGE 18

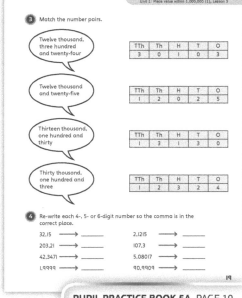

PUPIL PRACTICE BOOK 5A PAGE 19

Reflect

WAYS OF WORKING Independent thinking

IN FOCUS Here, children consider the different numbers that it could be depending on which digit is obscured. Can they find each possible answer?

ASSESSMENT CHECKPOINT Children should recognise that the number definitely has 5 ten thousands, 0 tens and 3 ones and give a possible number by assigning digits to the thousands and hundreds columns.

ANSWERS Answers for the **Reflect** part of the lesson can be found in the *Power Maths* online subscription.

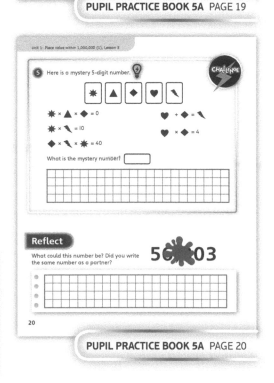

PUPIL PRACTICE BOOK 5A PAGE 20

After the lesson ⏸

- Can children correctly place the comma in numbers with four or more digits?
- Can children read numbers aloud correctly?
- Can children match numbers written in words to their numeric form?

Powers of 10

Learning focus

In this lesson, children focus on different powers of 10, such as 100 and 1,000, and consider how many other powers of 10 make up these numbers or any multiple of them. For example, they should already know that there are 10 hundreds in 1,000, and they will use this to find how many hundreds there are in a 4-digit number.

Before you teach ⏸

- Can children read 6-digit numbers aloud?
- Do children know the value of each digit in a 6-digit number?
- Do children know common equivalences, such as there being 10 tens in 100?

NATIONAL CURRICULUM LINKS

Year 5 Number – number and place value

Count forwards or backwards in steps of powers of 10 for any given number up to 1,000,000.

ASSESSING MASTERY

Children can identify how many 10s/100s/1,000s there are in different multiples of 100 and 1,000. They understand the relationship between 1s, 10s, 100s, 1,000s, 10,000s and 100,000s and can use this understanding to find other facts.

COMMON MISCONCEPTIONS

Children may think that numbers such as 14,000 can only be made using thousands because there are 14 thousands. They might think there are no hundreds in 14,000 because there is a zero in the hundreds column. Ask:

- *What does each zero mean? What are the zeros place holders for?*

STRENGTHENING UNDERSTANDING

Provide children with a place value grid and counters to make some of the smaller numbers and physically make exchanges to support them in working out their answers.

GOING DEEPER

Encourage children to explain the connection between their answers and use known facts to find other facts.

KEY LANGUAGE

In lesson: thousands (1,000s), hundreds (100s), tens (10s), ones (1s)

Other language to be used by the teacher: exchange, place value, ascending, descending

STRUCTURES AND REPRESENTATIONS

Place value grid, part-whole model

RESOURCES

Mandatory: place value grid, counters

Optional: laminated part-whole model, place value counters, play money (notes)

 In the eTextbook of this lesson, you will find interactive links to a selection of teaching tools.

Quick recap 🔁

Practise counting in 10s, 100s and 1,000s as a class. Start with the number 1,000 and count on in 1,000s for a while. Then start at 1,000 and count in 100s and then in 10s and 1s.

Discover

WAYS OF WORKING Pair work

ASK

- Question **1** a): *How much money does Luis have in total? What type of notes does he have? What calculation do you need to do to work out how many notes he has in total?*
- Question **1** b): *How many $1,000 notes make up a $10,000 note? How do you know? How many $100 notes make up a $1,000 note?*

IN FOCUS Here, children are introduced to powers of 10 in a common context that they are likely to be familiar with. Children will think about making different amounts of money using certain notes, and this could be scaffolded by linking to their own currency if needed.

PRACTICAL TIPS Provide children with paper money and ask them to make the amounts, counting aloud as they do so. Get them to ask a partner for a certain amount using only particular notes and work together to find how many notes they need.

ANSWERS

Question **1** a): Luis has 12 $10,000 notes.

Question **1** b): Bella has 120 $1,000 notes.

Powers of 10

Discover

1 a) Luis has $120,000 Singaporean dollars.
All his money is in $10,000 notes.
How many notes does he have?

b) Bella has the same amount of Singaporean dollars as Luis.
All her money is in $1,000 notes.
How many notes does she have?

28

PUPIL TEXTBOOK 5A PAGE 28

Share

WAYS OF WORKING Whole class teacher led

ASK

- Question **1** a): *Why does the place value grid show one 100,000 counter and two 10,000 counters? How many 10,000 counters did Dexter exchange his 100,000 counter for? Why? How did this help?*
- Question **1** b): *Why is 120,000 equal to 12 × 10,000 counters? Why do we multiply the 12 by 10? Can you show that this method is correct?*

IN FOCUS Children look at different methods for working out the number of notes needed. Ask them to discuss Flo's comment and work with a partner to show what she means. Ask them why this is more efficient and why they would not want to make that many exchanges physically.

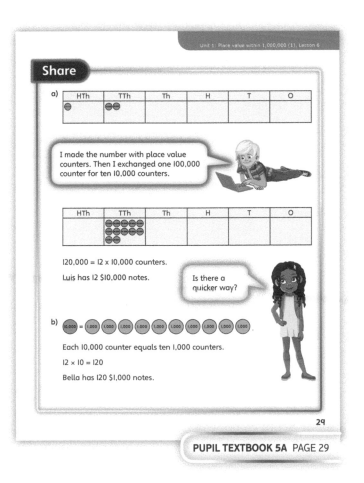

Share

a)

HTh	TTh	Th	H	T	O

I made the number with place value counters. Then I exchanged one 100,000 counter for ten 10,000 counters.

HTh	TTh	Th	H	T	O

120,000 = 12 × 10,000 counters.

Luis has 12 $10,000 notes.

Is there a quicker way?

b)

Each 10,000 counter equals ten 1,000 counters.

12 × 10 = 120

Bella has 120 $1,000 notes.

29

PUPIL TEXTBOOK 5A PAGE 29

Think together

WAYS OF WORKING Whole class teacher led (I do, We do, You do)

ASK

- Question **1** a): *How do you say Emma's number out loud? How does this help you find how many 1,000s there are?*
- Question **1** b) and c): *How many 100s are there in 1,000? How many 1,000s are there in 26,000? How can you use this to find how many 100s there are in 26,000?*

IN FOCUS In question **1**, children look at building a number using only 1,000s and then only 100s. They could make the numbers physically for question **1** a) but should recognise that this would take too long for question **1** b). Refer them back to Flo's method in **Share**. In question **2**, children use the fact that 10 hundreds are equal to 1 thousand to work out how many 100s there are in different numbers.

STRENGTHEN Provide children with a place value grid and counters when working with the smaller numbers to help them find the answers. Encourage them to spot patterns in their working and recognise why this would not be an efficient method for larger numbers, such as 93,000.

DEEPEN In question **3**, ask children to explain the connection between the different columns in Mo's table. What do they notice? Why does this happen? Ask them to explore what Ash and Flo are saying and try it for different numbers.

ASSESSMENT CHECKPOINT Use questions **1** and **2** to assess whether children can find the number of 1,000s and 100s in given numbers.

ANSWERS

Question **1** a): 2

Question **1** b): 6

Question **1** c): 0

Question **2** a): 4

Question **2** b): 9

Question **2** c): 0

Question **3** a):

100s	10s	1s
5,200	52,000	520,000

Think together

1 Emma has made the number 26,000.

TTh	Th	H	T	O

a) How many 10,000s are in 26,000?

b) How many 1,000s are in 26,000?

c) How many 100s are in 26,000?

2 How many 100s are there in each of these numbers?

a)

Th	H	T	O

b) 4,900

c) 93,000

There must be a way of working these out without using counters.

30

PUPIL TEXTBOOK 5A PAGE 30

3 I want to see how many 10s, 100s, 1,000s and 10,000s go into 6-digit numbers.

Mo

CHALLENGE

Mo makes this table.

Number	How many?				
	10,000s	1,000s	100s	10s	1s
520,000	52	520			

a) Complete Mo's table.

b) Discuss with a partner what you notice.

I wonder if this works for all 6-digit numbers.

What about 1,000,000?

31

→ Practice book 5A p21

PUPIL TEXTBOOK 5A PAGE 31

Practice

Independent thinking

IN FOCUS In question **1**, children use their knowledge of 10 hundreds being equal to 1,000 to find how many 100s there are in other numbers. In question **2**, they use the fact that 10 thousands are in 10,000 to find how many 1,000s there are in other numbers. In questions **3** and **4**, children look at building the same number using different counters and different exchanges. Question **5** encourages children to think about it the other way round; identifying the number from a given number of hundreds.

STRENGTHEN Provide children with a place value grid and counters when working with the smaller numbers to help them find the answers. In question **5**, they could draw or make the 100s and show the exchanges to support them.

DEEPEN Ask children to explain any patterns or connections they notice and why they happen. For example, there are ten times as many 100s as there are 10s in a number because there are 10 tens in 100.

ASSESSMENT CHECKPOINT Use questions **1** and **2** to assess whether children can work out the number of 10s and 1,000s in a number. Use question **3** to assess whether children can work out how many of any multiple of 10 there are in a number. Use question **5** to assess whether children can apply this in reverse to find a number, when given the number of 100s.

ANSWERS Answers for the **Practice** part of the lesson can be found in the *Power Maths* online subscription.

Reflect

Independent learning

IN FOCUS Here, children continue a sequence working with powers of 10. They should consider what is happening between each pair of numbers and use this to continue the sequence.

ASSESSMENT CHECKPOINT Children recognise that the numbers in the sequence are being multiplied by 10 each time. They use this information to find the missing numbers.

ANSWERS Answers for the **Reflect** part of the lesson can be found in the *Power Maths* online subscription.

After the lesson ⏸

- Can children identify the number of 1,000s that make up a number?
- Can children identify the number of 100s that make up a number?
- Can children identify the number of any power of 10 that make up a number?

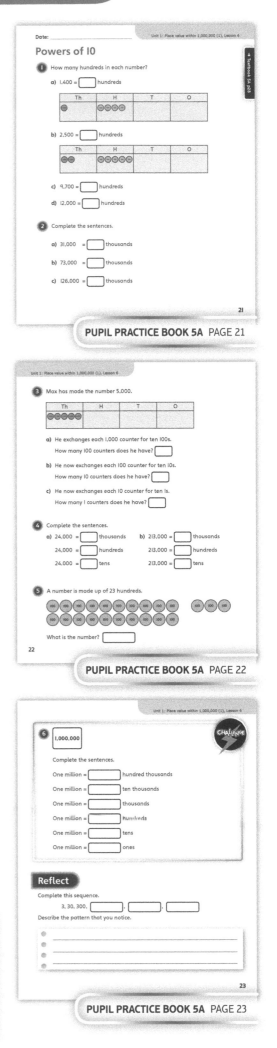

PUPIL PRACTICE BOOK 5A PAGE 21

PUPIL PRACTICE BOOK 5A PAGE 22

PUPIL PRACTICE BOOK 5A PAGE 23

10/100/1,000/10,000/100,000 more or less

Learning focus

In this lesson, children will develop their ability to count forwards and backwards in steps of 10, 100, 1,000 and 10,000.

Before you teach

- Which of the numbers in this lesson are likely to be tricky for some children?
- How will you scaffold their concrete understanding of those numbers?
- Can children name each column in the extended place value grid (up to 100,000s)?

NATIONAL CURRICULUM LINKS

Year 5 Number – number and place value

Count forwards or backwards in steps of powers of 10 for any given number up to 1,000,000.

ASSESSING MASTERY

Children can reliably and fluently count forwards and backwards in steps of 10, 100, 1,000 and 10,000 from a given number. Children can find 10, 100, 1,000 and 10,000 more or less than any given number, including the times where they may need to exchange (for example, 10 less than 7,000).

COMMON MISCONCEPTIONS

Children may struggle to bridge into the next place value column when adding 10, 100, 1,000 or 10,000. For example, when adding 100 to 900 they may say 'ten hundred'. Use base 10 equipment to demonstrate how the numbers change when adding 10, 100, 1,000 and 10,000. In particular, demonstrate how 10 hundreds is the same as 1 thousand. Ask:

- *What is the correct name of the number '10 hundreds'? What number comes after 9,999? What number do you get if you add 10,000 to 90,000?*

STRENGTHENING UNDERSTANDING

Provide children with place value counters and encourage them to make a number. Ask them to use the counters to add or subtract 10, 100, 1,000 or 10,000 to or from their number. Ask: *How did your number change? How did it stay the same? What is your new number?*

GOING DEEPER

Children could create inverse puzzles for each other. For example: *My final number was 550,000. To get to 550,000, I added 10,000, subtracted 100, added 1,000 and added another 10,000. What was my starting number?*

KEY LANGUAGE

In lesson: place value, sequence, table

Other language to be used by the teacher: ones (1s), tens (10s), hundreds (100s), thousands (1,000s), ten thousands (10,000s), hundred thousands (100,000s)

STRUCTURES AND REPRESENTATIONS

Place value grid

RESOURCES

Mandatory: place value counters

Optional: base 10 equipment, printed place value grids

 In the eTextbook of this lesson, you will find interactive links to a selection of teaching tools.

Quick recap

Ask children to make a 4-digit number in a place value grid using six counters, for example, 1,311. Give them one extra counter to add to the grid. Ask: *What new numbers can you make? What happens if you remove a counter? What numbers can you make now?*

Discover

WAYS OF WORKING Pair work

ASK

• Questions ➊ a) and b): *What number does Reena begin with?*

• Questions ➊ a) and b): *How will you know whether to add or subtract 10, 100, 1,000 or 10,000? What happens to a number when you add 10,000? What if both players' finishing numbers were 350,000?*

IN FOCUS This section gives children an opportunity to add and subtract 10, 100, 1,000 and 10,000 to and from a given number, within a context that they will be familiar with and find engaging. The learning is scaffolded by asking children to initially find only 10,000 more or less (in question ➊ a), before developing this to include finding 10, 100, 1,000 more or less (in question ➊ b).

PRACTICAL TIPS It would be an engaging introduction to this lesson if children were offered the opportunity to play a similar game. Using the grid in the picture, or by creating your own on paper, ask children to investigate different routes through the maze – which way leads to the greatest or smallest total?

ANSWERS

Question ➊ a): When Isla reaches the finish her score is 120,000.

Question ➊ b): When Reena reaches the finish her score is 241,980.

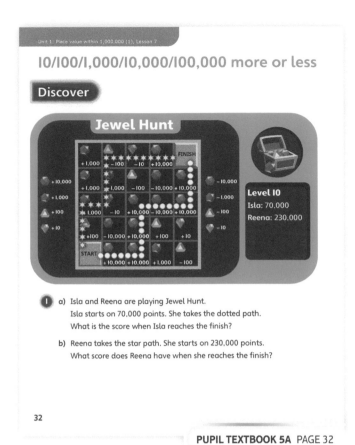

10/100/1,000/10,000/100,000 more or less

Discover

Jewel Hunt

Level 10
Isla: 70,000
Reena: 230,000

➊ a) Isla and Reena are playing Jewel Hunt. Isla starts on 70,000 points. She takes the dotted path. What is the score when Isla reaches the finish?

b) Reena takes the star path. She starts on 230,000 points. What score does Reena have when she reaches the finish?

32

PUPIL TEXTBOOK 5A PAGE 32

Share

WAYS OF WORKING Whole class teacher led

ASK

• Question ➊ a): *How did you make sure not to lose count? How could you represent how the number changed during the game?*

• Question ➊ b): *Which numbers were easier to add or subtract? Why?*

IN FOCUS It will be important to link children's understanding of place value with the concepts covered in this part of the lesson. Encourage children to demonstrate the addition and subtraction in question ➊ b) using place value counters in a place value grid, or with base 10 equipment.

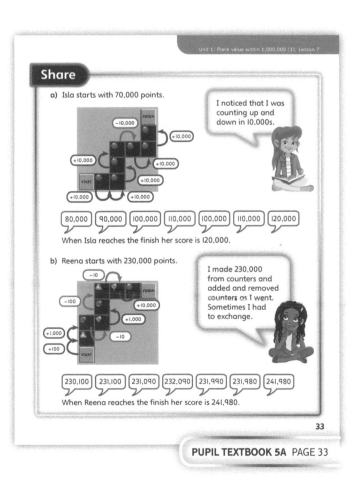

Share

a) Isla starts with 70,000 points.

I noticed that I was counting up and down in 10,000s.

80,000 90,000 100,000 110,000 100,000 110,000 120,000

When Isla reaches the finish her score is 120,000.

b) Reena starts with 230,000 points.

I made 230,000 from counters and added and removed counters as I went. Sometimes I had to exchange.

230,100 231,100 231,090 232,090 231,990 231,980 241,980

When Reena reaches the finish her score is 241,980.

33

PUPIL TEXTBOOK 5A PAGE 33

Think together

Whole class teacher led (I do, We do, You do)

ASK

- Question **1**: *What do you notice about how the numbers change? How will this help you find the missing numbers?*
- Question **2**: *How will the place value grid help you find the solution for each calculation?*
- Question **3** b): *Is there only one solution? How many successful routes can you discover?*

IN FOCUS Question **1** will help children to identify the patterns inherent in adding and subtracting multiples of 10, enabling them to become more fluent at counting in steps of 10, 100, 1,000 and 10,000. Question **2** demonstrates how adding different multiples of 10 will affect a number differently.

STRENGTHEN Provide place value counters to help complete the table in question **2**. Ask: *Show me 100 more using place value counters. What number do you have now? Explain how you know.*

DEEPEN For children mastering the concepts of the lesson, pose another challenge once they have solved question **3**. Ask: *If my final score was ⁻6,800, what was my route?*

ASSESSMENT CHECKPOINT Assess children's recognition of patterns when counting in steps of 10, 100, 1,000 and 10,000. Can they do so fluently? Check that children recognise the link between counting in steps of 10, 100, 1,000 and 10,000 and place value.

ANSWERS

Question **1** a): 72,000, 73,000, 74,000, **75,000, 76,000, 77,000**

Question **1** b): 272,700, 272,800, **272,900, 273,000**, 273,100, **273,200**

Question **1** c): 738,006, **638,006, 538,006**, 438,006, **338,006, 238,006**

Question **2**:

100,000 less	47,300	100,000 more	247,300
10,000 less	137,300	10,000 more	157,300
1,000 less	146,300	1,000 more	148,300
100 less	147,200	100 more	147,400
10 less	147,290	10 more	147,310

Question **3** a): Reena has a score of 400,500 by the end of the level. Children's explanation to detail how for every positive number Reena travels through, she also travels through a negative number cancelling out any points she scores.

Question **3** b): Look for children to find routes that total less than 500,000. An example is pictured below.

Think together

1 Work out the missing numbers in each sequence.

a) 72,000 , 73,000 , 74,000 , ☐ , ☐ , ☐

b) 272,700 , 272,800 , ☐ , ☐ , 273,100 , ☐

c) 738,006 , ☐ , ☐ , 438,006 , ☐ , ☐

2 Complete the table below for the number shown in the place value grid.

HTh	TTh	Th	H	T	O
1	4	7	3	0	0

100,000 less	47,300	100,000 more	247,300
10,000 less		10,000 more	
1,000 less		1,000 more	
100 less		100 more	
10 less		10 more	

34

PUPIL TEXTBOOK 5A PAGE 34

---

3 Isla and Reena are now on a different level of Jewel Hunt. CHALLENGE

a) Reena starts with 400,500 points. She takes this path.

What score does Reena have by the end of the level?
What do you notice?

b) Isla starts with 500,000 points.
What path could Isla take through the maze to finish with less than 500,000 points?

I will use trial and error and do it in my head.

I am going to look at the maze first and see if I notice anything.

35

→ Practice book 5A p24

PUPIL TEXTBOOK 5A PAGE 35

Practice

WAYS OF WORKING Independent thinking

IN FOCUS Question ❶ supports children's understanding of counting in steps of 10, 100, 1,000 and 10,000 from a given number, using the more concrete representation of place value counters in a place value grid. Children could replicate this representation using classroom resources to help them solve this question. Question ❷ offers children the opportunity to complete counts using simple numbers that are rounded to the place value that the count is in (for example, if they are counting in 1,000s, then the numbers do not have any 100s, 10s or 1s). This learning is extended in questions ❸, ❹ and ❺, which offer children an opportunity to continue counts with trickier numbers that are not conveniently rounded.

STRENGTHEN To strengthen understanding when counting on or back from the given numbers in questions ❻ and ❼, ask: *How could you represent the numbers in a way that would make it easier to see what happens when you count on or back?*

DEEPEN Question ❽ deepens children's reasoning and fluency. If children solve the given problem quickly, deepen their reasoning further by suggesting they create their own puzzle, like the one in the question, for a partner to solve.

ASSESSMENT CHECKPOINT Check that children can count on and back in steps of 10, 100, 1,000 and 10,000, fluently from a given number.

Assess children's recognition of place value, looking for clear understanding of how this is affected by counting on and back in steps of 10, 100, 1,000 and 10,000.

ANSWERS Answers for the **Practice** part of the lesson can be found in the *Power Maths* online subscription.

Reflect

WAYS OF WORKING Independent thinking

IN FOCUS This question will give an opportunity to assess children's fluency and reasoning with the numbers they have been learning about. They should be able to identify that counting in 10,000s will be quicker because the steps are much larger. Encourage children to prove their ideas in a concrete or pictorial way.

ASSESSMENT CHECKPOINT Look for recognition from children that they would need to count 1,000 hundreds but only 10 ten thousands.

ANSWERS Answers for the **Reflect** part of the lesson can be found in the *Power Maths* online subscription.

After the lesson ⏸

- Were all children equally able to count in all of the different steps confidently?
- How was children's conceptual understanding of the concepts developed in this lesson?
- Were children offered concrete, pictorial and abstract opportunities to access the learning?

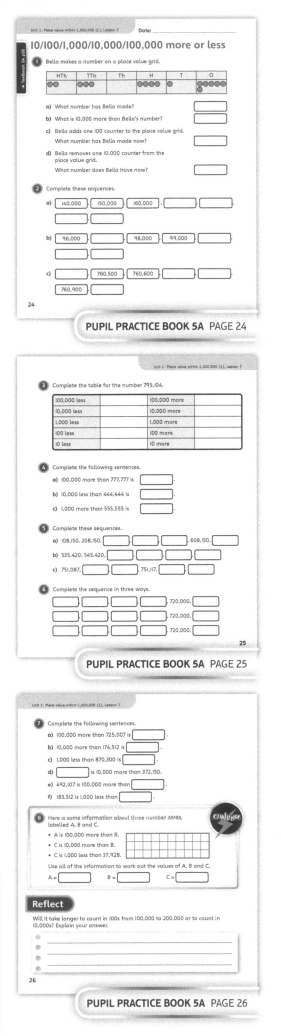

PUPIL PRACTICE BOOK 5A PAGE 24

PUPIL PRACTICE BOOK 5A PAGE 25

PUPIL PRACTICE BOOK 5A PAGE 26

Partition numbers to 1,000,000

Learning focus

In this lesson, children will use their understanding of the place value of numbers with up to 6 digits to partition and recombine numbers to solve number problems.

Before you teach

- Are there any misconceptions that you need to plan for, based on children's work in the previous lesson?
- How will you develop children's ability to generalise in this lesson?

NATIONAL CURRICULUM LINKS

Year 5 Number – number and place value

Read, write, order and compare numbers to at least 1,000,000 and determine the value of each digit.

ASSESSING MASTERY

Children can recognise, describe, read and write numbers with up to 6 digits. They can use what they know about place value to describe the value of each digit in a number and can solve problems using this understanding.

COMMON MISCONCEPTIONS

Children may assume that the only way to partition a number is into its place value headings (i.e. 136 can only be partitioned as 100, 30 and 6). Base 10 equipment or place value counters may help children to become more flexible in the way they partition. Ask:

- *Is that the only way you can break this number into partitions? Show me another way using a resource.*

STRENGTHENING UNDERSTANDING

Before starting the lesson, children could strengthen their fluency with 4-, 5- and 6-digit numbers and their understanding of place value by counting up and down in different steps from a given number. Ask: *Can you count up in 10,000s from 453? Can you count down in 100,000s from 876,232?*

GOING DEEPER

Encourage children to create their own missing number equalities to challenge their partner with. For example, 300,000 + ☐ + 6,000 + 40 + 5 = 376,045. Ask: *Create a challenge like this for your partner. How can you make your question easier or trickier?*

KEY LANGUAGE

In lesson: ones (1s), tens (10s), hundreds (100s), thousands (1,000s), ten thousands (10,000s), hundred thousands (100,000s), partition, place value, digit

Other language to be used by the teacher: re-combine

STRUCTURES AND REPRESENTATIONS

Place value grid

RESOURCES

Mandatory: place value counters

Optional: play money, base 10 equipment, whiteboards

 In the eTextbook of this lesson, you will find interactive links to a selection of teaching tools.

Quick recap 🔎

Ask children to write a number less than 100,000 on their whiteboards. Practise partitioning smaller numbers up to 100,000 and identify the value of each of the digits in the number.

Discover

Unit 1: Place value within 1,000,000 (1), Lesson 8

Partition numbers to 1,000,000

Discover

WAYS OF WORKING Pair work

ASK

- Question ❶ a): *How can you tell how much money is in the bank vault?*
- Question ❶ b): *What is the simplest way to partition £360,400? Have you found the same solution as your partner?*

IN FOCUS Question ❶ a) will give children an opportunity to begin recognising numbers when partitioned and recombining them into their total amount. Question ❶ b) reduces scaffolding by asking children to partition an abstract 6-digit number, without providing a pictorial representation.

PRACTICAL TIPS Encourage children to investigate different ways of making a given amount, using a variety of denominations of play money.

ANSWERS

Question ❶ a): In digits: £236,253 is in the bank vault.
In words: two hundred and thirty-six thousand, two hundred and fifty-three pounds.

Question ❶ b): The value of the 3 is 3 hundred thousands.
The value of the 6 is 6 ten thousands.
The value of the 4 is 4 hundreds.

❶ a) How much money is in the bank vault?

b) Another vault has £360,400 in it.
What is the value of each of the digits in this number?

36

PUPIL TEXTBOOK 5A PAGE 36

Share

WAYS OF WORKING Whole class teacher led

ASK

- Question ❶ a): *How does the place value grid match the bags of money? What amount of money do the 2 hundreds counters represent?*
- Question ❶ b): *How does the place value grid help you find the value of each digit?*

IN FOCUS Here, children use place value grids to support them in writing and interpreting numbers. They should be able to identify the value of any digit in a number using its place value.

Share

I used a place value grid to represent the amount of money in the bank vault.

a)

HTh	TTh	Th	H	T	O
●●	●●●	●●●●●●●●●●	●●	●●●●●	●●●

There is £236,253 in the bank vault.

b) There is £360,400 in the other bank vault.

HTh	TTh	Th	H	T	O
●●●	●●●●●●		●●●●		

The value of the 3 is 3 hundred thousands.
The value of the 6 is 6 ten thousands.
The value of the 4 is 4 hundreds.

I can also say that 3 is 300,000, 6 is 60,000 and 4 is 400.

37

PUPIL TEXTBOOK 5A PAGE 37

Think together

Think together

WAYS OF WORKING Whole class teacher led (I do, We do, You do)

ASK

- Question ❶: *How many counters are there in each column? What does each column represent? If there are 4 hundreds counters, what number does this represent?*
- Question ❷: *What do you do if there is a zero in the number?*

IN FOCUS In questions ❶ and ❷, children explore partitioning numbers by considering the place value of each digit. It is important that they fully understand place value, for example, in 58,415, the 8 does not represent 8, but instead 8,000. In question ❸, children explore this further by taking the given partition, finding the number and then exploring what happens if parts are removed.

STRENGTHEN Provide children with place value counters and a place value grid to make the numbers and strengthen their understanding. Encourage them to draw a part-whole model to support them in partitioning numbers.

DEEPEN In question ❸ b), ask children what other parts they could remove to leave zero as a place holder in a different column. Can they create a similar question for a partner that has a zero as a place holder in the tens column?

ASSESSMENT CHECKPOINT Use questions ❶ and ❷ to assess whether children can partition numbers in a standard way according to the place value of each digit. Use question ❸ a) to assess whether children can use the partitioned form to identify the number.

ANSWERS

Question ❶ a): 726,140 is 7 hundred thousands, 2 ten thousands, 6 thousands, 1 hundred and 4 tens.

Question ❶ b): 58,415: The value of the 5 is 5 ten thousands, the value of the 8 is 8 thousands, the value of the 4 is 4 hundreds, the value of the 1 is 1 ten and the value of the 5 is five ones.
604,003: The value of the 6 is 6 hundred thousands, the value of the 4 is 4 thousands and the value of the 3 is 3 ones.

Question ❷: 60,375 = 60,000 + 300 + 70 + 5
951,618 = 900,000 + 50,000 + 1,000 + 600 + 10 + 8
120,508 = 100,000 + 20,000 + 500 + 8

Question ❸ a): 57,312
300,562
104,500
26,503

Question ❶ b): 54,076
60,000

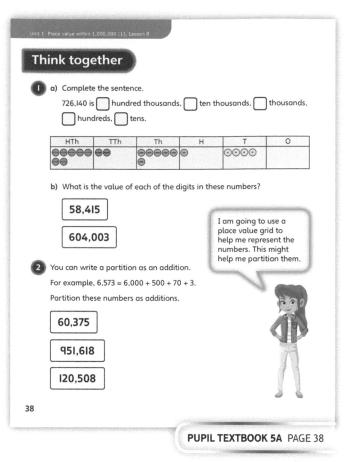

PUPIL TEXTBOOK 5A PAGE 38

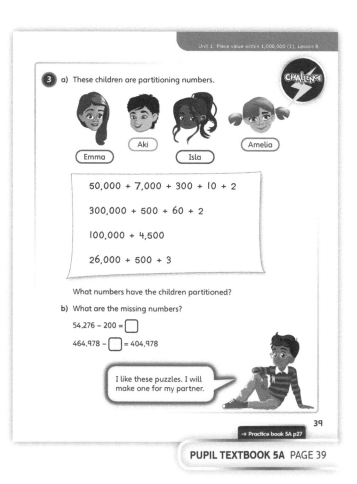

PUPIL TEXTBOOK 5A PAGE 39

Practice

WAYS OF WORKING Independent thinking

IN FOCUS Question ❶ provides an opportunity for children to recognise a partitioned number within a different context and independently recombine the number to find the total. Question ❺ has been designed to allow children to develop their fluency when partitioning and recombining 4-, 5- and 6-digit numbers recorded in an abstract way. It is scaffolded to enable children to work through parts a) to g) independently.

STRENGTHEN To strengthen understanding when recombining the numbers shown in question ❻, ask: *What structures or representations have you used previously to show a number? How could you use them here to help you?*

DEEPEN Question ❼ will test children's fluency with partitioned numbers in subtracting 10s, 100s, 1,000s and 10,000s from 6-digit numbers.

ASSESSMENT CHECKPOINT Can children fluently partition and recombine numbers with up to 6-digits?

ANSWERS Answers for the **Practice** part of the lesson can be found in the *Power Maths* online subscription.

Reflect

WAYS OF WORKING Independent thinking

IN FOCUS Here, children write numbers that meet certain criteria. This will check their understanding of the place value of digits and the position of these digits within a number.

ASSESSMENT CHECKPOINT Children correctly identify two 5-digit numbers and two 6-digit numbers that each have a 4 in the hundreds place.

ANSWERS Answers for the **Reflect** part of the lesson can be found in the *Power Maths* online subscription.

After the lesson

- How fluent were children at recognising that numbers can be partitioned in multiple ways?
- Were children equally confident in partitioning and recombining numbers? If not, how will you support their understanding in the concept they find trickier?
- Can children use both words and digits to write out the value of each digit of a 6-digit number?

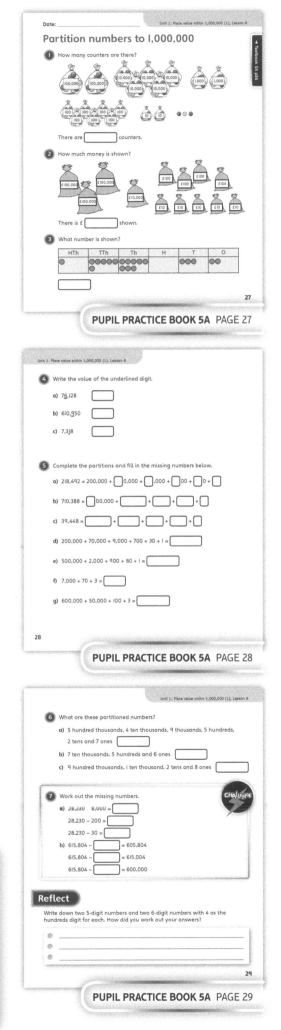

PUPIL PRACTICE BOOK 5A PAGE 27

PUPIL PRACTICE BOOK 5A PAGE 28

PUPIL PRACTICE BOOK 5A PAGE 29

End of unit check

Don't forget the unit assessment grid in your *Power Maths* online subscription.

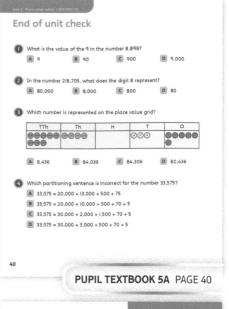

PUPIL TEXTBOOK 5A PAGE 40

WAYS OF WORKING Group work adult led

IN FOCUS

- In question ③, children check the value of each digit in abstract numbers to find the one that matches the pictorial representation, being careful not to make place value errors where numbers possess similar digits.
- In question ⑤, children have to identify which digit in 367,180 represents the thousands, and add 1 to that digit.
- Question ⑦ is a SATs-style question on Roman numerals.

ANSWERS AND COMMENTARY Children who have mastered the concepts in this unit will know the value of each digit in numbers up to 1,000,000. They can correctly identify numbers on a place value grid, from the ones column up to hundred thousands. They will flexibly partition numbers, appreciating that the combined parts must still be equivalent to the whole. Children will apply their knowledge of place value to help compare and order numbers. Children will convert between numbers and Roman numerals up to M.

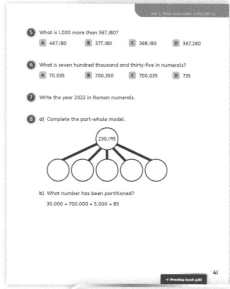

PUPIL TEXTBOOK 5A PAGE 41

Q	A	WRONG ANSWERS AND MISCONCEPTIONS	STRENGTHENING UNDERSTANDING
1	B	A, C and D suggest children do not know the value of digits within a 4-digit number.	Explore equivalence by building numbers with place value counters, identifying and writing the value of each digit as you go along. Put the counters into part-whole models, then move them to make a different number of parts by reducing or increasing the size of different parts. Build pairs of numbers with a varying number of digits using base 10 equipment or place value counters.
2	B	A, C and D suggest children do not know the value of digits within a 6-digit number.	
3	B	A indicates that children have not understood that zero is used as a place holder. C and D suggest that they are unsure of the value of each place.	
4	B	A suggests that children think there must be five parts for a 5-digit number. C indicates they think each part must be a multiple of a place value. D indicates that they do not understand equivalence.	
5	C	Any other answer suggests children have not correctly identified the number in the thousands place.	
6	C	A and D suggest that children may not be comparing the value of each digit, making decisions based on only the first two digits.	
7	MMXXII	Some children may put XXXXII because we often say the year as twenty twenty-two.	
8	a) 230,195 = 200,000 + 30,000 + 100 + 90 + 5, b) 735,085. If children answer incorrectly they may need assistance in understanding zeros as placeholders.		

My journal

WAYS OF WORKING Independent thinking

ANSWERS AND COMMENTARY

Children may describe the number 12,546 in many ways. For example:
- 12,546 is a 5-digit number because it has a digit in the 10,000s place
- 12,546 is 546 more than 12,000
- 12,546 is between the multiples 12,000 and 13,000
- 12,546 is a little more than half-way between 12,000 and 13,000
- 12,546 rounds to 13,000 to the nearest 1,000. 12,546 rounds down to 10,000 to the nearest 10,000.

If children need support using the key language, ask:
- *Is 12,546 closer to 12,000 or 13,000?*
- *How many 10,000s, 1,000s, 100s, 10s and 1s can you see in 12,546?*

Representations could include place value grids and partitioning in part-whole models, on number lines or as abstract number sentences.

If children need support representing the number in different ways, ask:
- *How have you represented numbers in this unit? What models have you used?*
- *How many columns will you need in a place value grid to represent 12,546?*

Power check

WAYS OF WORKING Independent thinking

ASK

- *What do you know about the different values of the digits in a 5-digit number?*
- *Can you say the multiples of 1,000 or 10,000 that a number lies between?*
- *Do you know how to round a number to the nearest 1,000?*
- *How confident do you feel ordering a group of 4- and 5-digit numbers?*

Power puzzle

WAYS OF WORKING Pair work

IN FOCUS Use this **Power puzzle** to assess whether children can recognise and follow number sequences. Understanding that number sequences follow a rule will help children to see that the numbers in a sequence will change in a regular way.

ANSWERS AND COMMENTARY Can children recognise consistent and regular number sequences in the given cards? The sequences are: 2, 8, 14, 20 and 16, 13, 10, 7, 4, 1 or 1, 4, 7, 10, 13, 16 and 20, 14, 8, 2.
If children need support, ask: *Which number is likely to be the first number In the sequence that gets bigger? Why?*

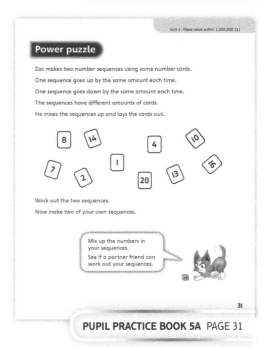

PUPIL PRACTICE BOOK 5A PAGE 30

PUPIL PRACTICE BOOK 5A PAGE 31

After the unit ⏸

- How well did the prompts and questions promote learning, and what were children's responses to them?
- Are children ready to move on with their learning? Following on from the end of unit assessment, how confident are they working with numbers within 1,000,000?

Strengthen and **Deepen** activities for this unit can be found in the *Power Maths* online subscription.

Unit 2
Place value within 1,000,000 ②

Mastery Expert tip! 'I made sure that children used their place value grids as much as possible in this unit to really embed the new place value columns they were working with. They used place value counters as well to help scaffold their understanding of the abstract numerals.'

Don't forget to watch the Unit 2 video!

WHY THIS UNIT IS IMPORTANT

In this unit, children will build on their understanding of numbers up to 1,000,000 by developing further fluency with these numbers by partitioning and recognising between which two multiples a number lies. Children will develop their understanding of number by recognising a number's position on a number line and identify numbers up to 1,000,000, going on to compare and order them. They will then develop their understanding of rounding numbers, with particular focus on 6-digit numbers. Throughout this unit, children will become more confident in dealing with much larger numbers and this will provide the foundation for work later in the year.

WHERE THIS UNIT FITS

→ Unit 1: Place value within 1,000,000 (1)
→ **Unit 2: Place value within 1,000,000 (2)**
→ Unit 3: Addition and subtraction

This unit builds on children's work in Unit 1, where they explored numbers to 1,000,000 and their representation on place value grids, as well as work on number lines from Year 4. It also builds on previous experience of comparing, ordering and rounding numbers from lower KS2. This unit provides preparation for using place value grids to add and subtract 5-digit numbers.

Before they start this unit, it is expected that children:
• can recognise 1s, 10s, 100s, 1,000s and 10,000s
• can use a place value grid to the 100,000s
• can recognise and use a number line
• can partition and recombine numbers.

ASSESSING MASTERY

Children who have mastered this unit will be able to demonstrate fluent understanding of place value of digits in numbers up to 1,000,000. They will continue to be able to say, partition and write numbers accurately and use this, and their understanding of place value, to compare and order numbers up to 1,000,000. Children will use number lines confidently. They will be able to use their understanding of place value to round up and down to a variety of different degrees of accuracy.

COMMON MISCONCEPTIONS	STRENGTHENING UNDERSTANDING	GOING DEEPER
Children may find it challenging to round large numbers to a smaller degree of accuracy, for example, rounding 176,315 to the nearest 10.	Encourage children to focus on the 10s and 1s. Explain that initially the other digits in the number are not as important. Ask: *Can you round 15 to the nearest 10?* This will help them round 176,315 to the nearest 10. Develop this by then getting children to consider the whole number, but to still just focus on the 10s column.	Ask: *Can you write down the lowest and greatest numbers that round to 170,000, to the nearest 10,000?* *Can you write down the lowest and greatest numbers if 170,000 is rounded to the nearest 1,000?*
Children may find it challenging to estimate the position of a number on a number line.	Begin with number lines divided into ten intervals and focus on smaller numbers, for example, 10,000 to 20,000. Ask: *Can you work out what the number line is going up in? Between which two numbers does the given number lie? Is it less than or greater than half-way between these two points?*	Ask children to estimate the position of a number on a blank number line. Look at the strategies that children adopt. Often children may split the number line into ten intervals to begin with.

UNIT STARTER PAGES

Use these pages to introduce the unit, with teacher-led discussion. Encourage children to think back to the previous unit and work out how the place value grid has changed. Challenge children to explain as many of the keywords as they can.

STRUCTURES AND REPRESENTATIONS

Place value grid: Place value grids are used in this unit to help children read numbers and recognise the value of each digit in numbers up to 1,000,000.

HTh	TTh	Th	H	T	O
●●●●●	●●● ●●●● ●		●●●●● ●● ●● ●		●●●●●

Number line: Number lines are used to help children plot numbers from 0 to 1,000,000.

0   10

KEY LANGUAGE

There is some key language that children will need to know as part of the learning in this unit:

→ place value
→ ones (1s), tens (10s), hundreds (100s), thousands (1,000s), ten thousands (10,000s), hundred thousands (100,000s), million (1,000,000)
→ partition, partitioning
→ number line
→ round, rounding, round up, round down
→ compare
→ order
→ ascending, descending
→ nearest
→ less than (<), greater than (>)

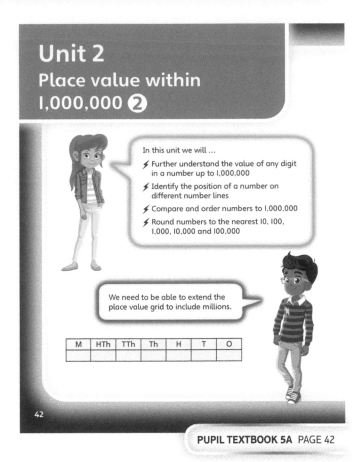

PUPIL TEXTBOOK 5A PAGE 42

PUPIL TEXTBOOK 5A PAGE 43

Number line to 1,000,000

Learning focus

In this lesson, children will estimate and accurately identify where numbers to 1,000,000 would lie on a number line. They will use their understanding of place value to help them achieve this.

Before you teach

- How confident are children in their use of number lines?
- Can children count fluently in 1,000s, 10,000s and 100,000s?
- What other real-life contexts can you apply to number lines in the lesson?

NATIONAL CURRICULUM LINKS

Year 5 Number – number and place value

Read, write, order and compare numbers to at least 1,000,000 and determine the value of each digit.

ASSESSING MASTERY

Children can use their understanding of place value to help them accurately identify, or estimate, where a number would lie on a number line for numbers up to 1,000,000.

COMMON MISCONCEPTIONS

Children may miscalculate, and so misinterpret, the unlabelled divisions on a number line. Ask:
- *If you count in those steps, does your number line work? Show me.*

STRENGTHENING UNDERSTANDING

Provide children with number cards up to 1,000,000. Encourage them to peg the cards along a piece of string in order of size, starting with the smallest. Ask: *How do you know which number comes first?*

Discuss with children whether the gap between each number should be the same. If not, why not? To show this practically, use four or five numbers (to 50) along a bead string. Label the appropriate beads and discuss why the gaps between them are different sizes.

GOING DEEPER

Children can draw their own number lines to 1,000,000 and place arrows along them. Suggest they challenge a partner to identify what number each arrow is pointing to.

KEY LANGUAGE

In lesson: intervals, **million**, hundred thousands (100,000s), ten thousands (10,000s), thousands (1,000s), hundreds (100s), tens (10s), ones (1s)

Other language to be used by the teacher: estimate, pattern

STRUCTURES AND REPRESENTATIONS

Number line

RESOURCES

Optional: bead strings, number cards, drywipe number lines

 In the eTextbook of this lesson, you will find interactive links to a selection of teaching tools.

Quick recap

Give children a ten-interval, wipe clean number line (or ask them to draw one). Ask them to mark the start number as 100 and the end number as 200. Ask: *What does the line go up in?*

Do the same for a number line with a start number of 1,000 and end number of 2,000. Ask: *What does the line go up in?* Ask children to mark certain numbers on each line too.

Discover

Unit 2: Place value within 1,000,000 (2), Lesson 1

WAYS OF WORKING Pair work

ASK

- Question ① a): *What is each interval worth? How do you know? What is different about the positions of the minimum price and maximum price? Which is easier to read and why?*
- Question ① b): *Why is it tricky to find £720,000 on this number line?*

IN FOCUS Question ① a) encourages children to explain how a number line is organised and ensures children develop their ability to recognise and calculate the value of unlabelled intervals. Question ① b) encourages children to accurately estimate where numbers appear between unlabelled intervals along a number line.

PRACTICAL TIPS When introducing this concept, provide children with number cards labelled with numbers up to 1,000,000. Ask children to order these numbers along a blank number line, placing them at regular intervals (for example, going up in steps of 100,000). This could be made trickier by including some numbers that do not fit the pattern.

ANSWERS

Question ① a): The minimum house price is £200,000.
The maximum house price is £850,000.

Question ① b):

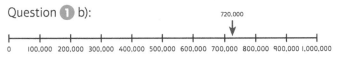

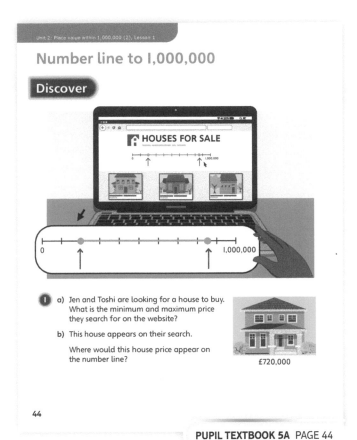

Number line to 1,000,000

Discover

① a) Jen and Toshi are looking for a house to buy. What is the minimum and maximum price they search for on the website?

b) This house appears on their search.

Where would this house price appear on the number line?

£720,000

44

PUPIL TEXTBOOK 5A PAGE 44

Share

WAYS OF WORKING Whole class teacher led

ASK

- Question ① a): *How can you test that the number line goes up in intervals of 100,000? How can you be sure that the maximum price is £850,000?*
- Question ① b): *Can you use this method to find £480,000?*

IN FOCUS At this point in the lesson, it will be important to ensure that children can calculate unlabelled intervals along a number line, and question ① a) will support this learning. Question ① b) will then develop children's learning as they visualise and calculate unlabelled intervals.

Share

a) The number line goes from 0 to 1,000,000.

1,000,000 is also known as one million.

There are 10 intervals.

The line goes up in intervals of 100,000s.

I can see that the maximum house price lies half-way between 800,000 and 900,000.

The minimum house price they search for on the website is £200,000.

The maximum house price is £850,000.

b) The house costs £720,000.

£720,000 is greater than £700,000, but less than £800,000.

720,000

45

PUPIL TEXTBOOK 5A PAGE 45

Think together

WAYS OF WORKING Whole class teacher led (I do, We do, You do)

ASK

- Question ❶: *How can you work out what to count up in? How can you check how accurate you have been?*
- Question ❷: *How can you work out what numbers the arrows are pointing to?*
- Question ❸: *How are the number lines the same? How are they different? How will the position of the number change in each number line? Is there any way of predicting where the number will be on each number line before you work it out?*

IN FOCUS Question ❶ will help children to develop their ability to use number lines, recognise patterns and use them to calculate unlabelled intervals. Question ❷ has been included to develop children's fluency with number lines. They will need to calculate what the unlabelled intervals represent and then use this to find A, B and C along the number line.

STRENGTHEN To strengthen understanding when recognising the missing numbers in question ❶, suggest that children read aloud the numbers that are already on the number line. Ask: *What patterns can you spot that will help you to find the missing numbers?*

DEEPEN Question ❸ deepens children's understanding as it demonstrates how changing the start and end point of a number line can change where a number will be placed along it.

ASSESSMENT CHECKPOINT All questions in this section assess children's reasoning and fluency when finding numbers along a number line. Check children are counting labelled intervals on the number line accurately and using any patterns they spot to calculate the unlabelled intervals.

ANSWERS

Question ❶ a): Missing numbers: 100,000, 200,000, 300,000, 400,000, 600,000, 700,000, 900,000

Question ❶ b): Missing numbers: 210,000, 230,000, 240,000, 250,000, 260,000, 270,000, 290,000

Question ❶ c): Missing numbers: 418,100, 418,200, 418,300, 418,400, 418,500, 418,600, 418,700, 418,800, 418,900

Question ❷: A = 353,000
B = 355,000
C = 359,000

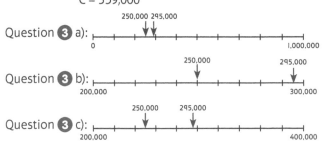

Question ❸ a):

Question ❸ b):

Question ❸ c):

Think together

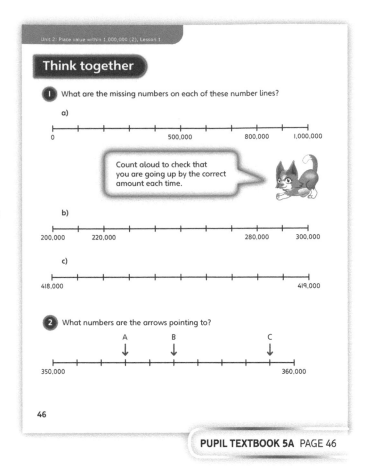

❶ What are the missing numbers on each of these number lines?

a)

Count aloud to check that you are going up by the correct amount each time.

b)

c)

❷ What numbers are the arrows pointing to?

46

PUPIL TEXTBOOK 5A PAGE 46

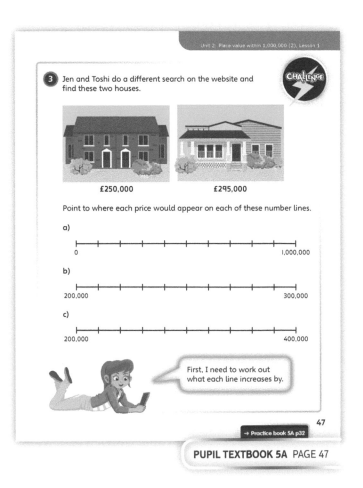

❸ Jen and Toshi do a different search on the website and find these two houses.

CHALLENGE

£250,000   £295,000

Point to where each price would appear on each of these number lines.

a)

b)

c)

First, I need to work out what each line increases by.

47

→ Practice book 5A p32

PUPIL TEXTBOOK 5A PAGE 47

Practice

WAYS OF WORKING Independent thinking

IN FOCUS In questions ❶ and ❷, children will practise calculating unlabelled intervals and finding missing numbers on a number line. Question ❸ offers children an opportunity to estimate where numbers lie along a number line.

STRENGTHEN If children are finding it difficult to estimate the location of the numbers in question ❸, ask: *Is it useful for you to find any particular points first? Explain which, and why. How does knowing half-way help you estimate the locations of the numbers?*

DEEPEN Question ❺ will enhance children's understanding of the similarities and differences between number lines. Extend this by asking: *Which number line is it easiest to plot 330,000 on? Why? Describe a number line where this number would not be found. Explain what makes the number line you described different from those in the question.*

THINK DIFFERENTLY Question ❹ develops children's ability to find solutions using number lines. It is a two-step problem as they will need to find out the value of point A and point B and then use this knowledge to identify which numbers from the list are found between the two points. Ask: *What is stopping you from identifying the numbers straight away?*

ASSESSMENT CHECKPOINT Check that children are accurately calculating the unlabelled intervals on a number line before they begin to work out any missing numbers. In question ❸, assess children's ability to recognise where numbers are positioned along a number line, looking in particular for an understanding of their proportional size based on where they are along the line. In question ❺, check that children recognise that the start and end point of a number line will determine where a number is positioned along the line.

ANSWERS Answers for the **Practice** part of the lesson can be found in the *Power Maths* online subscription.

Reflect

WAYS OF WORKING Pair work

IN FOCUS This **Reflect** question allows children to demonstrate the depth of their understanding by asking them to explain to a partner how they divided a given number line into intervals.

ASSESSMENT CHECKPOINT Assess whether children have divided the number line into an appropriate number of intervals and that the intervals are of equal size. Also, check that they have positioned both numbers in approximately the right place.

ANSWERS Answers for the **Reflect** part of the lesson can be found in the *Power Maths* online subscription.

After the lesson ⏸

- Were children able to plot numbers fluently on a number line with no marked intervals? Were they able to clearly explain ways of making this possible?
- How was children's understanding of place value developed in this lesson? Was the link between place value and number lines made explicit?

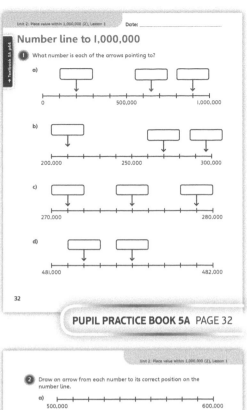

PUPIL PRACTICE BOOK 5A PAGE 32

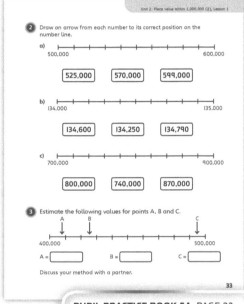

PUPIL PRACTICE BOOK 5A PAGE 33

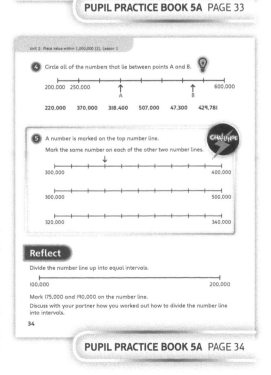

PUPIL PRACTICE BOOK 5A PAGE 34

Compare and order numbers to 100,000

Learning focus

In this lesson, children will compare and order numbers to 100,000 using what they know about place value and will identify which digits they need to compare first each time, explaining what to do when the digits are the same. They will use the signs < and > to show comparisons and order.

Before you teach

- Can children compare 3- and 4-digit numbers, explaining which is larger and why?
- Can they order a set of 3- or 4-digit numbers by comparing numbers within the set?

NATIONAL CURRICULUM LINKS

Year 5 Number – number and place value

Read, write, order and compare numbers to at least 1,000,000 and determine the value of each digit.

ASSESSING MASTERY

Children can compare pairs of numbers up to 100,000, explaining which is larger and why. They can write or represent a group of 4- and 5-digit numbers in ascending or descending order.

COMMON MISCONCEPTIONS

When comparing whole numbers, children may not check the number of digits first and look only at the size of the first digit. For example, when comparing the numbers 23,200 and 5,345, children may think that 5,345 is larger because 5 is larger than 2. Ask:

- *How would you represent these numbers on a place value grid?*
- *Which number has the digit in the position with the largest value?*
- *How does that help you decide which number is larger?*

STRENGTHENING UNDERSTANDING

Ask children to roll five dice (or the same dice five times) to create a 5-digit number. Ask them to arrange the digits to make the largest and then the smallest possible number, explaining their choices. Then ask them to make another number that is larger than the smallest number and another smaller than the largest, and finally a number that will go in the middle of the four numbers. Ask: *Why are the numbers now in order?*

GOING DEEPER

Give children measurements such as length or distance, mass or capacity to compare and order, asking them to explain how they can use what they know about ordering numbers to help them. Include some examples with different units, for example 34 km and 23,000 m so that children reason about why 34 km is further even though 23,000 is larger than 34.

KEY LANGUAGE

In lesson: compare, order, ascending, descending, less than (<), more than (>), smallest, largest, greatest, smaller, greater, larger, highest, lowest

STRUCTURES AND REPRESENTATIONS

Number line, place value grid

RESOURCES

Optional: place value counters, dice

 In the eTextbook of this lesson, you will find interactive links to a selection of teaching tools.

Quick recap

Ask: *How can you compare the numbers 145 and 156? How would you compare 176 and 172? Which is the larger number?* Discuss what methods children use. Where do they start comparing?

Discover

Compare and order numbers to 100,000

WAYS OF WORKING Pair work

ASK

- Question ① a): *What do you notice about the digits in all of the scores? Can you find some digits that have the same value in different scores?*
- Question ① a): *How do you know immediately that Amal did not achieve the highest score?*
- Question ① b): *How do you know that Holly was not last? Which digit in the number tells you this?*

IN FOCUS Question ① a) and b) use the familiar context of a game show for children to compare and order 5-digit numbers, each containing similar digits, but in different positions within the numbers. The scores have been chosen so that only two scores need to be compared more closely in question ① a), as 56,725 and 5,276 have a smaller number of 10,000s.

PRACTICAL TIPS Provide pairs or small groups of children with a large place value grid and sufficient place value counters to represent all of the scores in the picture. When they have made all of the scores, they can use the place value counters to prove how they know that Amal did not have the highest score.

ANSWERS

Question ① a): Amal achieved the lowest score of 5,276.

Question ① b): In ascending order, the scores are 5,276, 56,725, 65,272, 65,575.

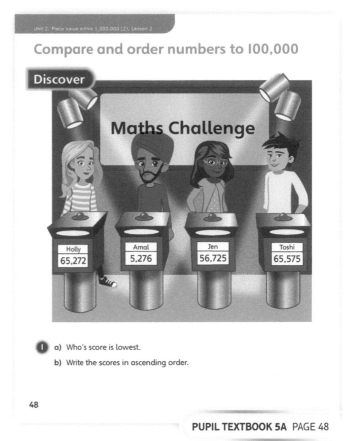

Discover

① a) Who's score is lowest.

   b) Write the scores in ascending order.

48

PUPIL TEXTBOOK 5A PAGE 48

Share

WAYS OF WORKING Whole class teacher led

ASK

- Question ① a): *Why is it important to check how many digits each number has before comparing digits? Can you suggest a pair of 5-digit numbers where you need to check the 1,000s but not the 100s? What would you do if the number of 100s in the two scores had also been the same?*
- Question ① b): *Why do you only need to compare the numbers 65,272 and 65,575 now? How does representing the two numbers with place value counters help you compare them? How do the signs between the numbers tell you that they are in ascending order? What does descending mean? How could you write the numbers in descending order?*

IN FOCUS Question ① b) introduces the term 'ascending' to specify how the numbers should be ordered. The two higher scores have been chosen to have the same number of 10,000s and 1,000s, so children are required to compare the 100s.

STRENGTHEN Ensure that children are confident to make sense of and use of the word 'ascending', recognising that 'descending' means in order from largest to smallest.

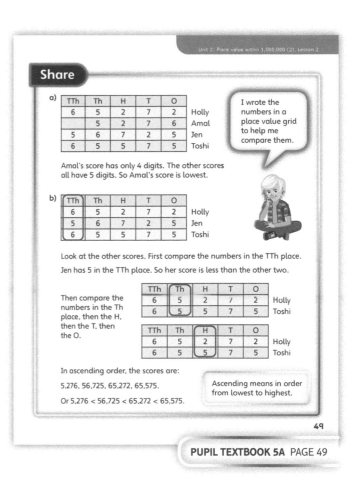

Share

a)

TTh	Th	H	T	O	
6	5	2	7	2	Holly
	5	2	7	6	Amal
5	6	7	2	5	Jen
6	5	5	7	5	Toshi

I wrote the numbers in a place value grid to help me compare them.

Amal's score has only 4 digits. The other scores all have 5 digits. So Amal's score is lowest.

b)

TTh	Th	H	T	O	
6	5	2	7	2	Holly
5	6	7	2	5	Jen
6	5	5	7	5	Toshi

Look at the other scores. First compare the numbers in the TTh place.

Jen has 5 in the TTh place. So her score is less than the other two.

Then compare the numbers in the Th place, then the H, then the T, then the O.

TTh	Th	H	T	O	
6	5	2	7	2	Holly
6	5	5	7	5	Toshi

TTh	Th	H	T	O	
6	5	2	7	2	Holly
6	5	5	7	5	Toshi

In ascending order, the scores are:

5,276, 56,725, 65,272, 65,575.

Or 5,276 < 56,725 < 65,272 < 65,575.

Ascending means in order from lowest to highest.

49

PUPIL TEXTBOOK 5A PAGE 49

Think together

WAYS OF WORKING Whole class teacher led (I do, We do, You do)

ASK

- Question ❶: *Grid B has more counters, but why might it not have the larger value?*
- Question ❶: *Which counters have the largest value? Why do you need to compare the numbers of 1,000s next?*
- Question ❷ b): *One of the numbers starts with a larger digit. Does this mean that it must be the larger number? Why?*
- Question ❷ b): *How do you immediately know that the second number is the smallest?*

IN FOCUS Question ❶ has been designed so there are more counters in grid B but the actual value of the grid is less. Question ❷ reinforces the use of the signs < and >. In question ❷ a), the numbers contain the same digits but in a different order. In question ❷ b), the numbers have been chosen to include a 4-digit and 5-digit number with the 4-digit number starting with a larger digit, requiring children to check the number of digits first.

STRENGTHEN To strengthen decision making in question ❷, ask children to represent each pair of numbers on a place value grid, using place value counters. Ask: *Can you explain which number is larger? Why?*

DEEPEN In question ❸, ask children to explore further examples with a partner. Ask: *Is it always true that a number starting with a 6 is larger than a number starting with a 4?*

ASSESSMENT CHECKPOINT Use questions ❶ and ❷ to assess whether children can compare 5-digit numbers. Check that they understand the use of the signs < and >. Use question ❷ b) to assess whether children can order 4- and 5-digit numbers. Use question ❸ to check that they recognise when a number with a higher first digit will be larger and when it will not, they should be able to give examples and counter examples.

ANSWERS

Question ❶: Grid A has the greater value because 23,110 > 22,512.

Question ❷ a): 34,790 < 43,970
21,033 > 8,968

Question ❷: b) 20,932 > 20,923 > 8,560

Question ❸: Sometimes true – it will depend on the number of digits and therefore the place value of the digits 9 and 5. Children should show this using a range of examples for which it is true (for example, 93,245 and 56,278) or not true (for example, 9,375 and 54,267).

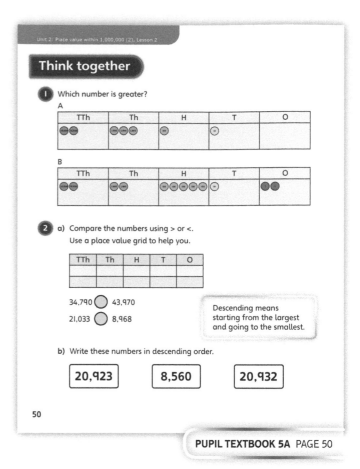

PUPIL TEXTBOOK 5A PAGE 50

PUPIL TEXTBOOK 5A PAGE 51

Practice

WAYS OF WORKING Independent thinking

IN FOCUS In question **1**, the numbers have been chosen so that the value of the 10,000s, 1,000s and 100s are the same so children must check the 10s. Question **6** involves comparing a 4-digit number with a 5-digit number, requiring children to check the number of digits first.

Question **8** requires children to reason about the positions of the missing digits. The question is made more challenging as there are no 'lower' digit cards to make the 1,000s position in the first number less than 6,000.

STRENGTHEN Encourage children to use place value counters to check any comparisons or ordering of numbers they are not sure about. Look together at the use of the < and > signs to ensure that children are confident to use these appropriately and recognise why they can also be used for showing order.

DEEPEN Ask children to make up some true or false statements (similar to question **6**) for a partner to solve.

ASSESSMENT CHECKPOINT Use questions **1**, **2**, **3** and **4** to assess whether children can compare and order 5-digit numbers. Use question **6** to assess whether children can compare 4- and 5-digit numbers, checking the number of digits first. Check that children understand the terminology 'ascending' and 'descending'. Use questions **7** and **9** to check that children can apply their understanding of order to measurement and money. In question **8**, check that children understand the use of the sign <.

ANSWERS Answers for the **Practice** part of the lesson can be found in the *Power Maths* online subscription.

Reflect

WAYS OF WORKING Independent thinking

IN FOCUS This reflection point draws attention to the common misconception that to compare two numbers you only have to look at the first digit in a number. This only works if each number has the same amount of digits.

ASSESSMENT CHECKPOINT Children should be able to explain clearly why 9,623 is smaller than 12,345 even though it starts with a 9. They should use correct vocabulary to explain that 9,623 has 9 thousands whereas 12,345 has 2 thousands but also 1 ten thousand.

ANSWERS Answers for the **Reflect** part of the lesson can be found in the *Power Maths* online subscription.

After the lesson 🕘

- Can children explain which number is larger in pairs of 5-digit and 4-digit numbers?
- Can they order numbers in ascending or descending order?
- Can they explain why 9,450 is smaller than 34,212 even though it starts with the digit 9?

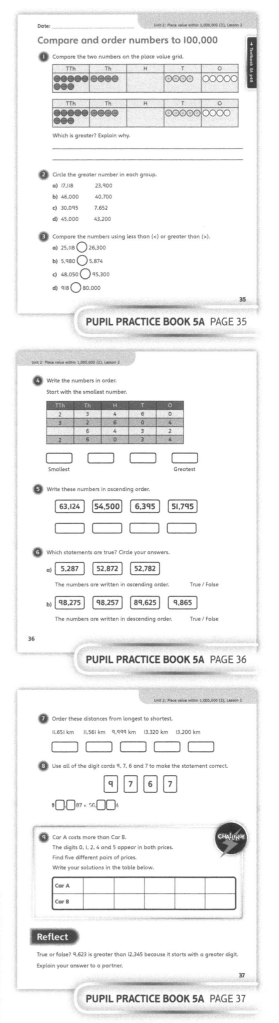

PUPIL PRACTICE BOOK 5A PAGE 35

PUPIL PRACTICE BOOK 5A PAGE 36

PUPIL PRACTICE BOOK 5A PAGE 37

Compare and order numbers to 1,000,000

Learning focus

In this lesson, children will use their understanding of place value and numbers up to 1,000,000 to compare and order numbers.

Before you teach ⏸

- Are all children comfortable using the new column, hundred thousands (HTh, 100,000s), in the place value grid?

NATIONAL CURRICULUM LINKS

Year 5 Number – number and place value

Read, write, order and compare numbers to at least 1,000,000 and determine the value of each digit.

ASSESSING MASTERY

Children can use their understanding of place value and numbers to 1,000,000 to explain how they know a number is greater or less than another, make accurate comparisons between numbers and order them correctly.

COMMON MISCONCEPTIONS

Children may order numbers incorrectly, for example, ordering in descending order instead of ascending order as the question may dictate. Ask:
- *How will you prove that you have put the numbers in that order?*

STRENGTHENING UNDERSTANDING

Children should be given opportunities to compare and order numbers up to 100,000. This will give children the chance to remind themselves of the process of comparing and ordering numbers before they move on to bigger numbers that include more place value headings.

GOING DEEPER

Children can play a game in groups of three or four. In turn, each child picks a number of digit cards and places them anywhere they like in a place value grid (up to 100,000s), with the aim of making the largest number (or smallest, depending on the rules) compared to the rest of their group. (Provide enough digit cards for each child to make a number in the 100,000s.) Ask: *Who has the biggest number? Who has the smallest? Use a number line to order the numbers.*

KEY LANGUAGE

In lesson: ascending, descending, compare, greater than (>), less than (<), largest, smallest, order, place value

Other language to be used by the teacher: equal to, most, number line, ones (1s), tens (10s), hundreds (100s), thousands (1,000s), ten thousands (10,000s), hundred thousands (100,000s), partition, partitioning

STRUCTURES AND REPRESENTATIONS

Place value grid

RESOURCES

Mandatory: place value counters

Optional: digit cards, dice

 In the eTextbook of this lesson, you will find interactive links to a selection of teaching tools.

Quick recap 🔄

Ask children to draw four boxes to put numbers in. They then roll a dice four times. Each time, children should put one of the numbers from the dice into their boxes. Once they have written a number in a box, they cannot change it. Ask the children to compare numbers with a partner. Ask: *Who has made the bigger number?* Order three or four of the numbers as a class. Repeat several times and extend to a 5-digit number.

Discover

WAYS OF WORKING Pair work

ASK

- Question ❶ a): *What have you learnt in previous lessons that can help you to compare these numbers? What is the same about the two numbers? What is different?*
- Question ❶ b): *What does 'ascending' mean? How could a place value grid help you to order these numbers?*

IN FOCUS In question ❶ a), children begin to investigate comparative sizes of numbers, and it will be important to link this to their prior learning about place value. Question ❶ b) requires children to use their understanding of numbers to arrange them in ascending order for the first time in the lesson.

PRACTICAL TIPS Suggest children use digit cards or place value arrow cards to create the numbers shown. Encourage them to discuss which numbers are bigger, which are smaller and how they can tell.

ANSWERS

Question ❶ a): Oxford has more 10,000s and so has a greater population.
162,200 > 123,900

Question ❶ b): The populations in ascending order are Durham, Cambridge, Oxford, Sunderland and Bristol.

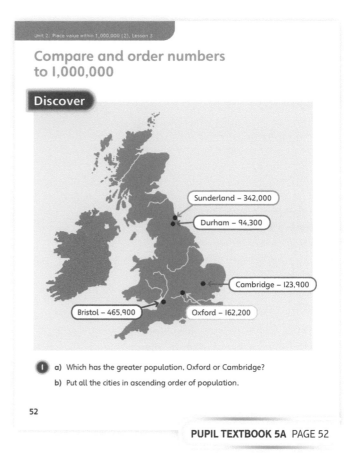

Compare and order numbers to 1,000,000

Discover

a) Which has the greater population, Oxford or Cambridge?
b) Put all the cities in ascending order of population.

52

PUPIL TEXTBOOK 5A PAGE 52

Share

WAYS OF WORKING Whole class teacher led

ASK

- Question ❶ a): *Show the number using a place value grid or place value counters. What other ways could you use to show the value of each number?*
- Question ❶ b): *How does the place value grid make the comparison clearer? What order did you put the populations into? Why? What end of the place value grid did you compare first? Why?*

IN FOCUS For question ❶ b), it is important to make explicit how the place value grid can help with comparing the value of numbers and how it can be used to order numbers easily. Make sure children recognise how to work through the place value grid, comparing the larger columns first.

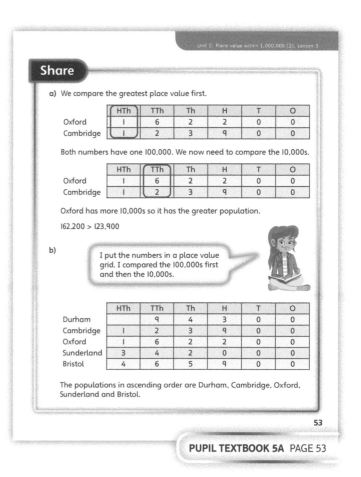

Share

a) We compare the greatest place value first.

	HTh	TTh	Th	H	T	O
Oxford	1	6	2	2	0	0
Cambridge	1	2	3	9	0	0

Both numbers have one 100,000. We now need to compare the 10,000s.

	HTh	TTh	Th	H	T	O
Oxford	1	6	2	2	0	0
Cambridge	1	2	3	9	0	0

Oxford has more 10,000s so it has the greater population.

162,200 > 123,900

b)

I put the numbers in a place value grid. I compared the 100,000s first and then the 10,000s.

	HTh	TTh	Th	H	T	O
Durham		9	4	3	0	0
Cambridge	1	2	3	9	0	0
Oxford	1	6	2	2	0	0
Sunderland	3	4	2	0	0	0
Bristol	4	6	5	9	0	0

The populations in ascending order are Durham, Cambridge, Oxford, Sunderland and Bristol.

53

PUPIL TEXTBOOK 5A PAGE 53

Think together

WAYS OF WORKING Whole class teacher led (I do, We do, You do)

ASK

- Question **1** a): *What will help you to compare these two numbers? What place value column should you begin your comparison at?*
- Question **1** b): *How is this question different from those you have seen before? What will you need to find before you can find the greatest population?*
- Question **2**: *How will you compare the numbers that have been written in different ways?*
- Question **3**: *What will you do to begin investigating this question?*

IN FOCUS Question **1** a): will develop children's fluency when comparing 6-digit numbers. Question **1** b) will then take this a step further by requiring children to compare more than two 5- and 6-digit numbers.

STRENGTHEN In question **2**, where children are asked to compare the numbers represented in different ways, encourage children to think about another way that they could represent the numbers to make them easier to compare. Ask: *What did you learn previously about the written names of numbers? Can you convert them to numerals?*

DEEPEN Deepen children's reasoning around comparing and ordering numbers in question **3** by offering them a question with multiple solutions. Ask: *How many solutions are there to this question? How can you prove that you have found them all?*

ASSESSMENT CHECKPOINT Check that children are able to compare two or more 6-digit numbers and use the information to order them correctly. Are children confident with the key language from the lesson, such as 'compare' and 'ascending'?

ANSWERS

Question **1** a): 311,899 < 312,785
Doncaster has the smaller population.

Question **1** b): Glasgow has the greatest population.

Question **2**: 195,311, 99,999, 308,000, seventy-nine thousand, two hundred

Question **3**: Look for any combination that works.
For example:
72,500, 126,091, 126,470, 133,904, 133,912
72,500, 126,191, 127,470, 133,904, 133,952

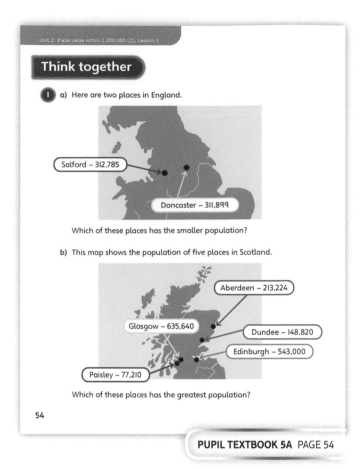

PUPIL TEXTBOOK 5A PAGE 54

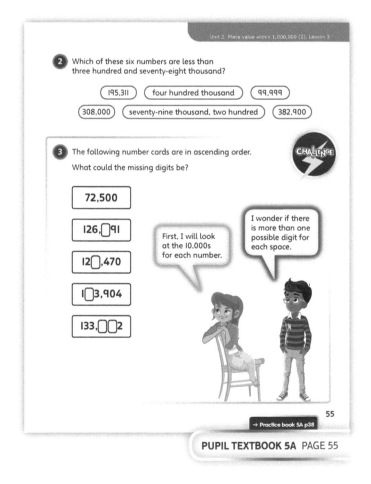

PUPIL TEXTBOOK 5A PAGE 55

Practice

Independent thinking

IN FOCUS Question ① offers children an opportunity to begin comparing numbers and identifying the largest, linking concrete representations of number to children's understanding of comparing and ordering numbers. Question ③ links children's use of the mathematical notation for comparison to their learning in this lesson. Question ④ requires children to order numbers in descending order for the first time in the lesson.

STRENGTHEN Provide place value grids if extra help is needed to compare pairs of numbers in question ②. Ask: *Show me how you would use a place value grid to compare the two numbers.*

DEEPEN Question ⑥ allows children to develop their understanding of the concepts of the lesson by requiring them to use reasoning to work out the missing digit in each case, looking at the largest place value first. Question ⑦ offers children the opportunity to investigate different solutions for the same question. Ask: *Is there only one solution to each statement? How can you prove that you have found all of the possible solutions?*

ASSESSMENT CHECKPOINT Can children compare numbers and order them correctly? Are they using the place value grid correctly to support them? In question ③, check children are correctly using the mathematical notation for number comparison. Question ④ assesses children's ability to compare and order numbers in descending order. Check that children have correctly started with the largest number.

ANSWERS Answers for the **Practice** part of the lesson can be found in the *Power Maths* online subscription.

Reflect

Independent thinking

IN FOCUS This **Reflect** question offers a good opportunity to assess children's understanding of how to compare numbers accurately. It may be interesting to ask children to compare their methods. What is the same and different about their approaches?

ASSESSMENT CHECKPOINT Children may refer to their use of the structures and representations covered in this unit, for example, the place value grid. Look for recognition of why it is important to begin comparing at the larger end of the place value grid.

ANSWERS Answers for the **Reflect** part of the lesson can be found in the *Power Maths* online subscription.

After the lesson ⏸

- Were children able to fluently and accurately order numbers in both ascending and descending order?
- What opportunities can you offer children to use these concepts in other areas of the curriculum?
- How many children made good progress in this lesson? What support will you offer to those children who did not?

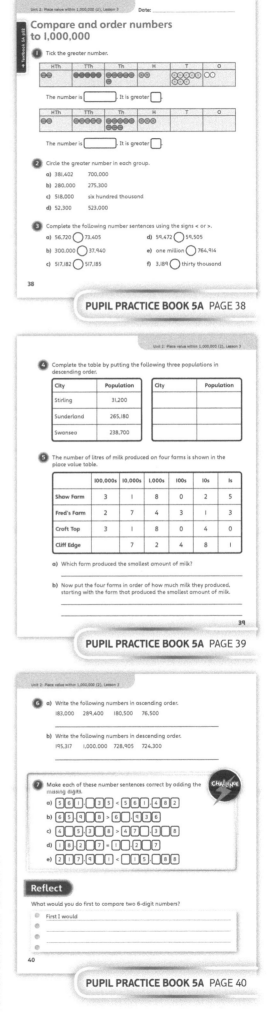

PUPIL PRACTICE BOOK 5A PAGE 38

PUPIL PRACTICE BOOK 5A PAGE 39

PUPIL PRACTICE BOOK 5A PAGE 40

Round numbers to the nearest 100,000

Learning focus

In this lesson, children learn how to round numbers to the nearest 100,000, through the real life context of the distance between the moon and earth and other planets. Children will use number lines to help them decide which 100,000 a number is closest to before rounding.

Before you teach

- Are children confident using number lines? Can they place numbers on a number line and work out intervals?
- Are children secure with place value of numbers up to 1,000,000? Can they read and write 6-digit numbers?

NATIONAL CURRICULUM LINKS

Year 5 Number – number and place value

Round any number up to 1,000,000 to the nearest 10, 100, 1,000, 10,000 and 100,000.

ASSESSING MASTERY

Children can round numbers up to 1,000,000 to the nearest 100,000. They can round 6-digit numbers to the nearest 100,000 and also smaller numbers such as 73,503.

COMMON MISCONCEPTIONS

Children may look at the wrong place value column when rounding. For example, when rounding 235,999, they may round up to 300,000 because the number ends with 9s. Ask:
- *Have you looked at the correct place value column?*
- *What could you use to check your answer?*

STRENGTHENING UNDERSTANDING

Give children access to wipe clean number lines so that they can use them throughout the lesson. You could also provide a template like in question 5 in the *Practice Book* to ensure they think about the previous and next 100,000 before they round.

GOING DEEPER

Ask children to explore the greatest and lowest possible numbers that round to a specific number of hundred thousands. They could also explore numbers with missing digits, for example, what could 3☐5,184 round to, to the nearest 100,000? What about this number 39☐,184? Ask: *Does it matter what the missing digit is?*

KEY LANGUAGE

In lesson: round, multiple

STRUCTURES AND REPRESENTATIONS

Number line

RESOURCES

Optional: digit cards

 In the eTextbook of this lesson, you will find interactive links to a selection of teaching tools.

Quick recap Q

Draw a number line on the board from 0 to 1,000,000, with all the 100,000s numbers marked. Ask children to estimate where the number 168,115 could be placed. Repeat for other numbers. Ask children to write down a number between 300,000 and 400,000. Discuss what children notice.

Discover

Pair work

ASK

- Question **1** a): *What is the smallest distance the moon could be from the sun? What is the greatest distance the moon could be from the sun? How many intervals will your number line need?*

IN FOCUS This context helps children visualise long distances and gives a reason for why distances might need to be rounded. Children begin to think about where the distances would lie on a number line between 0 and 1,000,000.

PRACTICAL TIPS Have a number line from 0 to 1,000,000 with intervals of 100,000 available and also a number line from 200,000 to 300,000 with intervals of 10,000.

ANSWERS

Question **1** a): Children should draw a number line from 0 to 1,000,000 with multiples of 100,000; with 200,000 and 300,000 labelled as the previous and next 100,000 for both numbers.

Question **1** b): 225,623 rounds to 200,000; 252,088 rounds to 300,000.

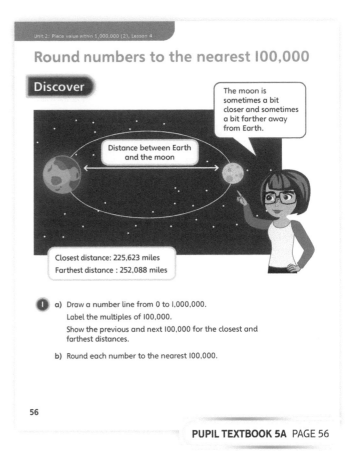

Round numbers to the nearest 100,000

Discover

The moon is sometimes a bit closer and sometimes a bit farther away from Earth.

Distance between Earth and the moon

Closest distance: 225,623 miles
Farthest distance : 252,088 miles

1 a) Draw a number line from 0 to 1,000,000.
Label the multiples of 100,000.
Show the previous and next 100,000 for the closest and farthest distances.

b) Round each number to the nearest 100,000.

56

PUPIL TEXTBOOK 5A PAGE 56

Share

Whole class teacher led

ASK

- Question **1** a): *Can you point to the approximate position of 225,623 and 252,088 on the number line?*
- Question **1** b): *Is 225,623 closer to 200,000 or 300,000? What about 252,088?*

IN FOCUS Children zoom in on a number line to see which multiple of 100,000 the numbers are closer to. This will help them round the numbers. Talk about which place value columns are most important when rounding to the nearest 100,000. Ask: *Does it matter what the digit is in the ones column? What about the tens column?*

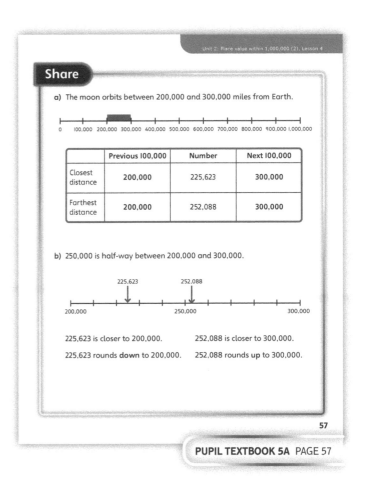

Share

a) The moon orbits between 200,000 and 300,000 miles from Earth.

0   100,000 200,000 300,000 400,000 500,000 600,000 700,000 800,000 900,000 1,000,000

	Previous 100,000	Number	Next 100,000
Closest distance	200,000	225,623	300,000
Farthest distance	200,000	252,088	300,000

b) 250,000 is half-way between 200,000 and 300,000.

225,623   252,088

200,000   250,000   300,000

225,623 is closer to 200,000.   252,088 is closer to 300,000.

225,623 rounds **down** to 200,000.   252,088 rounds **up** to 300,000.

57

PUPIL TEXTBOOK 5A PAGE 57

Think together

WAYS OF WORKING Whole class teacher led (I do, We do, You do)

ASK

- Question **1** a): *How did you choose your numbers? Are you likely to choose one that is the same as someone else's?*
- Question **1** b): *What does the word previous mean? Will your answers be the same as someone else's this time?*

IN FOCUS In Question **1** a), children have the freedom to think about their own number for each section of the number line. In question **1** b), children need to think about the previous and next multiples of 100,000 for each of their numbers. This helps children round in question **2**. Draw children's attention to the place value columns in question **3**.

STRENGTHEN Provide children with printed number lines so that they can focus on rounding rather than worrying about drawing accurate number lines.

DEEPEN Ask: *How would you round 999,999 to the nearest 100,000?*

ASSESSMENT CHECKPOINT Children can place numbers on a number line, identify the previous and next 100,000 of the number and round numbers to the nearest 100,000.

ANSWERS

Question **1** a): Answers will vary but the number for A must be between 0 and 100,000, the number for B between 300,000 and 400,000 and the number for C between 800,000 and 900,000.

Question **1** b): A: 0 and 100,000
B: 300,000 and 400,000
C: 800,000 and 900,000

Question **2** a): 450,000

Question **2** b): Children should point to where 403,511, 449,789 and 470,000 are on the number line.

Question **2** c): 403,511 rounds to 400,000; 449,789 rounds to 400,000; 470,000 rounds to 500,000.

Question **3** a): Answers will vary but should include looking at the first digit to identify the previous and next 100,000 and using the other digits to round up or down.

Question **3** b): 268,200 rounds to 300,000; 409,975 rounds to 400,000; and 700,500 rounds to 700,000.

Question **3** c): 50,000 rounds to 100,000; 350,000 rounds to 400,000; and 950,000 rounds to 1,000,000.

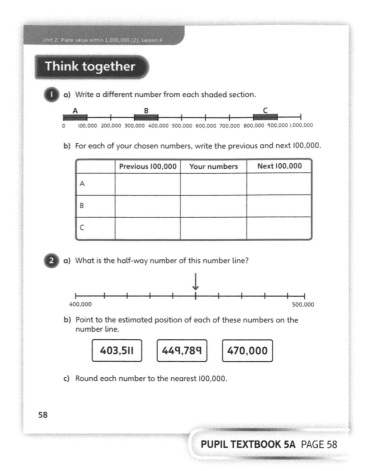

PUPIL TEXTBOOK 5A PAGE 58

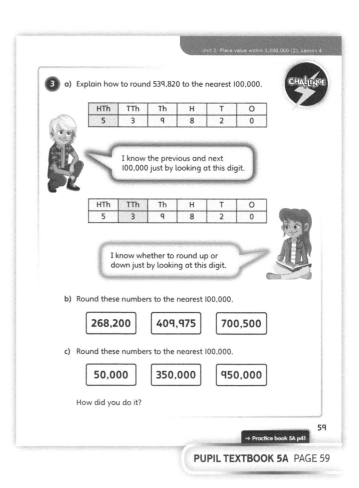

PUPIL TEXTBOOK 5A PAGE 59

Practice

WAYS OF WORKING Independent thinking

IN FOCUS The exercises build from children simply writing 6-digit numbers to rounding numbers to the nearest 100,000. The first part of question ④ links exactly to question ③ to help build confidence.

STRENGTHEN Provide children with wipe clean number lines to help them. You could provide blank number lines with ten intervals or add the start and end values. If you add the start and end values, you will need one from 0 to 1,000,000 then one for each 100,000, for example, 0 to 100,000, 100,000 to 200,000, 200,000 to 300,000, etc. Provide digit cards for question ⑤ so that children can physically manipulate them.

DEEPEN Ask children to identify which digits are relevant when rounding to the nearest 100,000 in each of these numbers: 268,912, 68,912, 8,912. Ask: *What is the greatest number that still rounds to 0 when rounding to the nearest 100,000?*

ASSESSMENT CHECKPOINT Children can place numbers on a number line, identify the previous and next multiple of 100,000 of the number and round numbers to the nearest 100,000.

ANSWERS Answers for the **Practice** part of the lesson can be found in the *Power Maths* online subscription.

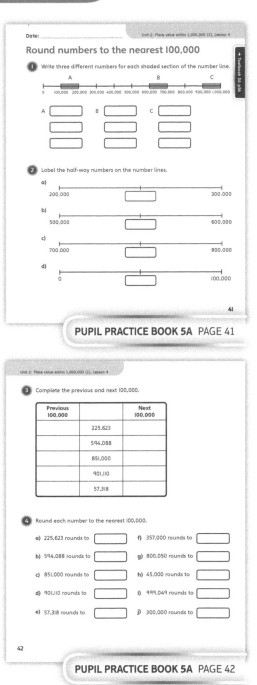

PUPIL PRACTICE BOOK 5A PAGE 41

PUPIL PRACTICE BOOK 5A PAGE 42

Reflect

WAYS OF WORKING Independent thinking

IN FOCUS Children can describe a method for rounding to the nearest 100,000. You would expect them to talk about identifying the previous 100,000 and the next 100,000 of a number and then decide which it is closest to.

ASSESSMENT CHECKPOINT Children can identify the important digits in a number when rounding to the nearest 100,000.

ANSWERS Answers for the **Reflect** part of the lesson can be found in the *Power Maths* online subscription.

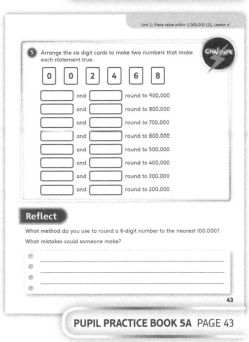

After the lesson ⏸

- Can children explain which digit should be checked each time in a number to be rounded to 100,000?
- Can children reason about numbers that do or do not round to a given multiple?
- Do children know which multiple of 100,000 comes after 900,000?

PUPIL PRACTICE BOOK 5A PAGE 43

Round numbers to the nearest 10,000

Learning focus

In this lesson, children will learn to round to the nearest 10,000 and identify the next and previous multiple, reasoning about which digit to check in a number to help make decisions on rounding.

Before you teach

- Can children work out intervals on a number line?
- Can children estimate where a number lies on a number line?
- Can children say the two multiples of 10,000 that a number lies between?

NATIONAL CURRICULUM LINKS

Year 5 Number – number and place value

Round any number up to 1,000,000 to the nearest 10, 100, 1,000, 10,000 and 100,000.

ASSESSING MASTERY

Children can round numbers up to 1,000,000 to the nearest 10,000. They can round 5-digit numbers to the nearest 10,000 and also smaller numbers such as 3,503.

COMMON MISCONCEPTIONS

Children may look at the wrong place value column when rounding. For example, when rounding 34,999, they may round up to 40,000 because the number ends with 9s. Ask:
- *Have you looked at the correct place value column?*
- *What could you use to check your answer?*

STRENGTHENING UNDERSTANDING

Give children access to wipe clean number lines so that they can use them throughout the lesson. You could also provide a template like the one in the **Share** section in the **Textbook** to ensure they think about the previous and next 10,000 before they round.

GOING DEEPER

Ask children to explain why, for example, 95,300 rounds to the same 10,000 as 100,000. Encourage them to find other numbers that round in a similar way. Challenge them to make generalisations about numbers that round in this way.

KEY LANGUAGE

In lesson: round, multiple, digit, ten thousands (10,000s), nearest, next, previous, closer to

Other language to be used by the teacher: positive (⁺)

STRUCTURES AND REPRESENTATIONS

Place value grid, number lines

RESOURCES

Mandatory: digit cards

Optional: place value counters

 In the eTextbook of this lesson, you will find interactive links to a selection of teaching tools.

Quick recap

Give children digits cards/fan. Ask them to make a 4-digit number using their cards/fan.

Children should then ask a partner to round their number to the nearest 1,000.

Ask children to make numbers that round to 5,000 to the nearest 1,000. Ask: *What strategies did you use?*

Discover

Pair work

ASK

- Question ① a): *Which pool holds the most water? Which pool holds the least water? How many intervals will the number line have? Why is the position of each number an estimate?*

IN FOCUS In question ① a), children consider where each 5-digit number lies on a number line from 0 to 100,000. In question ① b), they consider which multiple of 10,000 each number is closest to and then round to the nearest 10,000.

PRACTICAL TIPS Ask children to read all of the numbers aloud and explain the value of each digit. Practise counting in steps of 10,000 on to 100,000 and back again. Discuss some other numbers that will or will not round to the same numbers and use place value counters to help visualise the values of numbers.

ANSWERS

Question ① a): Children should draw a number line from 0 to 100,000 with multiples of 10,000 labelled and with estimates for 41,300, 77,735 and 98,275 marked.

Question ① b): 41,300 rounds to 40,000; 77,735 rounds to 80,000 and 98,275 to 100,000.

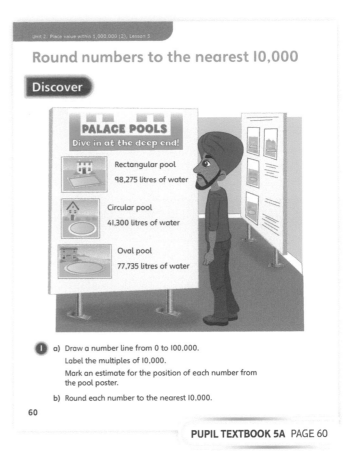

Round numbers to the nearest 10,000

Discover

① a) Draw a number line from 0 to 100,000.
Label the multiples of 10,000.
Mark an estimate for the position of each number from the pool poster.

b) Round each number to the nearest 10,000.

60

PUPIL TEXTBOOK 5A PAGE 60

Share

Whole class teacher led

ASK

- Question ① a): *Which multiples of 10,000 does each number lie between?*
- Question ① b): *How does the place value grid help you round the numbers? What does each digit in the number tell you? Which digits are the most important when rounding to the nearest 10,000?*

IN FOCUS In question ① a) children are reminded about the use of number lines and the position of numbers between two multiples to help make decisions about rounding up or down. Question ① b) focuses more on the digits that should be checked each time to help round, developing the ability to round without a number line. The number line is provided for support, but the reasoning is based on the digit position and place value.

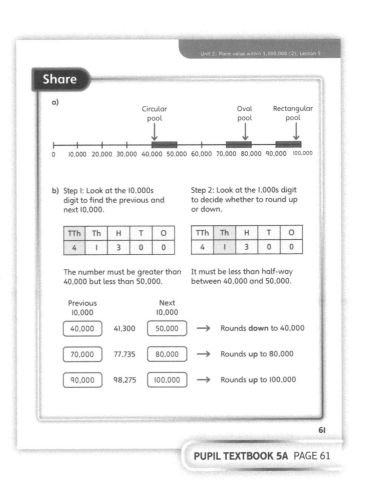

Share

61

PUPIL TEXTBOOK 5A PAGE 61

Think together

WAYS OF WORKING Whole class teacher led (I do, We do, You do)

ASK

- Question **1**: *What is the previous multiple of 10,000 and the next multiple of 10,000 for each number? Which place value column should you consider when deciding whether to round up or down?*
- Question **2**: *How does the number line help you decide which numbers round to 30,000? What is the smallest number that will round to 30,000? What is the greatest number that will round to 30,000?*

IN FOCUS Place value grids are used in question **1** to draw children's attention to the digit in the thousands column. This is the digit that tells them whether to round up or down. The fourth example has been chosen carefully to ensure children are exposed to examples that round to 0.

STRENGTHEN Provide place value grids and number lines for children so that they can recreate each question practically.

DEEPEN Encourage children to think about the smallest and greatest numbers that round to a certain multiple of 10,000. Ask: *What is the greatest number that would round to 0 when rounded to the nearest 10,000? What is the smallest number that would round to 70,000 when rounded to the nearest 10,000?*

ASSESSMENT CHECKPOINT Children can round numbers to the nearest 10,000 and talk about why place value grids and number lines are useful ways to explain and justify their answers.

ANSWERS

Question **1**: 26,291 rounds to 30,000; 63,059 rounds to 63,000; 89,001 rounds to 90,000 and 4,275 rounds to 0.

Question **2**: 27,700 and 33,501

Question **3** a): Children should recognise that Danny is looking at the 100,000s not the 10,000s and is incorrect. Ebo is on the right track in counting in 10,000s but needs to identify the 10,000 before 426,835 and the 10,000 after. Kate is incorrect and Lexi is correct. 426,835 rounded to the nearest 10,000 is 430,000.

Question **3** b): 151,380 rounds to 150,000; 199,250 rounds to 200,000; 277,907 rounds to 280,000; and 5,001 rounds to 10,000.

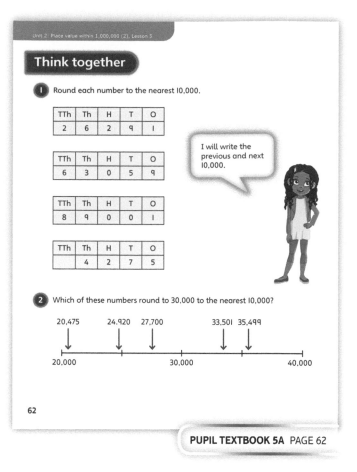

PUPIL TEXTBOOK 5A PAGE 62

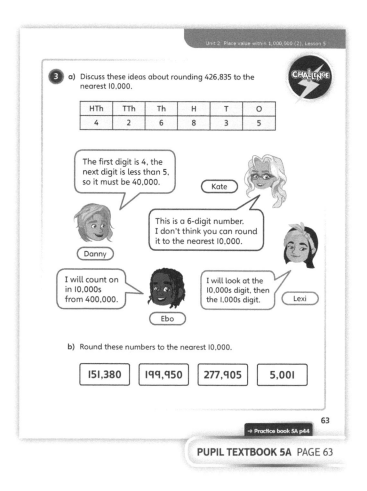

PUPIL TEXTBOOK 5A PAGE 63

Practice

WAYS OF WORKING Independent thinking

IN FOCUS The exercises build from children simply writing 5-digit numbers to rounding numbers to the nearest 10,000.

STRENGTHEN Provide children with wipe clean number lines to help them. You could provide blank ones with ten intervals or add the start and end values. If you add the start and end values, you will need one from 0 to 100,000 then one for each 10,000, for example, 0 to 10,000, 10,000 to 20,000, 20,000 to 30,000, etc.

DEEPEN Ask: *Which digits are relevant when rounding to the nearest 10,000 in each of these numbers: 68,912, 8,912, 912?* Ask: *What is the greatest number that still rounds to 0 when rounding to the nearest 10,000?*

THINK DIFFERENTLY Question **5** makes children aware of another real-life context for negative numbers. This question will challenge children's thinking as the 0 point (sea level) is an imaginary line, not actually below the mountain itself but some way up it.

ASSESSMENT CHECKPOINT Children can place numbers on a number line, identify the previous and next multiple of 10,000 of the number and round numbers to the nearest 10,000.

ANSWERS Answers for the **Practice** part of the lesson can be found in the *Power Maths* online subscription.

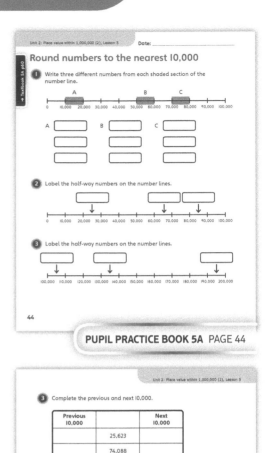

PUPIL PRACTICE BOOK 5A PAGE 44

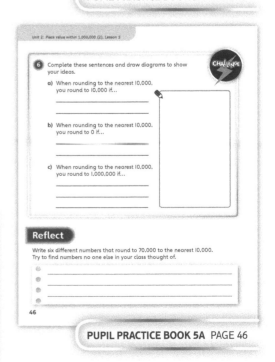

PUPIL PRACTICE BOOK 5A PAGE 45

Reflect

WAYS OF WORKING Independent thinking

IN FOCUS Children are asked to write down six different numbers that round to 70,000. Some children may go for six consecutive numbers starting with the highest that rounds to 70,000, for example, 69,999, 69,998, 69,997, 69,996, 69,995, 69,994. This is an interesting and systematic approach which should be celebrated. Children could then be challenged to choose a number with a 7 in the thousands column.

ASSESSMENT CHECKPOINT Children can identify the important digits in a number when rounding to the nearest 10,000.

ANSWERS Answers for the **Reflect** part of the lesson can be found in the *Power Maths* online subscription.

After the lesson ⏸

- Can children explain which digit should be checked when rounding to the nearest 10,000? Can they compare their method of rounding to the nearest 10,000 and rounding to the nearest 100,000?
- Can children reason about numbers that do or do not round to a given multiple?
- Do children know which multiple of 10,000 comes after 90,000?

PUPIL PRACTICE BOOK 5A PAGE 46

Round numbers to the nearest 10, 100 and 1,000

Learning focus

In this lesson, children will use their understanding of place value to help them round numbers to the nearest 10, 100 and 1,000. They will discuss when rounding is appropriate and which multiple of 10 to round to in a given context.

Before you teach

- Can children recall the rules for rounding?
- Are children confident when using a number line?
- How will you help children understand the importance of place value when rounding?

NATIONAL CURRICULUM LINKS

Year 5 Number – number and place value

Round any number up to 1,000,000 to the nearest 10, 100, 1,000, 10,000 and 100,000.

ASSESSING MASTERY

Children can recognise when to round numbers and explain how to do so fluently. They can apply their understanding of rounding to larger numbers and can reliably round up and down to the nearest 10, 100, 1,000, 10,000 and 100,000.

COMMON MISCONCEPTIONS

Children may assume that rounding only affects the place value column it references, for example that rounding to the nearest 100 will only affect the digits in the hundreds, tens and ones columns. Ask:
- *If you round 1,972 to the nearest 100, what happens?*
- *Show me how the number will change using a picture or resources.*
- *What happens when you count on one more hundred from 900?*

STRENGTHENING UNDERSTANDING

Before the lesson, children who may need more help should be encouraged to recap their understanding of rounding. Encourage children to round numbers to the nearest 10,000. Ask: *What does it mean to round a number? What numbers can you round to? Try showing me how to round x to the nearest x?*

GOING DEEPER

Provide children with six digit cards. Ask: *What numbers can you make with these digit cards? What numbers in the 100,000s can you make that will round down to the nearest 1,000? What numbers in the 100,000s can you make that will round up to the nearest 1,000? Is there more of one type than the other? Why?*

KEY LANGUAGE

In lesson: compare, greater, greatest, place value, round, trial and error

Other language to be used by the teacher: round up, round down

STRUCTURES AND REPRESENTATIONS

Number line, place value grid

RESOURCES

Optional: digit cards

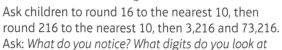 In the eTextbook of this lesson, you will find interactive links to a selection of teaching tools.

Quick recap ↻

Ask children to round 16 to the nearest 10, then round 216 to the nearest 10, then 3,216 and 73,216. Ask: *What do you notice? What digits do you look at each time?*

Ask children to round 562 to the nearest 100, then the same for 1,562, 31,562 and 931,562. Ask: *What do you notice?*

Discover

WAYS OF WORKING Pair work

ASK

- Questions ① a): *How can you tell that Jamie has not rounded the number correctly?*
- Questions ① a): *Has Danny rounded the number to the nearest 100?*
- Question ① b): *How will you know whether to round up or down? Which digit will you need to look at when rounding to the nearest 100?*

IN FOCUS Children have the opportunity to explore some common misconceptions as a class when rounding to the nearest 100. Jamie has rounded to the nearest 10 and Danny has rounded to the nearest 100,000.

PRACTICAL TIPS Provide children with digit cards, as in the picture. Suggest they investigate the different numbers that can be made with the same five or six digits. Challenge them to find out what numbers they can make that can be rounded to the same 100, 1,000, 10,000 and so on.

ANSWERS

Question ① a): Jamie has rounded to the nearest 10.

Question ① b): 124,600

Round numbers to the nearest 10, 100 and 1,000

Discover

I think she made a mistake. It should be 100,000.

I rounded your number.

124,580

1 2 4 5 7 8

Jamie   Danny

① a) What nearest multiple has Jamie rounded Danny's number to?

b) Round Danny's number to the nearest 100.

64

PUPIL TEXTBOOK 5A PAGE 64

Share

WAYS OF WORKING Whole class teacher led

ASK

- Question ① a): *What digit did you look at to know how to round the number? Why? How does the number line show this?*
- Question ① b): *How could you prove the number rounded to 124,600? What rules for rounding helped you identify what numbers would and would not work?*

IN FOCUS For questions ① a) and b), it is important to make sure children are given the opportunity to identify the rules that apply to rounding to the nearest 100. Remind children to look at the place value column before the one they are rounding to, for example the tens column if they are rounding to the nearest 100. If the number is less than 5, they round down. If the number is 5 or more, they round up.

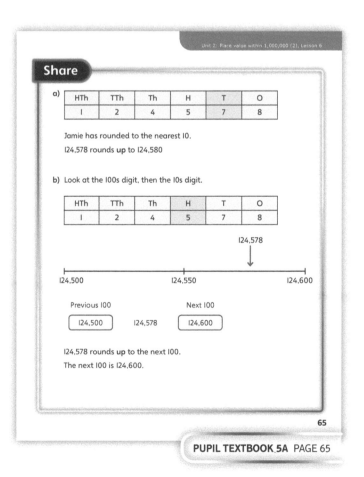

Share

a)

HTh	TTh	Th	H	T	O
1	2	4	5	7	8

Jamie has rounded to the nearest 10.
124,578 rounds up to 124,580

b) Look at the 100s digit, then the 10s digit.

HTh	TTh	Th	H	T	O
1	2	4	5	7	8

124,578

124,500 ———— 124,550 ———— 124,600

Previous 100   Next 100

124,500   124,578   124,600

124,578 rounds up to the next 100.
The next 100 is 124,600.

65

PUPIL TEXTBOOK 5A PAGE 65

Think together

Whole class teacher led (I do, We do, You do)

ASK

- Question **1**: *Will any of your answers be the same? Which place value column do you need to look at in each case?*
- Question **2**: *What changed each time? What stayed the same?*
- Question **3**: *Why have the usual rounding rules been ignored for the t-shirts?*

IN FOCUS Question **1** provides an opportunity for children to investigate how rounding to a different multiple of 10 can affect a number in different ways, helping children to strengthen their understanding that a number can round up or down depending on the value of each digit.

STRENGTHEN Provide place value grids and number lines for children so that they can re-create each question practically.

DEEPEN Question **3** gives children the opportunity to explore where rounding is used in real-life contexts and that sometimes this means going against the usual rounding rules.

ASSESSMENT CHECKPOINT Are children able to use a number line to support them when rounding up or down? Are they confident when deciding whether to round up or down? Check that children understand the rule for rounding. Use question **1** to assess whether children can correctly round a number up or down to the nearest 10, 100, 1,000, 10,000 and 100,000.

ANSWERS

Question **1**: 100,000; 130,000; 128,000; 127,900; 127,850

Question **2**: 60; 960; 1,960; 21,960; 521,960
You always look at the 1s digit and the last two digits always round to 60. It does not matter about the other digits when rounding to the nearest 10.

Question **3** a): The music promoter rounded up to the next 100 to make sure that there were t-shirts for everyone.
The engineer has rounded up to the nearest 1,000 but it is not safe to do this since this is more than the maximum load of 24,150 kg.
The cost of the microscope has been rounded to the nearest £50,000. This will ensure that there is more than enough money to pay for this but they could have rounded to the nearest £10,000 and still had enough funds.

Question **3** b): All the newspaper headlines have likely been rounded up or down, probably up in order to grab people's attention with a more dramatic storyline.

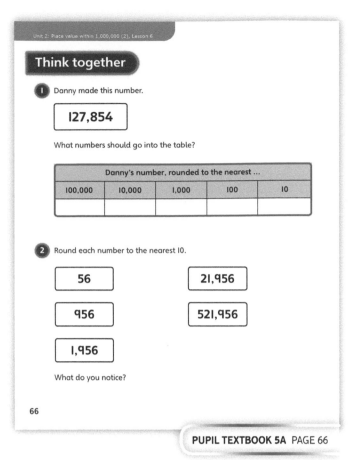

PUPIL TEXTBOOK 5A PAGE 66

PUPIL TEXTBOOK 5A PAGE 67

Practice

Independent thinking

IN FOCUS Question ❶ gives children plenty of practice at rounding a number to a given multiple of 10. Question ❷ allows children to see that the same rounding rules apply even if the question is in a context such as money. Question ❹ provides the opportunity to think about maximum and minimum values within the context of mass.

STRENGTHEN To strengthen understanding in question ❷, you could first present the question without the units. Ask: *Do the units make a difference?*

DEEPEN Question ❻ deepens understanding by asking children to think about numbers that fulfil the given criteria and those that do not. Being able to explain what something is and what something is not shows a deeper understanding of a concept.

ASSESSMENT CHECKPOINT Check children are rounding to the nearest 10, 1,000, 10,000 and 100,000 accurately. For question ❹, are children able to state the maximum and minimum values? Are children able to round accurately when the questions are presented in a context?

ANSWERS Answers for the **Practice** part of the lesson can be found in the *Power Maths* online subscription.

Reflect

WAYS OF WORKING Independent thinking

IN FOCUS This question encourages children to generalise about the properties of numbers and how these properties can influence how a number changes when rounded. Ask: *What do you notice about numbers where all the digits change? Does this apply when rounding to other multiples of 10? Explain your answer.*

ASSESSMENT CHECKPOINT Assess children's recognition of how the properties of a number can affect how it is rounded, in particular how multiples of 9 in a number can produce a chain, crossing the multiples of 10.

ANSWERS Answers for the **Reflect** part of the lesson can be found in the *Power Maths* online subscription.

After the lesson

- Are children able to clearly explain the rules of rounding?
- Which did the class find most challenging; rounding to the nearest 100,000, 10,000, 1,000, 100 or 10?

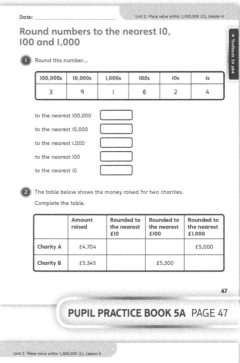
PUPIL PRACTICE BOOK 5A PAGE 47

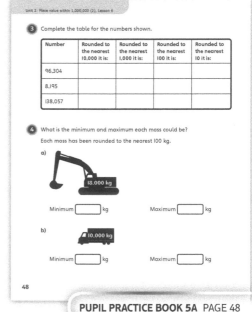
PUPIL PRACTICE BOOK 5A PAGE 48

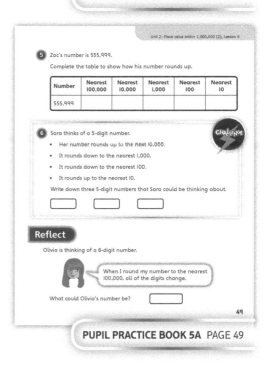
PUPIL PRACTICE BOOK 5A PAGE 49

End of unit check

> **Don't forget the unit assessment grid in your *Power Maths* online subscription.**

WAYS OF WORKING Group work teacher led

IN FOCUS These questions are designed to draw out misconceptions or misunderstandings.

- Question ❶ assesses children's recognition of the place value of digits in a number up to 1,000,000.
- Question ❸ assesses children's ability to recognise numbers up to 1,000,000 along an unlabelled number line.
- Question ❻ is a SATs-style question and requires children to use their understanding of the rules for rounding and how these are linked to the properties of numbers.
- Question ❼ is a SATs-style question that looks at focusing on rounding a given number to different degrees of accuracy.

ANSWERS AND COMMENTARY Children who have mastered the concepts in this unit will be able to demonstrate fluency in place value within numbers up to 1,000,000. They will name, partition and write the names of numbers accurately and compare and order numbers up to 1,000,000 with confidence. They will complete partial number lines, using the unlabelled marks to help them calculate what each interval is worth. They will round up and down to the nearest multiple of 10 and will identify and use the rules that govern number sequences.

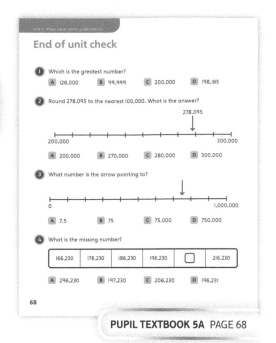

PUPIL TEXTBOOK 5A PAGE 68

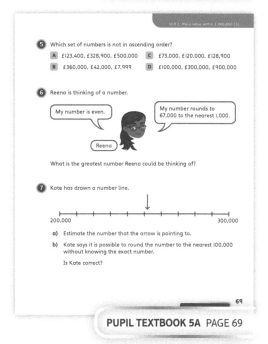

PUPIL TEXTBOOK 5A PAGE 69

Q	A	WRONG ANSWERS AND MISCONCEPTIONS	STRENGTHENING UNDERSTANDING
1	C	Incorrect choices indicate children have not mastered the unit.	Allow children to build numbers using manipulatives including place value grids and place value counters.
2	D	C suggests that children have rounded to the nearest 10,000 instead of 100,000.	
3	D	A, B or C indicate mistakes when recognising the position of numbers between two points on an unlabelled number line.	
4	C	A and B suggest recognising the repetition of 230 without knowing the reason. D suggests counting on 1.	
5	B	Incorrect answers suggest that children do not understand what the term ascending means.	
6	**67,498**. An incorrect answer will indicate children are not clear on the rules for rounding.		
7	Part a) **any number between 250,001 and 253,000**. Part b) **Yes**. Incorrect answers suggest that children are not clear on the rules for rounding.		

My journal

WAYS OF WORKING Independent thinking

ANSWERS AND COMMENTARY

A number between 250,000 and 350,000.	Solutions such as: 315,689, 315,896, 316,589, …
A number that has a smaller number of 100s than 10,000s.	Solutions such as: 896,153, 968,153, 695,813, …
The greatest even number that can be made.	985,316
A number that rounds to 600,000 to the nearest 100,000.	Any number between 550,000 and 649,999
The smallest number that rounds to 600,000 to the nearest 100,000.	561,389
The number that is 10,000 less than 875,913.	865,913

To strengthen children's understanding, ask:
- If the answer must be between 250,000 and 350,000, what does that mean for the 100,000s digit?
- Are there any clues where there is only one solution? Can you prove it?

Power check

WAYS OF WORKING Independent thinking

ASK

- How confident are you with the place value of numbers up to 1,000,000?
- Could you explain to someone how to round numbers up to 1,000,000?
- Are you confident to compare and order numbers to 1,000,000?

Power play

WAYS OF WORKING Pair work

IN FOCUS Use this **Power play** to assess whether children can complete the rounding tasks according to the number they roll on the dice.

ANSWERS AND COMMENTARY Can children round correctly according to the number they land on and the number they roll? Can they answer the place value questions correctly?

If children need support, ask: *Can you count on using a number line?*

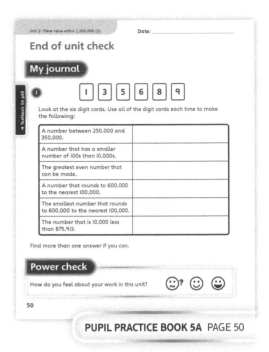

PUPIL PRACTICE BOOK 5A PAGE 50

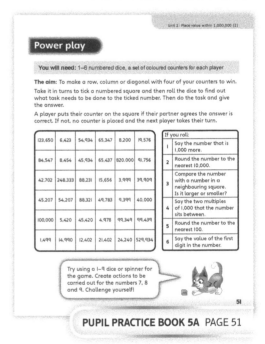

PUPIL PRACTICE BOOK 5A PAGE 51

After the unit ⏸

- How confident are you that children have a deep conceptual understanding of the place value of 6-digit numbers?
- What one strategy or teaching approach worked particularly well in this unit? How can you apply the same strategy in other areas of your mathematics teaching?

Strengthen and **Deepen** activities for this unit can be found in the *Power Maths* online subscription.

Unit 3
Addition and subtraction

Mastery Expert tip! 'To build confidence and recall with addition and subtraction methods, I encourage adding and subtracting in different areas of the curriculum throughout the year. It definitely helps children become more familiar and confident with the method.'

Don't forget to watch the Unit 3 video!

WHY THIS UNIT IS IMPORTANT

This unit is important because it allows children to apply the formal written methods of addition and subtraction to numbers with up to five digits. The range of problem-solving questions involving adding and subtracting, including mentally, will develop confidence and flexibility when exploring the most efficient ways to add and subtract.

WHERE THIS UNIT FITS

→ Unit 2: Place value within 1,000,000 (2)
→ **Unit 3: Addition and subtraction**
→ Unit 4: Multiplication and division (1)

This unit builds on children's work in Year 4, where they learnt to add and subtract 4-digit numbers. It will extend their knowledge of addition and subtraction using formal written methods for numbers with up to five digits. It will give children the opportunity to build confidence with problem solving and to explore efficient methods for addition and subtraction calculations, including those that can be solved mentally.

Before they start this unit, it is expected that children:
- can add and subtract numbers with up to four digits
- are able to solve sum and difference word problems
- have experience of using inverse operations to check calculations
- are able to round numbers to the nearest 10, 100, 1,000 and 10,000.

ASSESSING MASTERY

Children who have mastered this unit can add and subtract numbers with up to five digits using a variety of methods, including formal written methods and mental methods. They can confidently apply their knowledge of addition and subtraction to solving word problems.

COMMON MISCONCEPTIONS	STRENGTHENING UNDERSTANDING	GOING DEEPER
When adding or subtracting numbers with a different number of digits, children may line the numbers up from the left, not the right.	Provide place value grids and counters to help children visualise where the digits should be and when they will need to exchange.	Compare calculations laid out correctly and incorrectly. What is the difference between the correct and incorrect answer? Why?
When solving a problem, children may choose the wrong calculation. For example, adding when they should subtract.	Use a bar model to help children understand whether a calculation requires addition or subtraction.	Ask increasingly difficult word problems, including ones that bring in other areas of the mathematics curriculum.
Children may think they need to rely on written methods, even when a mental method might be more efficient.	Ensure that children know the bonds to any number within 10 to help them mentally add and subtract quickly.	Ask children to justify the most efficient way of solving addition and subtraction problems. This is not always the column method.

Unit 3: Addition and subtraction

UNIT STARTER PAGES

Go through the unit starter pages of the textbook with the whole class. Talk through the key learning points that the characters mention and the key vocabulary. Discuss Dexter's comment about the importance of setting out calculations in columns neatly.

STRUCTURES AND REPRESENTATIONS

Place value grid: This model uses counters to show the value of each digit in a number, which supports the column method layout.

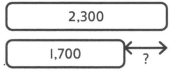

TTh	Th	H	T	O

Column addition and column subtraction: This model demonstrates the place value of each digit in addition and subtraction calculations and shows exchanges between columns.

TTh	Th	H	T	O
1	6	9	9	8
+	2	1	5	6
1	9	1	5	4
		1	1	1

TTh	Th	H	T	O
⁷8̶	¹2	⁶7̶	¹0	6
− 3	9	4	1	5
4	3	2	9	1

Bar model: This model can be used to represent the calculation needed in some addition and subtraction word problems.

2,300

1,700	?

Part-whole model: This model can be used to partition a number into two parts. It can also be used to help solve addition and subtraction problems.

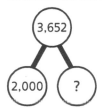

KEY LANGUAGE

There is some key language that children will need to know as part of the learning in this unit.

→ add, subtract
→ ones (1s), tens (10s), hundreds (100s), thousands (1,000s), ten thousands (10,000s)
→ inverse
→ round
→ mentally
→ estimate
→ distance chart

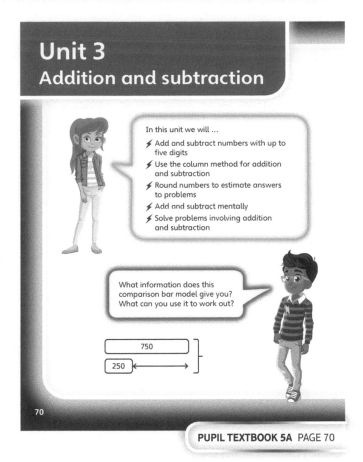

PUPIL TEXTBOOK 5A PAGE 70

PUPIL TEXTBOOK 5A PAGE 71

Mental strategies (addition)

Learning focus

In this lesson, children will learn how to add whole numbers mentally by choosing the most efficient method from a variety of strategies.

Before you teach

- Do children know how to partition a number into 1s, 10s, 100s, and so on?
- Can they add multiples of 10, 100 and 1,000 mentally?
- Can children round to the nearest 10, 100 and 1,000?

NATIONAL CURRICULUM LINKS

Year 5 Number – addition and subtraction

Add and subtract numbers mentally with increasingly large numbers.

ASSESSING MASTERY

Children can understand which mental methods can be used to efficiently add whole numbers.

COMMON MISCONCEPTIONS

Children may miscalculate when working out mentally. For example, they think the answer to $53 + \square = 80$ is 37. Ask:
- *What method are you using? What will the next 10 be in this calculation? Are you sure?*

Children may ignore an exchange that is needed because they are working out mentally. Ask:
- *What is the place value of each digit in your calculation? Do you need to do an exchange here?*

STRENGTHENING UNDERSTANDING

Children should start by adding 1-digit and then 2-digit numbers with no exchanges. Build up to calculations that require an exchange and calculations with 3-digit numbers. Encourage children to use a number line to visualise the calculation in a pictorial way.

GOING DEEPER

Ask children to add mentally more than two whole numbers together. Represent the calculations in different ways, for example on a part-whole model or bar model, and ask them to work out the totals or missing parts mentally.

KEY LANGUAGE

In lesson: method, add, subtract

Other language to be used by the teacher: total, round, exchange, ones (1s), tens (10s), hundreds (100s)

STRUCTURES AND REPRESENTATIONS

Number lines, part-whole models

RESOURCES

Optional: place value counters

 In the eTextbook of this lesson, you will find interactive links to a selection of teaching tools.

Quick recap

Ask children to add multiples of 100 mentally, such as 200 + 400. How quickly can they get the answer? If children are confident, move on to examples where the 100s cross the 1,000, for example 700 + 600. Briefly discuss their methods.

Discover

WAYS OF WORKING Pair work

ASK

- Question ❶ a): *What do you need to think about if you are doing this mentally? What is the same and what is different in this pair?*
- Question ❶ b): *Can you partition these numbers to help you? What is the same and what is different in this pair?*

IN FOCUS Question ❶ a) requires children to use number facts they know to calculate mentally with greater numbers. Pair A is an addition of two 4-digit numbers and an addition of two 5-digit numbers. Also, in question ❶ b) children are given two numbers and asked to work out the total mentally. Pair B is an addition of two 2-digit numbers and an addition of two 3-digit numbers.

PRACTICAL TIPS Make flashcards of the calculations in the **Discover** picture. Ask children to read each calculation aloud and arrange them in pairs, as in the picture. Can they explain why the calculations have been paired like this?

ANSWERS

Question ❶ a): 2,000 + 7,000 = **9,000**

40,000 + 30,000 = **70,000**

Question ❶ b): 45 + 23 = **68**

450 + 230 = **680**

Mental strategies (addition)

Discover

Work these out mentally.

Pair A

Pair B

2,000 + 7,000 = ☐

45 + 23 = ☐

40,000 + 30,000 = ☐

450 + 230 = ☐

❶ a) Work out the answers to Pair A in your head.

Now explain to a partner how you worked them out.

b) Work out the answers to Pair B in your head.

Now explain to your partner how you worked them out.

72

PUPIL TEXTBOOK 5A PAGE 72

Share

WAYS OF WORKING Whole class teacher led

ASK

- Question ❶ a): *What is 2 + 7? How can you use this to work out 2,000 + 7,000?*
- Question ❶ a): *What is 4 + 3? How can you use this to work out 40,000 + 30,000?*
- Question ❶ b): *How does partitioning the numbers help you add them?*

IN FOCUS Children are encouraged to use known facts to work out additions with greater numbers. In question ❶ a), children only need to consider one step, whereas, in question ❶ b), children will need to think about partitioning before adding. Ensure children are fluent in examples like those in ❶ a) before tackling examples like those in ❶ b).

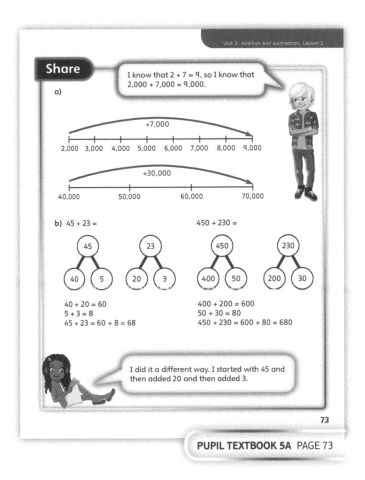

Share

I know that 2 + 7 = 9, so I know that 2,000 + 7,000 = 9,000.

a)

+7,000

2,000 3,000 4,000 5,000 6,000 7,000 8,000 9,000

+30,000

40,000   50,000   60,000   70,000

b) 45 + 23 =   450 + 230 =

45 → 40, 5   23 → 20, 3   450 → 400, 50   230 → 200, 30

40 + 20 = 60   400 + 200 = 600
5 + 3 = 8   50 + 30 = 80
45 + 23 = 60 + 8 = 68   450 + 230 = 600 + 80 = 680

I did it a different way. I started with 45 and then added 20 and then added 3.

73

PUPIL TEXTBOOK 5A PAGE 73

Think together

Whole class teacher led (I do, We do, You do)

ASK

- Question ❶: *What is each number made up of? How can you add these mentally? What do you need to do with your answers?*
- Question ❷: *What is the same and what is different? Can you add each place value separately?*
- Question ❸: *How much more has Andy added on each time? How will this affect the answer? What will you do now to get to the correct answer?*

IN FOCUS Questions ❶ and ❷ look at adding whole numbers together mentally by adding each place value separately and combining the answers. Some of the questions require an exchange, which children can make mentally. Encourage children to discuss and explain their methods. Question ❸ allows children to explore different strategies for doing mental calculations. Encourage children to realise the benefit of this mental strategy when numbers are close to a multiple of 100 or 1,000. Ensure children are confident in choosing which numbers to add on, i.e. rounding to the nearest 100 or 1,000 and subtracting the difference.

STRENGTHEN To support understanding, encourage children to draw their own number lines and to think about how much they should add on to get to, for example, the next 10.

DEEPEN Give children different calculations and ask them to explain which mental strategy would be most beneficial. For example, to work out 32 + 98 it may be easier to add 100 then subtract 2, but to work out 34 + 65 it may be easier to add the 10s then the 1s and combine them. Encourage children to explain why they chose each method.

ASSESSMENT CHECKPOINT Can children mentally add whole numbers using an appropriate strategy?

ANSWERS

Question ❶ a): 50,000 + 30,000 = 80,000

Question ❶ b): 70,000 + 60,000 = 130,000

Question ❶ c): 54 + 35 = 89

Question ❶ d): 540 + 350 = 890

Question ❷ a): 24 + 69 = 93

Question ❷ b): 240 + 690 = 930

Question ❸ a): Andy needs to add 200 then subtract 2 to find the first answer and add 200 then subtract 3 to find the second answer.
324 + 198 = **522**
324 + 197 = **521**

Question ❸ b): 672 + 99 = **771**
426 + 397 = **823**
296 + 3,147 = **3,443**
7,608 + 1,998 = **9,606**
18,790 + 39,990 = **58,780**

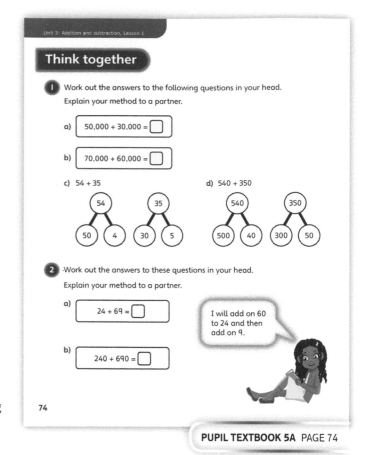

PUPIL TEXTBOOK 5A PAGE 74

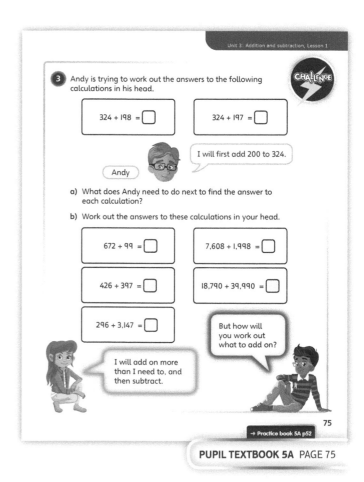

PUPIL TEXTBOOK 5A PAGE 75

Practice

WAYS OF WORKING Independent thinking

IN FOCUS Questions ❶ and ❷ aim to consolidate children's understanding of place value in additions. Question ❸ encourages children to partition numbers before adding them. In question ❹, each addition builds on the previous one to encourage children to make connections between calculations. Question ❺ is more open and encourages children to write their own strategies for mental calculations. They may choose to add the 1s and add the 10s or to add on, for example, 200 and subtract.

STRENGTHEN Encourage children to use place value counters to help them partition the numbers and add the 1s and the 10s separately.

DEEPEN Question ❻ involves mentally adding more than two whole numbers. This can be explored further by giving children other missing number problems with more than two whole numbers. Say numbers aloud and ask children to add them mentally.

ASSESSMENT CHECKPOINT By the end of the practice section, children should be confident in choosing the most useful method for adding whole numbers mentally. Question ❺ gives children an opportunity to show how well they have grasped mental addition, through the working out they show.

ANSWERS Answers for the **Practice** part of the lesson can be found in the *Power Maths* online subscription.

Reflect

WAYS OF WORKING Independent learning

IN FOCUS This **Reflect** activity checks that children can explain how to mentally add two whole numbers. Encourage children to explain carefully how they would carry out each calculation, and why, as well as actually answering it. Look out for children who need support to do this.

ASSESSMENT CHECKPOINT For each calculation, make sure children can explain how to choose the most efficient method for finding the total of two whole numbers mentally.

ANSWERS Answers for the **Reflect** part of the lesson can be found in the *Power Maths* online subscription.

After the lesson ⏸

- Can children add two whole numbers mentally?
- Do they understand that there are several mental methods to choose between?
- Are there any children who are not able to explain why they would choose one method over another?

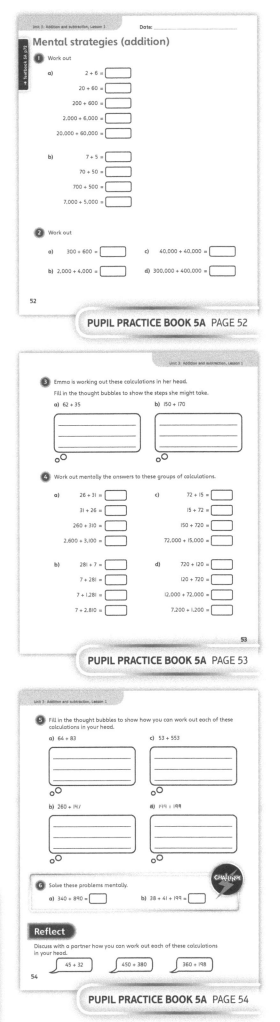

PUPIL PRACTICE BOOK 5A PAGE 52

PUPIL PRACTICE BOOK 5A PAGE 53

PUPIL PRACTICE BOOK 5A PAGE 54

Mental strategies (subtraction)

Learning focus

In this lesson, children will learn how to mentally subtract whole numbers by choosing the most efficient method from a variety of strategies.

Before you teach

- Do children know how to partition a number into 100s, 10s and 1s?
- Can children add and subtract multiples of 10, 100 and 1,000 mentally?

NATIONAL CURRICULUM LINKS

Year 5 Number – addition and subtraction

Add and subtract numbers mentally with increasingly large numbers.

ASSESSING MASTERY

Children can understand what mental methods can be used to efficiently subtract whole numbers.

COMMON MISCONCEPTIONS

Children may miscalculate when doing a subtraction in their head, for example, when mentally calculating 82 – 45, children may do 80 – 40 = 40 and 5 – 2 = 3 and get an answer of 43. Ask:
- *Are you confident of your answer? Is there another method you could use?*

Ask children if there is a count on method that they could use. For example: *What do you need to add to 45 to make 82?*

STRENGTHENING UNDERSTANDING

Children should use a number line to aid their understanding, and start by subtracting 1- and 2-digit numbers with no exchanges, building up to calculations that require an exchange. Also ask children how they can check their answer.

GOING DEEPER

Give children questions that involve both mental addition and subtraction. Ask questions verbally for children to work out mentally. Introduce a context to the question, for example, give children the cost of some items and ask them to mentally work out the total cost or the difference in price between two items.

KEY LANGUAGE

In lesson: mentally, subtract, partition, tens (10s), hundreds (100s), number line, method

Other language to be used by the teacher: addition, exchange

STRUCTURES AND REPRESENTATIONS

Number lines, part-whole models

RESOURCES

Optional: stopwatch or egg timer, place value counters

 In the eTextbook of this lesson, you will find interactive links to a selection of teaching tools.

Quick recap

Ask children to subtract multiples of 100 mentally, such as 800 – 300. Give them examples where they have to cross the 1,000, such as 1,200 – 600. Discuss their methods.

Discover

WAYS OF WORKING Pair work

ASK

- Question ❶ a): *What can you think about if you are doing this calculation in your head and trying to do it quickly?*
- Question ❶ b): *Could you use a similar method to Ebo to work out this calculation mentally?*

IN FOCUS Question ❶ a) will help children to see why it can be useful to work out calculations mentally rather than always using a written method. Question ❶ b) builds on this, with children working out a subtraction calculation mentally.

PRACTICAL TIPS Ask children to model the written method that they think Lexi might have tried to use. Time them working it out and point out how long it takes. Ask: *Do you think you can do it any quicker?* A similar context to the **Discover** picture could be carried out in class. Give two children a calculation to work out. One child must use a written method and the other a mental method.

ANSWERS

Question ❶ a): Ebo used a mental method. He counted on from 1,995 to 2,002.

Question ❶ b): 700 − 200 = 500
60 − 50 = 10
760 − 250 = 510

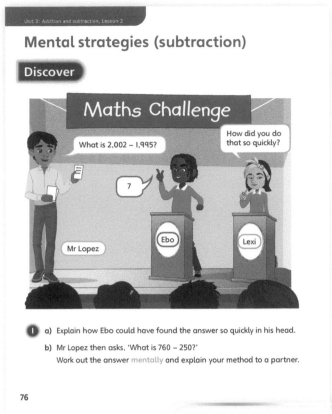

Share

WAYS OF WORKING Whole class teacher led

ASK

- Question ❶ a): *What is the next 1,000 after 1,995? How many 1s do you need to add to get to 2,000? How many do you need to add to 2,000 to get to 2,002? What should you do with these two answers?*
- Question ❶ b): *What is 760 made up of? What is 250 made up of? Can you subtract the 100s and the 10s separately? What should you do with these two answers?*

IN FOCUS For question ❶ a), show the number line from 1,995 to 2,000 and 2,002. Can children explain why we first add 5 ones and then 2 ones? Discuss how, overall, we have 5 + 2 = 7. Can children see why this mental method is useful and how Ebo got his answer so quickly? For question ❶ b), discuss what the numbers 760 and 250 are made up of. Use correct mathematical language, such as how 760 is 7 hundreds and 6 tens so 760 = 700 + 60. Encourage children to subtract the 100s and the 10s separately, and then combine their answers. Discuss an alternative strategy, for example, subtracting 200 from 760 then subtracting the 50. Ask children if they can think of any other suitable mental methods.

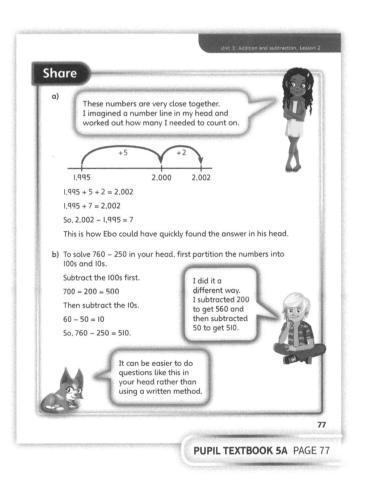

Think together

Whole class teacher led (I do, We do, You do)

ASK

- Question ❶ a): *What do you notice about your answers? Is there anything you can say about the number of 0s in the questions and the number of 0s in each answer?*
- Question ❶: *Did the questions get harder or were they all a similar difficulty?*
- Question ❸ a): *How does the number line help you answer the question? Why can you count on even though it is a subtraction?*
- Question ❸ b): *Can you sketch a number line to show how you worked it out?*

IN FOCUS In question ❸, children work out subtractions that cross a 10 or 100 boundary by counting on. Help children recognise that this is a more efficient method than formal written methods for questions like this. They should see that, even though the calculations look quite complex, they are easy to perform mentally.

STRENGTHEN Provide children with blank number lines, so that they can sketch how they are counting on to calculate answers.

DEEPEN Give children a start number and an answer and ask them what the question is. For example, the start number is 3,201 and the answer is 8, so the question must be 3,201 − 3,193. How was their thought process different?

ASSESSMENT CHECKPOINT Children can work out subtractions mentally using a counting on strategy. Ideally, they can visualise the number line in their head, however, some children may prefer to sketch it.

ANSWERS

Question ❶: 7 − 2 = **5**
70 − 20 = **50**
700 − 200 = **500**
7,000 − 2,000 = **5,000**
70,000 − 20,000 = **50,000**
700,000 − 200,000 = **500,000**

Question ❶ b): 750 − 240 = 510

Question ❷: 76 − 40 matches 76 − 40 = 36
76 − 42 matches 76 − 40 = 36, 36 − 2 = 34
70 − 40 = 30, 6 − 2 = 4,
30 + 4 = 34
72 − 46 matches 72 − 40 = 32, 32 − 2 = 30,
30 − 4 = 26

Question ❸ a): 498 + 2 = 500, 500 + 6 = 506,
2 + 6 = 8, so 506 − 498 = 8

Question ❸ b): 710 − 697 = **13**
4,302 − 4,299 = **3**
10,005 − 9,987 = **18**

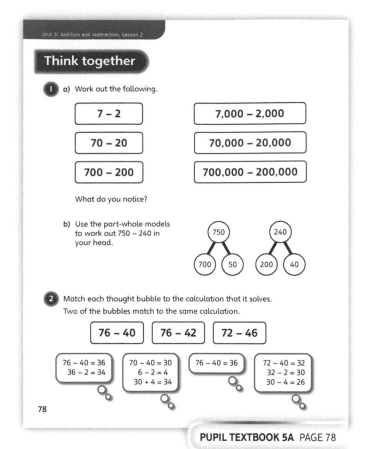

Think together

❶ a) Work out the following.

7 − 2	7,000 − 2,000
70 − 20	70,000 − 20,000
700 − 200	700,000 − 200,000

What do you notice?

b) Use the part-whole models to work out 750 − 240 in your head.

750 → 700, 50   240 → 200, 40

❷ Match each thought bubble to the calculation that it solves. Two of the bubbles match to the same calculation.

| 76 − 40 | 76 − 42 | 72 − 46 |

76 − 40 = 36
36 − 2 = 34

70 − 40 = 30
6 − 2 = 4
30 + 4 = 34

76 − 40 = 36

72 − 40 = 32
32 − 2 = 30
30 − 4 = 26

78

PUPIL TEXTBOOK 5A PAGE 78

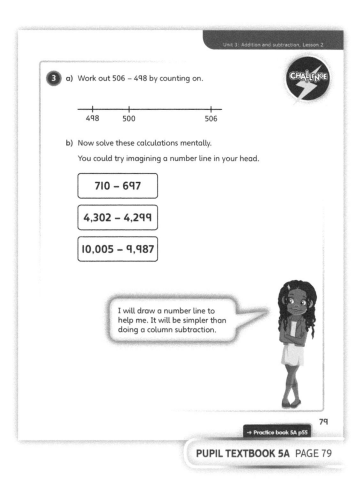

❸ a) Work out 506 − 498 by counting on.

CHALLENGE

498   500   506

b) Now solve these calculations mentally. You could try imagining a number line in your head.

| 710 − 697 |
| 4,302 − 4,299 |
| 10,005 − 9,987 |

I will draw a number line to help me. It will be simpler than doing a column subtraction.

79

→ Practice book 5A p55

PUPIL TEXTBOOK 5A PAGE 79

Practice

WAYS OF WORKING Independent thinking

IN FOCUS Question ❶ requires children to use two different mental methods to work out a calculation. Encourage children to discuss the benefits and difficulties of both methods even if they have a preferred one. For question ❷, children can use their preferred method, or try each of the two methods with different subtractions. Each subtraction builds upon the previous one, to encourage children to make connections between calculations. Encourage children to compare their answer for each question and discuss how the answers have changed. In questions ❹ and ❺, children must think carefully about which strategy they choose for each calculation. They may realise that adding on is quicker for some questions because the numbers are quite close together.

STRENGTHEN Encourage children to use a number line to help with adding on and place value counters to help with partitioning.

DEEPEN Question ❻ can be explored further by giving children other missing number problems with more than two whole numbers, for example, 740 – 260 – ☐ = 190. Give numbers verbally for children to add mentally. Include numbers that are not as close together or that feature a mixture of addition and subtraction.

THINK DIFFERENTLY Question ❸ asks children to fill in the missing numbers, to describe the method and to show this on a number line, making the link between subtraction and adding on.

ASSESSMENT CHECKPOINT Can children confidently subtract whole numbers mentally, choosing from a variety of strategies and identifying the most appropriate strategy each time?

ANSWERS Answers for the **Practice** part of the lesson can be found in the *Power Maths* online subscription.

Reflect

WAYS OF WORKING Independent thinking

IN FOCUS This **Reflect** activity checks that children can explain what method to use to mentally subtract two whole numbers. Encourage children to carefully explain how they would carry out the calculation as well as actually answering it. Encourage children to imagine or sketch a number line, if necessary.

ASSESSMENT CHECKPOINT Can children explain how to subtract with two whole numbers mentally and why a particular mental method is appropriate?

ANSWERS Answers for the **Reflect** part of the lesson can be found in the *Power Maths* online subscription.

After the lesson ⏸

- Can children subtract two whole numbers mentally?
- Can children identify the various mental subtraction methods that are available?
- Do children know what features to look for to choose the most sensible method?

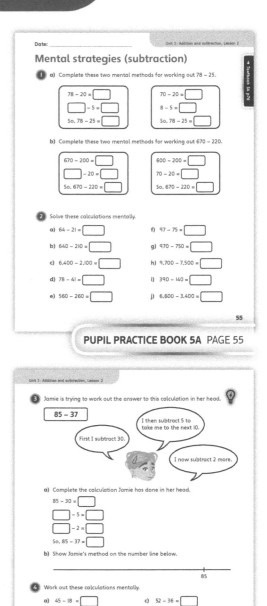

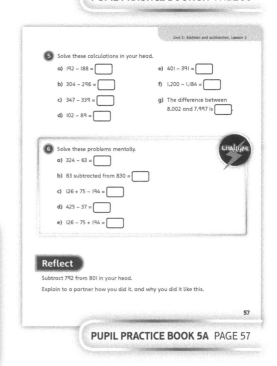

Add whole numbers with more than 4 digits ❶

Learning focus

In this lesson, children will use the formal written method to add whole numbers with more than four digits, recognising the importance of place value.

Before you teach

- Do children know how to add 2- and 3-digit numbers?
- Do children know how to make an exchange when using column addition?

NATIONAL CURRICULUM LINKS

Year 5 Number – addition and subtraction

Add and subtract whole numbers with more than 4 digits, including using formal written methods (columnar addition and subtraction).

ASSESSING MASTERY

Children can use the formal written method of column addition to add whole numbers with more than four digits.

COMMON MISCONCEPTIONS

Children may not know which place value column to start with when adding two whole numbers together, using column addition. Ask:
- *Why do you always start by adding the column that has the smallest place value in this method?*

Children may not understand the concept of exchanging between columns. Ask:
- *How might a place value grid and counters help you to see what is happening?*

STRENGTHENING UNDERSTANDING

Children should first practise adding whole numbers with two or three digits before moving on to whole numbers with more than four digits. Encourage them to clearly describe the place value of each column, and make sure they understand the importance of this, in particular when making an exchange.

GOING DEEPER

Give children a total and ask how many different ways they can make the total, for example, ☐ + ☐ = 8,876. This could also be represented on a part-whole model to help children see the link between adding and subtracting.

KEY LANGUAGE

In lesson: add, total, digit, column, place value

Other language to be used by the teacher: exchange, ones (1s), tens (10s), hundreds (100s), thousands (1,000s), ten thousands (10,000s)

STRUCTURES AND REPRESENTATIONS

Place value grid, column addition

RESOURCES

Mandatory: place value counters

 In the eTextbook of this lesson, you will find interactive links to a selection of teaching tools.

Quick recap 🔎

Recall addition of numbers within 1,000 or 10,000. Give children quick examples to remind them of the column addition method. Include examples where they have to exchange.

Discover

Unit 3: Addition and subtraction, Lesson 3

WAYS OF WORKING Pair work

ASK

- Question ① a): *How many views are there on Tuesday? How many views are there on Wednesday? What does 'total' mean? How could you add these amounts together?*
- Question ① b): *What method could you use to find which two days make this total?*

IN FOCUS Question ① a) requires children to identify two numbers and find the total. This calculation requires children to make one exchange when using column addition. Question ① b) gives children a total and asks them to work out which two numbers make this total.

PRACTICAL TIPS Make sure children understand that the boy is watching videos and can see the number of views for each video. Show a video on a real video sharing website and ask children to point out the number of views.

ANSWERS

Question ① a): The total number of video views for Tuesday and Wednesday is 39,328.

Question ① b): Wednesday and Friday have the total views of 37,592.

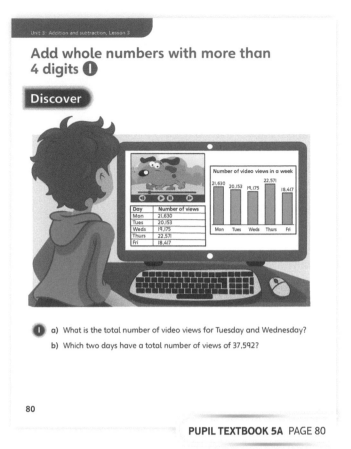

Add whole numbers with more than 4 digits ①

Discover

① a) What is the total number of video views for Tuesday and Wednesday?

b) Which two days have a total number of views of 37,592?

80

PUPIL TEXTBOOK 5A PAGE 80

Share

WAYS OF WORKING Whole class teacher led

ASK

- Question ① a): *What method can you use to add the two numbers together?*
- Question ① a): *Which place value column do you need to start with? Do you need to make an exchange?*
- Question ① b): *What method could you use to work out which two numbers make that total? Is there a quicker way?*

IN FOCUS For question ① a), take the opportunity to discuss how the word 'total' leads us to carry out an addition for this question and make sure children know which place value column we begin with when adding. Check that children are able to identify and say the larger numbers correctly. The place value grids can be used to reinforce the place value of each digit when carrying out the calculation, for example, ask: *What is 3 ones add 5 ones? What is 5 tens add 7 tens?* Demonstrate why this is important when children are required to carry out one exchange, of 10 tens for 1 hundred.

Discuss the use of the trial and improvement method in question ① b), emphasising the need to not miss any calculations out. Draw out that this will be time-consuming with such big numbers and encourage children to think flexibly about using a different strategy. For example, Flo's method of adding just the 1s of each number instead to see which gives a total ending with 2.

Share

a) Add the number of video views for Tuesday and Wednesday.

I used counters to help me. I set out the work in columns and added them together. I started with the column of least place value.

The total number of video views for Tuesday and Wednesday is 39,328.

b) I used trial and improvement to find the correct two days. I took my time and was careful not to miss any.

The last digit of Wednesday is 5.
The last digit of Friday is 7.

5 + 7 = 12

Wednesday and Friday have a total number of views of 37,592.

I simply added the last digits.

81

PUPIL TEXTBOOK 5A PAGE 81

Think together

Think together

WAYS OF WORKING Whole class teacher led (I do, We do, You do)

ASK

• Question ❶: *How many views are there on Thursday? How many views are there on Friday? What method can you use to add these amounts together?*
• Question ❷: *Do you need to make any exchanges for these additions? How should we approach question ❷ c)?*
• Question ❸: *How do you set numbers out in columns when they have a different number of digits?*

IN FOCUS In question ❶ and question ❷, children practise column addition with differing numbers of exchange. In questions ❷ b) and d), children see that they will need to set the problem out themselves. Check that children align the digits correctly. In question ❸, children may choose to add two numbers that do not have the same number of digits. Make sure they lay out the column addition correctly. When working out which two numbers have made a given total, encourage children to look at the last digit in each number instead of carrying out the full calculation.

STRENGTHEN Support understanding in question ❸ by representing calculations using counters on a place value grid.

DEEPEN For question ❸, ask children to work out the total for other combinations of numbers. Encourage them to add numbers that have a different number of digits.

ASSESSMENT CHECKPOINT Can children use the formal written method to add whole numbers with four or more digits where one or more exchanges are required? Make sure they pay attention to laying it out neatly and accurately and identifying the importance of the place value of each column.

ANSWERS

Question ❶: 22,571 + 18,417 = 40,988
The total number of views is 40,988.

Question ❷ a): 26,915 + 30,241 = 57,156

Question ❷ b): 37,418 + 4,157 = 41,575

Question ❷ c): 1,564 + 18,417 = 19,981

Question ❷ d): 28,019 + 4,096 = 32,115

Question ❸: Children should work out any two from the following:
34,171 + 61,426 = 95,597
34,171 + 5,458 = 39,629
34,171 + 1,023 = 35,194
61,426 + 5,458 = 66,884
61,426 + 1,023 = 62,449
5,458 + 1,023 = 6,481

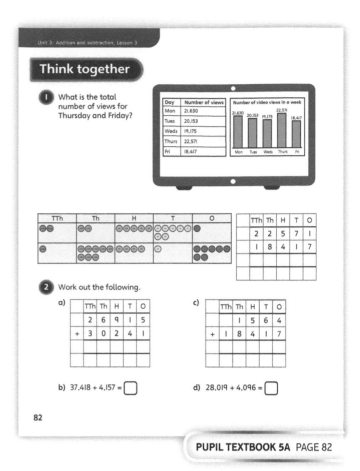

PUPIL TEXTBOOK 5A PAGE 82

PUPIL TEXTBOOK 5A PAGE 83

Practice

WAYS OF WORKING Independent thinking

IN FOCUS Questions **1** to **3** consolidate understanding of adding two whole numbers using column addition where the information is represented with counters on a place value grid, in a column and abstractly. Question **4** asks children to problem solve and work out missing digits in addition calculations while linking to subtraction.

Question **6** introduces a context for adding two whole numbers that have a different number of digits.

STRENGTHEN Encourage children to use counters on a place value grid to support understanding and, when the calculation is not given in a column layout, encourage them to write it in columns.

DEEPEN Explore question **4** in more depth by giving other missing number problems. Question **5** can be explored further by saying numbers for children to add together mentally, rather than using a written method.

THINK DIFFERENTLY In question **5**, numbers are given in words and children need to write these as numerals in order to work out the total. Some children may be able to work out the answers without using column addition.

ASSESSMENT CHECKPOINT Children are confident in using column addition to add whole numbers with four or more digits.

ANSWERS Answers for the **Practice** part of the lesson can be found in the *Power Maths* online subscription.

Reflect

WAYS OF WORKING Pair work

IN FOCUS This **Reflect** activity checks understanding of adding two whole numbers with four or more digits. Encourage children to explain how they would carry out the calculation as well as actually answering it. Look for children who are able to do this without any support.

ASSESSMENT CHECKPOINT Assess if children can correctly explain how to find the total of two whole numbers, emphasising the importance of the place value of each column, and identifying the need to make exchanges.

ANSWERS Answers for the **Reflect** part of the lesson can be found in the *Power Maths* online subscription.

After the lesson ⏸

- Can children show how to use column addition to add whole numbers with four or more digits?
- Do children understand the importance of a neat and accurate layout for this method?
- Which children needed to use counters on a place value grid for support?

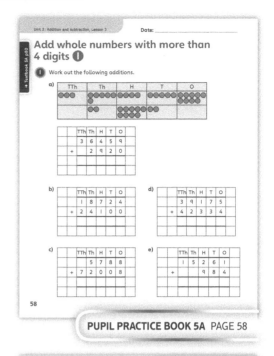

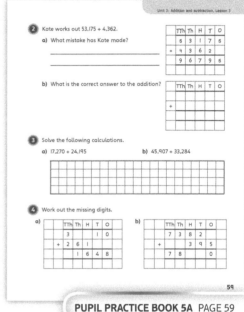

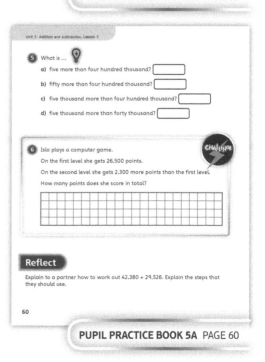

Add whole numbers with more than 4 digits ➋

Learning focus

In this lesson, children will identify large numbers in the context of distance and will use the formal written method to add two or more whole numbers with more than four digits.

Before you teach

- Do children know how to make an exchange when using column addition?
- Can children set out a column addition where the two numbers have a different number of digits?

NATIONAL CURRICULUM LINKS

Year 5 Number – addition and subtraction

Add and subtract whole numbers with more than 4 digits, including using formal written methods (columnar addition and subtraction).

ASSESSING MASTERY

Children can identify, compare and add two or more whole numbers with more than four digits using the formal written method, in a real-life context.

COMMON MISCONCEPTIONS

Children may not set out the column addition correctly when adding numbers with a different number of digits. Ask:
- *What is the value of each digit? Are you sure each digit in the same column has the same place value?*

STRENGTHENING UNDERSTANDING

Encourage children to use counters on a place value grid to help them see and understand things in a concrete way. This may be done alongside the column addition so children see how the concrete method links to the abstract method. Children are encouraged more to write out their own column additions to increase familiarity with the place value columns.

GOING DEEPER

Give children information in different forms such as tables and charts where they must first read and extract the information before adding numbers together. Provide missing number problems, for example, 15,239 + ☐ = 29,034, and encourage children to make the link between addition and subtraction when solving this type of calculation.

KEY LANGUAGE

In lesson: distance chart, add, total, distance, kilometres (km), metres (m), column, shortest, digit

Other language to be used by the teacher: place value, ones (1s), tens (10s), hundreds (100s), thousands (1,000s), ten thousands (10,000s)

STRUCTURES AND REPRESENTATIONS

Column addition, distance chart

RESOURCES

Mandatory: place value counters

 In the eTextbook of this lesson, you will find interactive links to a selection of teaching tools.

Quick recap

Write on the board the digits 2, 4, 5, 7 and 9.

Ask children to write down a 5-digit number using these digits. Add it to their partner's number. The answer closest to 100,000 wins. Repeat, changing the target number each time.

Discover

WAYS OF WORKING Pair work

ASK

- Question **1** a): *What information does the pilot's chart show us?*
- Question **1** a): *Which column represents London? Which row represents Sydney? Where do they meet? Can you find 9,385 km on the table?*
- Question **1** b): *What is the distance between London and Auckland? What is the distance between Auckland and Dubai?*

IN FOCUS Question **1** a) helps to ensure that children can use a distance chart to find relevant information. Question **1** b) asks children to identify numbers and find the total. This calculation requires children to make one exchange.

PRACTICAL TIPS Make sure children are confident in using a distance chart to find and use information. If necessary, provide similar charts with fewer, and smaller, numbers, so that children can practise looking across the rows and down the columns to identify the numbers they need.

ANSWERS

Question **1** a): The distance between London and Sydney is 16,998 km.
Shanghai and Auckland are 9,385 km apart.

Question **1** b): 18,360 + 14,212 = 32,572 km
Holly flies 32,572 km in total.

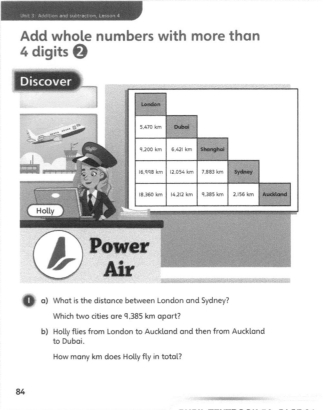

Add whole numbers with more than 4 digits ❷

Discover

1 a) What is the distance between London and Sydney?

Which two cities are 9,385 km apart?

b) Holly flies from London to Auckland and then from Auckland to Dubai.

How many km does Holly fly in total?

84

PUPIL TEXTBOOK 5A PAGE 84

Share

WAYS OF WORKING Whole class teacher led

ASK

- Question **1** a): *Look down the column for London. What number do you need to look at so you are in the row for Sydney? Which row and which column is 9,385 km in? Which two cities does this represent?*
- Question **1** b): *What method will you use to work out the total distance? Do you need to make an exchange?*

IN FOCUS Question **1** a) presents information in a real-life context. Children will need to understand the significance of each column and row in the table and use this to draw out the relevant information. Can children explain how they know that the distance between the two cities is 16,998 km and which two cities are 9,385 km apart?

For question **1** b), show children the column addition. They should be able to explain which place value column to start with when adding numbers together and why. Use this question to reinforce the place value of each digit when carrying out the calculation and discuss the exchange that needs to take place, of 10 thousands for 1 ten thousand. Emphasise the use of the correct units of measurement (km) in this context.

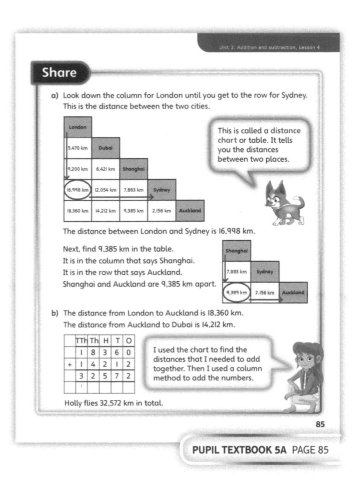

Share

a) Look down the column for London until you get to the row for Sydney. This is the distance between the two cities.

This is called a distance chart or table. It tells you the distances between two places.

The distance between London and Sydney is 16,998 km.

Next, find 9,385 km in the table.
It is in the column that says Shanghai.
It is in the row that says Auckland.
Shanghai and Auckland are 9,385 km apart.

b) The distance from London to Auckland is 18,360 km.
The distance from Auckland to Dubai is 14,212 km.

	TTh	Th	H	T	O
	1	8	3	6	0
+	1	4	2	1	2
	3	2	5	7	2
	1				

I used the chart to find the distances that I needed to add together. Then I used a column method to add the numbers.

Holly flies 32,572 km in total.

85

PUPIL TEXTBOOK 5A PAGE 85

Think together

WAYS OF WORKING Whole class teacher led (I do, We do, You do)

ASK

- Question ❶: *What is the distance from London to Sydney? What is the distance from Sydney to Auckland? What method will you use to add these distances together?*
- Question ❷: *How will you find the total distance from London to Auckland for each route? How do you know which route is the shortest distance?*
- Question ❸: *Can you add all three prices using column addition? If you add the first two prices then add on the final price afterwards, do you still get the same answer?*

IN FOCUS Question ❶ looks at adding together whole numbers that require exchanges. Encourage children to use a column addition and ensure they lay it out correctly as the two numbers have a different number of digits. Question ❷ requires children to add two whole numbers together involving more than one exchange, and then to compare their answers to say which is smaller. In question ❸, children can explore adding more than two whole numbers together. Children may choose to add all three numbers in one go or they may add two together and then add on the final number to get the total. Encourage children to try both methods, ensuring they lay out the column addition correctly. Discuss which method children prefer.

STRENGTHEN To support understanding, represent each calculation using counters on a place value grid and display this alongside the abstract calculations.

DEEPEN For question ❸, add in painting B and ask children to work out the total price now. Encourage children to find the answer using more than one method.

ASSESSMENT CHECKPOINT Can children explore the best method to use to add two or more whole numbers with four or more digits, including those with more than one exchange? Can they compare their answers to explain which is bigger and which is smaller?

ANSWERS

Question ❶: Mo flies 19,154 km in total.

Question ❷: The total distance of Route 1 is 19,682 km. The total distance of Route 2 is 18,585 km. David should choose Route 2.

Question ❸ a): The total cost of paintings C and D is £472,629.

Question ❸ b): The total cost of paintings C, D and A is £511,379.

Question ❸ c): This will be each child's personal choice.

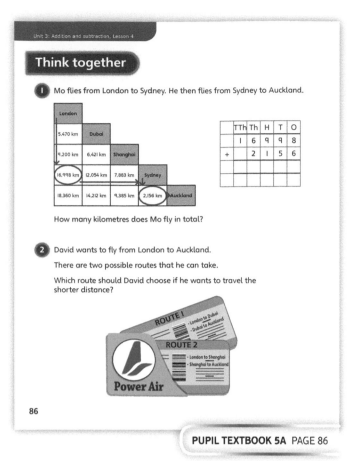

PUPIL TEXTBOOK 5A PAGE 86

PUPIL TEXTBOOK 5A PAGE 87

Practice

WAYS OF WORKING Independent thinking

IN FOCUS Questions ❶ and ❷ aim to consolidate children's understanding of adding two whole numbers together using column addition, where the information is represented in a column with place value headings and abstractly. In question ❸, children will need to use their preferred method to add together three whole numbers and then decide if the total is less than or more than 6,000 m. Question ❹ requires children to spot a mistake in a column addition and to work out the correct answer.

STRENGTHEN Encourage children to use counters on a place value grid to support their understanding. They should write each calculation in columns, paying close attention to laying this out carefully and accurately. They can use this to help write out their own column additions.

DEEPEN Question ❻ is more open ended and allows children to look at different ways to make totals. Explore this further by giving other totals, but where the digit cards may be used more than once. Encourage children to start with the ones column and think about which digits they could put there to make the total of the ones column.

THINK DIFFERENTLY Question ❺ encourages children to problem solve and work out missing digits in addition calculations. They will need to make the link between addition and subtraction to solve them.

ASSESSMENT CHECKPOINT Are children confident in using column addition to add two or more whole numbers with four or more digits, including those with exchanges? For questions ❺ and ❻, do they understand how they can use the link between addition and subtraction to find missing numbers?

ANSWERS Answers for the **Practice** part of the lesson can be found in the *Power Maths* online subscription.

Reflect

WAYS OF WORKING Independent thinking

IN FOCUS This **Reflect** activity checks understanding of adding two whole numbers that both have five digits and that require two exchanges. Encourage children to explain how they know that their calculation will have exactly two exchanges and ask them to also work out the answer to the calculation. Identify children who still need support to do this.

ASSESSMENT CHECKPOINT Can children work out what information they will need to know to make sure they are writing an addition question that requires exactly two exchanges? Can they explain why the place value of each column is important for this? Can they lay out their calculation neatly and accurately and correctly find the answer to it?

ANSWERS Answers for the **Reflect** part of the lesson can be found in the *Power Maths* online subscription.

After the lesson

- Can children confidently use column addition to add two or more whole numbers with four or more digits?
- Can children identify calculations that require exchanges and correctly solve these?
- Are any children still overlooking the importance of place value in these calculations?

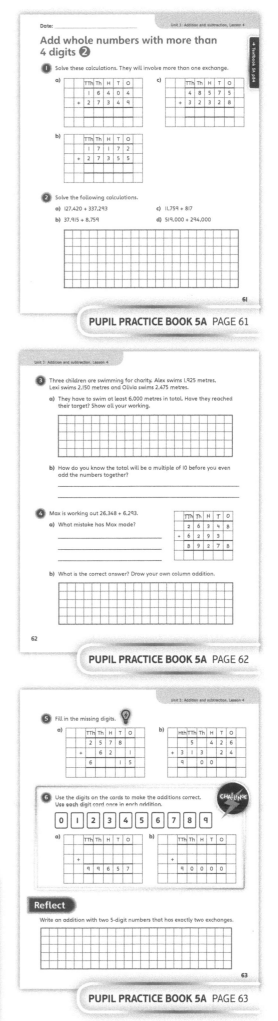

PUPIL PRACTICE BOOK 5A PAGE 61

PUPIL PRACTICE BOOK 5A PAGE 62

PUPIL PRACTICE BOOK 5A PAGE 63

Subtract whole numbers with more than 4 digits

Learning focus

In this lesson, children will use the formal written method to subtract whole numbers with more than four digits, in the context of taking away and of finding a difference. This includes examples where an exchange is required.

Before you teach ⏸

- Can children subtract with 2- and 3-digit numbers?
- Do children understand key vocabulary, such as 'greater than', 'less than' and 'difference'?

NATIONAL CURRICULUM LINKS

Year 5 Number – addition and subtraction

Add and subtract whole numbers with more than 4 digits, including using formal written methods (columnar addition and subtraction).

ASSESSING MASTERY

Children can recognise why exchanges are needed when subtracting whole numbers with more than four digits and can use the formal written method to find the answer. They can use comparison bar models to express a problem and show what they need to work out.

COMMON MISCONCEPTIONS

Children may just subtract the smaller digit from the bigger digit. For example, when working out 38,792 – 17,345 they may write 3 in the ones column as they have worked out 5 – 2 and not made an exchange. Ask:
- *Which digit do you need to subtract? What will you do if the digit you are subtracting is bigger?*

Children may not know how to correctly set out the column subtraction when subtracting two numbers that have a different number of digits. Ask:
- *What is the place value of each digit in these numbers? Why is it important to lay this out accurately?*

STRENGTHENING UNDERSTANDING

Children should first practise subtracting whole numbers with two or three digits before moving on to subtracting whole numbers with more than four digits.

GOING DEEPER

Give children an answer and ask how many different ways they can make this number using subtraction of a 4-digit number from a 5-digit number. For example, ☐ + ☐ = 34,728.

KEY LANGUAGE

In lesson: subtract, difference, exchange, greater than, more, less, digit, capacity

Other language to be used by the teacher: column, place value, ones (1s), tens (10s), hundreds (100s), thousands (1,000s), ten thousands (10,000s)

STRUCTURES AND REPRESENTATIONS

Place value grid, column subtraction, bar model

RESOURCES

Mandatory: counters

 In the eTextbook of this lesson, you will find interactive links to a selection of teaching tools.

Quick recap

Recall subtraction of numbers within 1,000 or 10,000. Give children quick examples to remind them of column subtraction. Include examples where they have to exchange and where the numbers have a different number of digits.

Discover

Subtract whole numbers with more than 4 digits ①

WAYS OF WORKING Pair work

Discover

ASK

- Question ① a): *What calculation do you need to do to work out how much greater the capacity of the velodrome is than the archery field?*
- Question ① b): *What is the capacity of the athletics stadium? How will you work out how many seats were empty?*

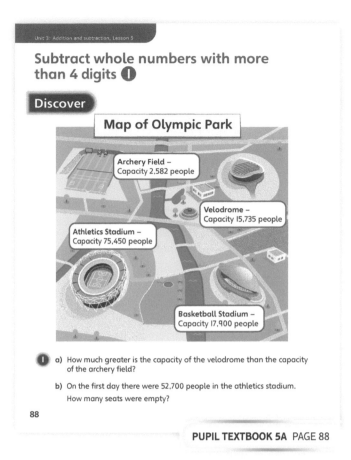

IN FOCUS Question ① a) presents subtraction in the context of a 'how much greater than' question and requires children to identify and subtract a 4-digit number from a 5-digit number using column subtraction and making one exchange. Question ① b) requires children to work out the difference between a given number and the total, by subtracting a 5-digit number from a 5-digit number with one exchange.

① a) How much greater is the capacity of the velodrome than the capacity of the archery field?

   b) On the first day there were 52,700 people in the athletics stadium. How many seats were empty?

88

PRACTICAL TIPS Explain that the word 'capacity' means how many people will fit in each area of the sports park. Find images and capacity information about local and national sports centres that children may be familiar with to help them visualise and understand the large numbers and the difference in capacity between these.

PUPIL TEXTBOOK 5A PAGE 88

ANSWERS

Question ① a): 15,735 – 2,582 = 13,153
The velodrome capacity is 13,153 greater than the archery field capacity.

Question ① b): 75,450 – 52,700 = 22,750
22,750 seats were empty in the athletics stadium.

Share

WAYS OF WORKING Whole class teacher led

ASK

- Question ① a): *What calculation will you do if the question is 'how much greater than'? What method will you use? Which place value column do you start with? Do you need to make an exchange?*
- Question ① b): *What type of calculation will this be? How do you lay this out as a column subtraction?*

IN FOCUS For question ① a), make sure children are able to correctly identify and say the given capacity of the velodrome and the archery field. Discuss how the words 'how much greater' lead us to carry out a subtraction. Show children the column subtraction and ensure they are confident with using this layout. Use this to reinforce the place value of each digit. Explore why it is not possible to subtract 8 tens from 3 tens and discuss the need to exchange 1 hundred for 10 tens. For question ① b), discuss which calculation is needed and show the bar model so children can see why they need to subtract. Emphasise the place value of each digit and the exchange of 1 thousand for 10 hundreds so they can easily subtract the 7 hundreds.

PUPIL TEXTBOOK 5A PAGE 89

Think together

Whole class teacher led (I do, We do, You do)

ASK

- Question **1**: *If you are taking away the number of people who leave, what calculation will you need to do? What method will you use?*
- Question **2**: *What is the capacity of the basketball stadium? What calculation will you do to work out how many more people could have watched the game?*
- Question **3**: *What information do you have? How does the bar model show this? How will you find the capacity of the hockey centre? What calculation will you do to work out how much greater this is than the velodrome?*

IN FOCUS Question **1** involves subtracting whole numbers without any exchanges. Encourage children to use column subtraction. In question **2**, children will subtract whole numbers with one exchange. Children can use the bar model to help them decide which calculation they need to do. To solve question **3**, children will need to problem solve and explore a two-step subtraction. Encourage children to look at the comparison bar model to help them decide which calculation and numbers to use.

STRENGTHEN To support understanding, children should represent each calculation using counters on a place value grid alongside the abstract calculations. Where there are exchanges, encourage them to make and move each number and to explain what is happening and why.

DEEPEN Extend question **3** by asking children to use subtraction to compare the capacities of other structures in the sports parks. For example, how much greater is the capacity of the hockey centre than the archery field?

ASSESSMENT CHECKPOINT Can children identify where a subtraction is needed to find an answer and relate this to finding a difference or taking away? Can they use column subtraction to subtract whole numbers with four or more digits, including those where an exchange is required?

ANSWERS

Question **1**: 15,735 – 3,620 = 12,115
There are 12,115 people left in the velodrome.

Question **2**: 17,900 – 10,840 = 7,060
7,060 more people could have watched the game.

Question **3**: 72,450 – 42,300 = 33,150
The capacity of the hockey centre is 33,150.
33,150 – 15,735 = 17,415
The capacity of the hockey centre is 17,415 greater than the capacity of the velodrome.

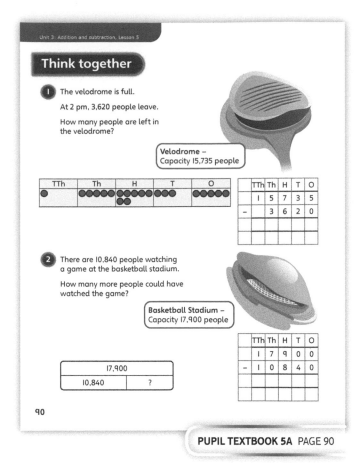

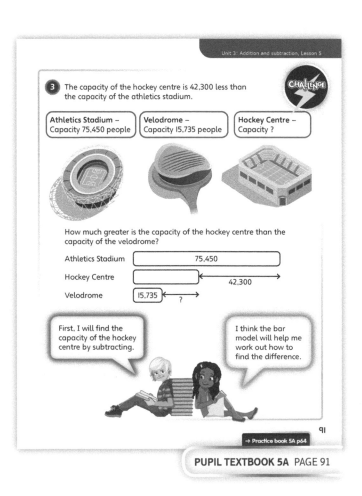

PUPIL TEXTBOOK 5A PAGE 90

PUPIL TEXTBOOK 5A PAGE 91

Practice

WAYS OF WORKING Independent thinking

IN FOCUS Questions ❶ and ❷ aim to consolidate children's understanding of subtracting two whole numbers using column subtraction where the information is represented with counters on a place value grid, in columns and as a bar model. Question ❸ uses the context of prices; children will need to identify that a subtraction is required. Encourage them to lay out the column subtraction correctly, paying attention to place value.

STRENGTHEN Encourage children to use counters on a place value grid to support their understanding. When the calculation is not given in columns, encourage them to lay it out carefully in columns, describing the place value of each column. Support children in drawing given numbers in a comparison bar model.

DEEPEN To extend question ❺, give examples of other problems that require more than one calculation. Ask a child to present some information in the form of a comparison bar model and then ask the other children to write a word problem to match it that they can solve using column subtraction.

THINK DIFFERENTLY Question ❹ encourages children to problem solve and work out missing digits in subtraction calculations by linking them to addition. This can be explored further by giving children other missing number problems, such as $13,565 - \boxed{} = 4,099$.

ASSESSMENT CHECKPOINT Are children confident in laying out the column subtraction and using it to subtract whole numbers with four or more digits, including where an exchange is needed?

ANSWERS Answers for the **Practice** part of the lesson can be found in the *Power Maths* online subscription.

Reflect

WAYS OF WORKING Pair work

IN FOCUS This **Reflect** activity checks children's understanding of column subtraction by asking them to explain how to use it to subtract two whole numbers that both have five digits. Encourage children to actually answer the calculation as well as explain the steps. Look out for children who do not address the importance of place value or who do not identify how to correctly carry out the exchange that is needed.

ASSESSMENT CHECKPOINT Can children explain how to use column subtraction to subtract two 5-digit whole numbers with one exchange? Have children clearly explained place value and the process of carrying out one exchange?

ANSWERS Answers for the **Reflect** part of the lesson can be found in the *Power Maths* online subscription.

After the lesson ⏸

- Can children lay out a column subtraction neatly and accurately?
- Can children identify which information in a problem shows that a subtraction is needed?
- Do children understand how comparison bar models can help them to work out missing information?

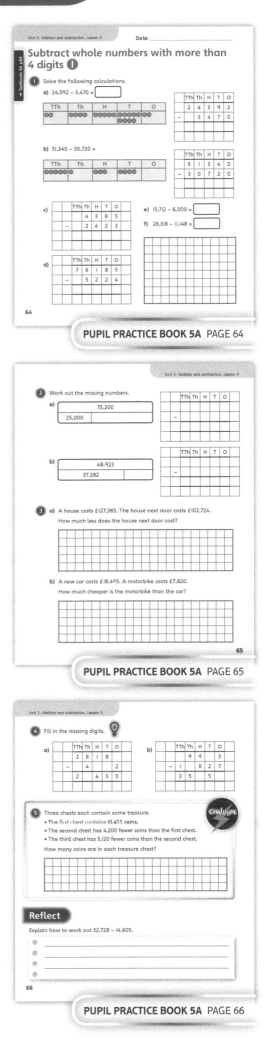

PUPIL PRACTICE BOOK 5A PAGE 64

PUPIL PRACTICE BOOK 5A PAGE 65

PUPIL PRACTICE BOOK 5A PAGE 66

Subtract whole numbers with more than 4 digits ②

Learning focus

In this lesson, children will explore how and why exchanges can occur in subtractions. They will use the formal written method to subtract whole numbers with more than four digits, including where exchanges are needed in some or all columns.

Before you teach

- Do children know how to accurately lay out a column subtraction?
- Do children know how to make an exchange when using column subtraction?

NATIONAL CURRICULUM LINKS

Year 5 Number – addition and subtraction

Add and subtract whole numbers with more than 4 digits, including using formal written methods (columnar addition and subtraction).

ASSESSING MASTERY

Children can lay out the formal written method neatly and accurately and can use it to subtract whole numbers with more than four digits, including needing to make multiple exchanges.

COMMON MISCONCEPTIONS

Children may make mistakes where an exchange is needed because there are 0s in the bigger number, for example, for 28,300 – 18,976, exchanging 1 hundred for 10 ones, not 1 hundred for 10 tens, and then 1 ten for 10 ones. Ask:
- *Can you make an exchange from this column? Why not? Which column can you exchange from? How much will you be exchanging?*

STRENGTHENING UNDERSTANDING

Encourage children to use counters on a place value grid to help them see the calculation in a concrete way. Do this alongside the column subtraction so that children can see how the concrete links to the abstract method. If children do not understand the concept of exchanges, focus on calculations that require just one exchange before moving on to calculations with more than one exchange.

GOING DEEPER

Give children two 5-digit numbers, for example, 89,343 and 17,824, and ask how many times they can subtract the smaller number from the greater number without going into negative numbers.

KEY LANGUAGE

In lesson: subtract, exchange, digit, column, place value

Other language to be used by the teacher: ones (1s), tens (10s), hundreds (100s), thousands (1,000s), ten thousands (10,000s), less than

STRUCTURES AND REPRESENTATIONS

Column subtraction

RESOURCES

Mandatory: digit cards 0–9, counters, place value grids, whiteboards

 In the eTextbook of this lesson, you will find interactive links to a selection of teaching tools.

Quick recap ↺

Ask children to write, on a whiteboard, their own column subtraction using numbers within 1,000. Ask them to rub out one digit from each column. Give it to their partner and see if they can work out the missing digits. Repeat several times, including with greater numbers.

Discover

ASK

- Question ① a): *What is the new calculation? What do you need to look at to decide if an exchange is needed?*
- Question ① b): *What has changed in each calculation? Can you see any new exchanges? How do you know?*

IN FOCUS Question ① a) requires children to identify when an exchange is needed and to use column subtraction to work out the answer to a subtraction with one exchange. Question ① b) introduces more subtractions where the same digits are rearranged, so that each question has a different number of exchanges in different place value columns.

PRACTICAL TIPS Ask children to arrange real digit cards 0–9 to make Max and Jamilla's subtraction from the picture. These can then be used to create subtractions throughout the lesson.

ANSWERS

Question ① a): One exchange is needed now, as there are not enough 1,000s to subtract from.
The answer to the new subtraction is 44,563.

Question ① b): 62,097 − 18,534 = **43,563**
62,037 − 18,594 = **43,443**
62,034 − 18,597 = **43,437**

PUPIL TEXTBOOK 5A PAGE 92

Share

ASK

- Question ① a): *In which column do you need to make an exchange? How do you know?*
- Question ① b): *Do you need to make more than one exchange in each of these calculations?*

IN FOCUS For question ① a), encourage children to write out the new calculation, or make it with digit cards, and to say the numbers correctly. Show the column subtraction and ensure children are confident in using this layout. Check that children can explain how we know that we need to make an exchange in the thousands column. Use a column subtraction to reinforce the place value of each digit when carrying out the calculation and explain that we always start by subtracting in the ones column in case we need to make an exchange. For question ① b), show children the new column subtractions and reinforce the place value of each digit. Highlight and describe any exchanges needed to ensure that children do not just subtract the smaller digit from the greater digit.

PUPIL TEXTBOOK 5A PAGE 93

129

Think together

Whole class teacher led (I do, We do, You do)

ASK

- Question ❶: *Which place value column should you start with? What is 6 ones subtract 5 ones? Do you need to make any exchanges? How do you know?*
- Question ❷: *What do you subtract from 3 ones to get 1 one? What is the missing digit in the ones column? Can you subtract from 0 tens to get 1 ten? What must have happened here?*
- Question ❸: *What is the same about the subtractions 27,910 – 15,462 and 27,900 – 15,462, and what is different? What will be different about the answers? Is there an easier way to find the answers than doing two column subtractions?*

IN FOCUS Question ❷ introduces problem solving to find missing digits. Encourage children to start with the ones column and to think about how to identify if an exchange has taken place. Encourage them to check that their calculation is correct by working it out and checking they get an answer of 61,611. In question ❸, encourage children to use column subtraction to begin with but to think about what is the same in each question (the 15,462) and what is different (such as, 27,900 is 10 less than 27,910, so the answer should be 10 less). Where there are more 0s in the numbers, there will also be more exchanges. Use 20,000 – 15,462 to demonstrate how to get rid of the 0s and the need for exchanges, by subtracting 1 from both numbers. The answer to 19,999 – 15,461 will still be the same. Children can still carry out the original column subtraction to check this.

STRENGTHEN To support understanding, children should represent the calculations using counters on a place value grid alongside the abstract calculations. For question ❸, ask children to make all four calculations using counters so they can easily see what is different in each question.

DEEPEN For question ❸, ask children to work out the answers to other calculations where they are always subtracting 15,462. Children could think of their own questions for this and then answer them.

ASSESSMENT CHECKPOINT Can children identify where exchanges are needed in subtractions and use column subtraction to subtract whole numbers with four or more digits that involve exchanges?

ANSWERS

Question ❶: 82,706 – 39,415 = **43,291**

Question ❷: 7**6**,503 – 1**4**,**892** = 61,611

Question ❸ a): 27,910 – 15,462 = **12,448**

Question ❸ b): 27,900 – 15,462 = **12,438**

Question ❸ c): 27,000 – 15,462 = **11,538**

Question ❸ d): 20,000 – 15,462 = **4,538**

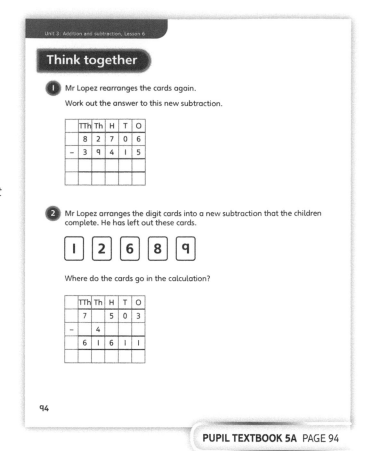

PUPIL TEXTBOOK 5A PAGE 94

PUPIL TEXTBOOK 5A PAGE 95

Practice

WAYS OF WORKING Independent thinking

IN FOCUS Questions ❶, ❷ and ❸ aim to consolidate children's understanding of subtracting two whole numbers using column subtraction, where the information is represented with counters on a place value grid, or abstractly in a column. Encourage children to choose their preferred method for question ❹. They may subtract 18,926 once to get an answer and then subtract 18,926 from this, and so on. Alternatively, they may choose to work out 18,926 + 18,926 + 18,926 and then subtract this answer from 76,350. Question ❺ encourages children to problem solve and work out missing digits in subtraction calculations by linking them to addition. Question ❻ involves a lot of 0s so will require multiple exchanges. Can children recall a strategy to avoid this?

STRENGTHEN Encourage children to use counters on a place value grid to support their understanding and, when a calculation is not given in columns, encourage them to write it out in columns. Children may find it useful to draw a bar model for question ❼, to help them understand which calculations they need to do.

DEEPEN Extend the context in question ❹ by asking children after how many hours will there be 646 left.

ASSESSMENT CHECKPOINT Are children confident in laying out the column subtraction neatly and accurately and using it to subtract whole numbers with four or more digits, including those with multiple exchanges? Do they use the link with addition, think flexibly to employ a variety of strategies and use bar models to help them work out what calculation is needed?

ANSWERS Answers for the **Practice** part of the lesson can be found in the *Power Maths* online subscription.

Reflect

WAYS OF WORKING Independent thinking

IN FOCUS This **Reflect** activity checks children's understanding of using column subtraction to subtract two 5-digit whole numbers with exactly two exchanges. Encourage children to explain how they know their subtraction has exactly two exchanges and encourage them to focus on using key vocabulary and the correct place value terms. Also, ask children to find the answer to their calculation.

ASSESSMENT CHECKPOINT Can children explain how they know that a calculation will have exactly two exchanges? Can they accurately use column subtraction to subtract the two whole numbers and find the correct answer to their calculation?

ANSWERS Answers for the **Reflect** part of the lesson can be found in the *Power Maths* online subscription.

After the lesson

- Can children confidently use column subtraction to subtract two whole numbers with four or more digits where there are multiple exchanges?
- Can children identify where exchanges will occur in subtractions?
- Do children employ a variety of strategies to solve problems?

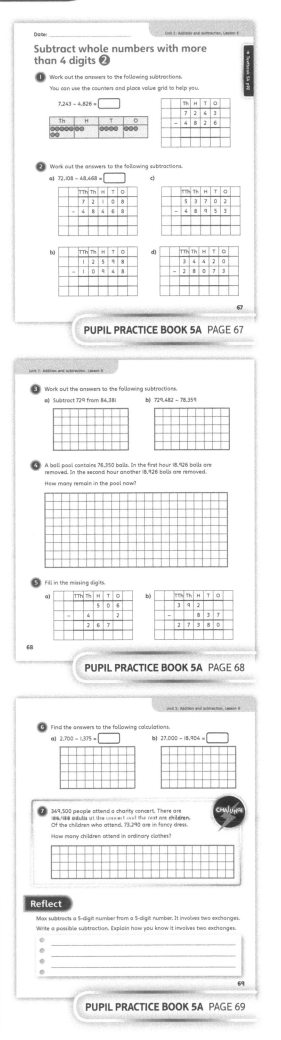

PUPIL PRACTICE BOOK 5A PAGE 67

PUPIL PRACTICE BOOK 5A PAGE 68

PUPIL PRACTICE BOOK 5A PAGE 69

Round to check answers

Learning focus

In this lesson, children will learn how to use rounding to help make estimates, identify sensible answers, find mistakes and check answers to calculations.

Before you teach

- Do children know how to round to the nearest 10,000, 1,000, 100 and 10?
- Are children confident in using column addition and subtraction?

NATIONAL CURRICULUM LINKS

Year 5 Number – addition and subtraction

Use rounding to check answers to calculations and determine, in the context of a problem, levels of accuracy.

ASSESSING MASTERY

Children can round numbers and use this to make estimates, find mistakes and check answers.

COMMON MISCONCEPTIONS

Children may not know what to round each number to. Ask:
- *Can you find the number on a number line? What multiple of 10/100/1,000 does the number round to?*

Some children may not round numbers appropriately. For example, rounding 39,921 + 15,839 to 39,920 + 15,840 which is still very close to the original. Ask:
- *Will that help you to easily check your answer? What would be a simpler estimate to calculate with?*

STRENGTHENING UNDERSTANDING

Encourage children to use number lines to help them identify what to sensibly round each number to before using the rounded numbers to make an estimate.

GOING DEEPER

Give children a rounded answer and ask what the question could have been. For example: *Two numbers are added together and an estimated answer is 12,000. What could the calculation have been?* For example, it could have been 7,136 + 4,834 which would round to 7,000 + 5,000.

KEY LANGUAGE

In lesson: round, estimate, check, difference, total, close to, sensible, reasonable

Other language to be used by the teacher: addition, subtraction, exchange, ones (1s), tens (10s), hundreds (100s), thousands (1,000s), ten thousands (10,000s)

STRUCTURES AND REPRESENTATIONS

Column addition, column subtraction, number line

RESOURCES

Optional: number lines

 In the eTextbook of this lesson, you will find interactive links to a selection of teaching tools.

Quick recap

Check children's ability to round a number to a given degree of accuracy. Write a 4-digit number on the board, such as 1,719. Ask: *What is this number rounded to the nearest 1,000? 100? 10?* Repeat for other 4- and 5-digit numbers.

Discover

Pair work

ASK

- Question **1** a): *What could you round each number to, to make a sensible estimate?*
- Question **1** b): *Do you think Bella has laid out the calculation carefully? What do you need to think about when laying out a column addition?*

IN FOCUS Question **1** a) helps children realise the importance of making estimates by rounding numbers to check if their answer is logical. Question **1** b) requires children to focus on the layout of a column addition in order to identify the mistake that has been made.

PRACTICAL TIPS Consider showing children some sample test papers similar to the one that Bella is doing. Explain the importance of checking your answers when doing a test.

ANSWERS

Question **1** a): 18,000 + 4,000 = 22,000. Bella's answer should be close to 22,000.

Question **1** b): Bella has lined up the numbers incorrectly in the column addition.
The 1,000s need to be lined up underneath the 1,000s, and so on.
The correct answer is 21,889.

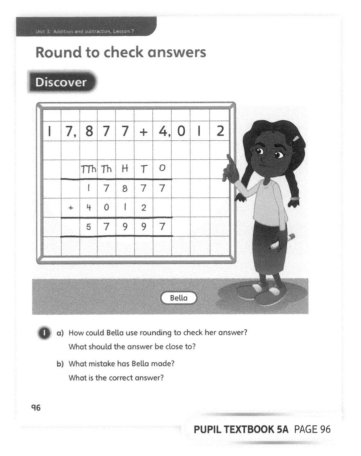

Share

Whole class teacher led

ASK

- Question **1** a): *What number is 17,877 close to? What number is 4,012 close to? Can you work out 18,000 add 4,000 mentally?*
- Question **1** b): *How do you know Bella has lined the numbers up incorrectly? Why is place value important?*

IN FOCUS In question **1** a), the number line can be used to prompt a discussion about which numbers are closest to 17,877 and 4,012. Encourage children to say the numbers they have rounded to correctly then add them together. Can children explain why the answer should be close to 22,000? Compare Bella's column addition to the correct column addition in question **1** b). Can children explain why Bella's answer must be incorrect? Emphasise the importance of lining the 1s up under the 1s, and so on. Discuss the exchange that needs to take place. Can children explain why the correct answer is 21,889? Emphasise why rounding is useful for checking answers by discussing how 21,889 and 22,000 are close to each other.

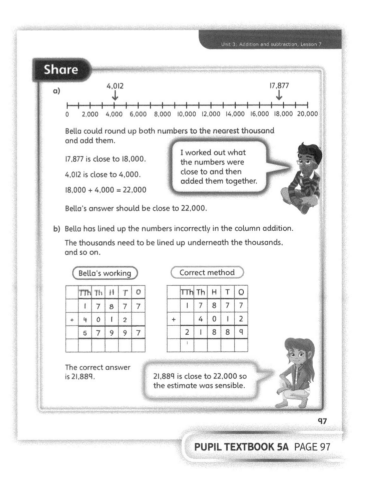

Think together

WAYS OF WORKING Whole class teacher led (I do, We do, You do)

ASK

- Question **1**: *What will you round 4,935 to? What will you round 322 to? Can you subtract mentally?*
- Question **2**: *How do you know what to round each number to? Can you tell if your estimates are sensible?*
- Question **3**: *What methods will you use to find the total and the difference? How will you check your answers?*

IN FOCUS Question **1** requires children to use rounding to check if an answer is correct and to explain the mistake that has been made. Encourage children to round each number to a suitable degree of accuracy. Children may round 4,935 to 5,000 or to 4,900. Discuss why 4,900 will give a more accurate answer. In question **2**, children use rounding to estimate answers and compare them to the correct answer. Encourage children to explain how they know if their estimate is sensible. Question **3** presents calculations in the context of prices and involves more than two numbers. Although the question does not ask children to use rounding, discuss the different strategies children could use to check their answers by rounding.

STRENGTHEN Children can use a number line to help them to decide what to round each number to. If necessary, encourage children to use the column method to work out their estimations rather than doing it mentally.

DEEPEN For question **3**, ask children to find the difference between the price of the car and the TV, and the difference between the price of the laptop and the TV, giving their answers both as an estimate and as an exact answer.

ASSESSMENT CHECKPOINT Can children round numbers to an appropriate degree of accuracy and use this to make sensible estimates, identify mistakes, and check answers when adding and subtracting?

ANSWERS

Question **1**: 4,935 is close to 4,900 or 5,000
322 is close to 300
4,900 – 300 = 4,600 or 5,000 – 300 = 4,700
Bella has laid out the calculation incorrectly. She has not lined up the 1s under the 1s, and so on.

Question **2** a): 17,240 rounds to 17,000
28,385 rounds to 28,000
17,000 + 28,000 = 45,000
17,240 + 28,385 = 45,625

Question **2** b): 7,010 rounds to 7,000
3,997 rounds to 4,000
7,000 – 4,000 = 3,000
7,010 – 3,997 = 3,013
The estimates were sensible.

Question **3** a): 12,795 + 1,199 + 298 = £14,292
The items cost £14,292 in total.

Question **3** b): 12,795 – 1,199 = £11,596
The difference in price between the car and the laptop is £11,596.

Unit 3: Addition and subtraction, Lesson 7

Think together

1 Bella works on the next question.

Use rounding to show that Bella's answer must be incorrect.

```
4, 9 3 5 – 3 2 2
```

	Th	H	T	O
	4	9	3	5
–	3	2	2	
	1	7	1	5

What mistake has Bella made?

2 Use rounding to estimate the answers to the following calculations.

a) 17,240 + 28,385

b) 7,010 – 3,997

Now work out the answers to each of the questions.

Were your estimates sensible?

98

PUPIL TEXTBOOK 5A PAGE 98

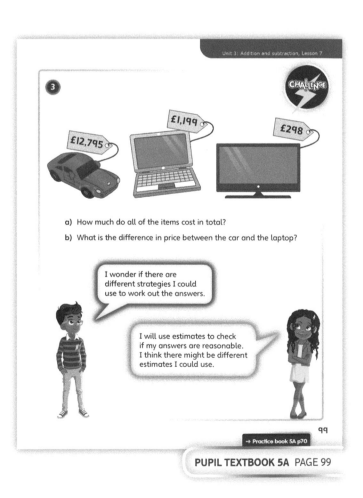

3

CHALLENGE

£12,795   £1,199   £298

a) How much do all of the items cost in total?

b) What is the difference in price between the car and the laptop?

I wonder if there are different strategies I could use to work out the answers.

I will use estimates to check if my answers are reasonable. I think there might be different estimates I could use.

99

→ Practice book 5A p70

PUPIL TEXTBOOK 5A PAGE 99

Practice

WAYS OF WORKING Independent thinking

IN FOCUS Questions ❶, ❷ and ❸ aim to consolidate children's understanding of using rounding to check answers and to compare actual answers to estimates. Question ❺ is a two-step problem involving both addition and subtraction. Discuss the different ways in which this can be calculated. Children may add the first two numbers and then subtract the third. Encourage children to compare their estimate to the actual answer and to decide if it is a sensible estimate.

STRENGTHEN In question ❹, support children in realising that it can be possible to round to different numbers and that this will affect the estimate. Encourage children to use a number line with 100s and 1,000s marked so they can see why 2,187 can round to both 2,000 and 2,200. Discuss which estimate is more accurate and why.

DEEPEN Ask children to come up with at least two different estimates for question ❶, question ❷ and question ❸.

ASSESSMENT CHECKPOINT Are children confident in using rounding to make sensible estimates and then using these estimates to check and correct answers? Can they identify the mistakes made and explain what should be changed?

ANSWERS Answers for the **Practice** part of the lesson can be found in the *Power Maths* online subscription.

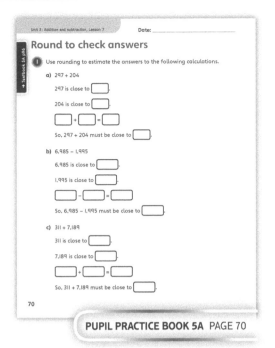

PUPIL PRACTICE BOOK 5A PAGE 70

PUPIL PRACTICE BOOK 5A PAGE 71

Reflect

WAYS OF WORKING Independent thinking

IN FOCUS This **Reflect** activity checks if children understand the importance of using rounding to make estimates. Children should be able to explain that rounding helps to find mistakes and to check if the answer to a calculation looks sensible and is likely to be correct.

ASSESSMENT CHECKPOINT Can children explain why rounding can be used to make estimates and check answers?

ANSWERS Answers for the **Reflect** part of the lesson can be found in the *Power Maths* online subscription.

After the lesson ⏸

- Can children round numbers to an appropriate degree of accuracy?
- Can children make sensible estimates and use these to find mistakes?
- Do children understand the importance of checking answers?

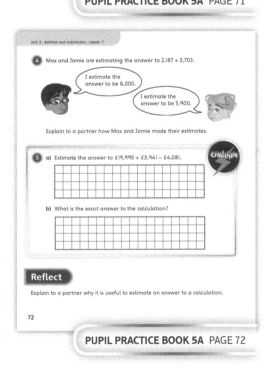

PUPIL PRACTICE BOOK 5A PAGE 72

Inverse operations (addition and subtraction)

Learning focus

In this lesson, children will learn how to use the inverse operation in order to check the answers to addition and subtraction calculations.

Before you teach

- Can children write out the four number sentences from a part-whole model?
- Can children add and subtract using the column method?

NATIONAL CURRICULUM LINKS

Year 5 Number – addition and subtraction

Estimate and use inverse operations to check answers to a calculation.

ASSESSING MASTERY

Children can use the inverse operations of addition and subtraction to check the answers to calculations.

COMMON MISCONCEPTIONS

Children may choose the wrong numbers when trying to identify the inverse calculation. For example, for 3,482 − 1,232 = 2,250, they do 3,482 + 2,250 instead of 1,232 + 2,250. Ask:

- *What type of calculation is this? What are you trying to check? Does your answer make sense? How could we represent this information as a bar model to help us?*

STRENGTHENING UNDERSTANDING

Children can use a part-whole model or a bar model to help them see why addition and subtraction are inverse operations of each other.

GOING DEEPER

Give children questions that involve both mental addition and subtraction and ask them to mentally check the answer to each calculation using the inverse operation.

KEY LANGUAGE

In lesson: inverse, addition, subtraction, check, equal, exchange, operation, fact family

STRUCTURES AND REPRESENTATIONS

Column addition, column subtraction, part-whole model

RESOURCES

Optional: place value counters, blank part-whole models

 In the eTextbook of this lesson, you will find interactive links to a selection of teaching tools.

Quick recap

Ask: *The answer to an addition is 4,000. What is the question?* Look at the strategies that children use to get to the answer. Repeat for different, more challenging numbers and also for subtractions.

Discover

WAYS OF WORKING Pair work

ASK

- Question ❶ a): *Which numbers do you need to add together? What do you think the answer might be?*
- Question ❶ b): *Why do you think this person has made this mistake?*

IN FOCUS Question ❶ a) is used to demonstrate why it is useful to use the inverse operation in order to check a calculation. Question ❶ b) builds on this and requires children to identify the mistake that has been made.

PRACTICAL TIPS Use place value counters to make the numbers in Reena and Lee's calculation. Use them to model why Lee's answer cannot be right.

ANSWERS

Question ❶ a): Reena: 2,355 + 5,191 = 7,546
Lee: 2,355 + 5,211 = 7,566
Reena is correct as 2,355 + 5,191 is equal to 7,546.

Question ❶ b): Lee subtracted the smaller digit from the greater digit in each column. He should have exchanged 1 hundred for 10 tens so that he could do the subtraction.

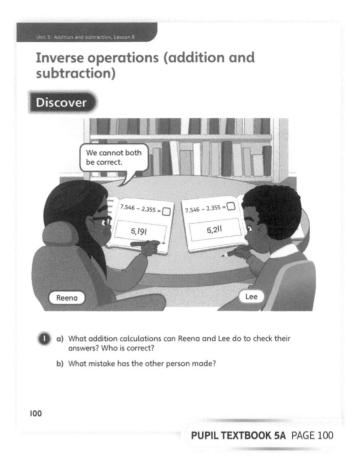

Inverse operations (addition and subtraction)

Discover

❶ a) What addition calculations can Reena and Lee do to check their answers? Who is correct?

b) What mistake has the other person made?

100

PUPIL TEXTBOOK 5A PAGE 100

Share

WAYS OF WORKING Whole class teacher led

ASK

- Question ❶ a): *Why should you use addition to check the calculation? What is the whole? What are the parts? What should you add together?*
- Question ❶ b): *Look at the place value for each column. Where has Lee made his mistake?*

IN FOCUS For question ❶ a), show children the part-whole model and discuss why we can use addition to check the calculation. Emphasise that addition is the inverse of subtraction. Show children Reena and Lee's column additions. Look at the place value of each digit. For question ❶ b), discuss with children if we can easily subtract 5 tens from 4 tens, identify the exchange that needs to take place, of 10 tens for 1 hundred, and encourage children to spot Lee's mistake here. Can children explain why the total is 7,546 and not 7,566?

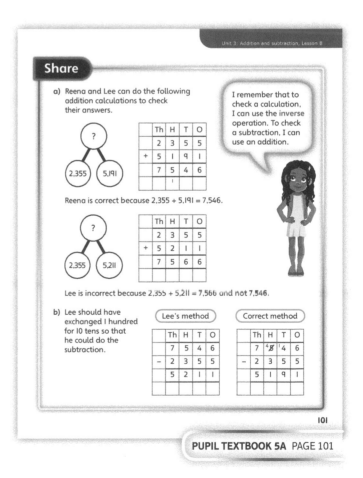

Share

a) Reena and Lee can do the following addition calculations to check their answers.

Reena is correct because 2,355 + 5,191 = 7,546.

Lee is incorrect because 2,355 + 5,211 = 7,566 and not 7,546.

b) Lee should have exchanged 1 hundred for 10 tens so that he could do the subtraction.

I remember that to check a calculation, I can use the inverse operation. To check a subtraction, I can use an addition.

101

PUPIL TEXTBOOK 5A PAGE 101

Think together

WAYS OF WORKING Whole class teacher led (I do, We do, You do)

ASK

- Question **1**: *What is the inverse operation of addition? Can you use this to check Reena and Lee's answers?*
- Question **2**: *Why do you use addition to check Reena and Lee's answers? Are there any other ways that you could check?*
- Question **3**: *What is the whole? What are the parts? Can you write more than one calculation?*

IN FOCUS Question **2** looks at using the inverse operation to check a calculation where two different mistakes have been made. Discuss if 11,315 is a sensible answer to 46,795 – 3,548 and link this to the lesson about using rounding to check answers. For question **3** b), discuss how it may be easier to subtract 1 from each number in the two subtractions so that we have 9,999 rather than 10,000 and do not have to deal with an exchange. Can children explain why subtracting 1 from each number will not affect the overall answer?

STRENGTHEN To support understanding, encourage children to use a part-whole model to show how addition and subtraction are inverse operations of each other and to help them identify which numbers they need to add or subtract. Write the number sentences that relate to the part-whole model, using blank boxes, with the addition, subtraction and equals signs in the correct place. Ask children to fill in the blank boxes with the correct numbers.

DEEPEN Question **3** can be explored further by giving children other part-whole models and asking them to write out all the possible number sentences, then using inverse operations to check they are correct. Give children more calculations that involve subtracting from numbers with lots of 0s, such as 1,000 and 10,000.

ASSESSMENT CHECKPOINT Can children identify and use the inverse operation, and other useful mental methods, to check the answers to calculations and find and correct mistakes?

ANSWERS

Question **1** a): The correct answer is 23,405 + 7,892 = 31,297, so Lee is correct. You can check the answer is correct by doing subtractions.

Question **1** b): Reena has not exchanged 10 hundreds for 1 thousand or 10 thousands for 1 ten thousand.

Question **2**: The correct answer is 46,795 – 3,548 = 43,247. Reena has not laid out the column subtraction correctly. The 8 ones need to be under the 5 ones, and so on.
Lee has not exchanged 1 ten for 10 ones and has just subtracted the smaller digit from the greater digit in the ones column.

Question **3** a): 770 + 230 = 1,000   230 + 770 = 1,000
1,000 – 770 = 230   1,000 – 230 = 770

Question **3** b) and c): The calculation is incorrect because
10,000 – 7,730 = 2,270 and
10,000 – 3,270 = 6,730.

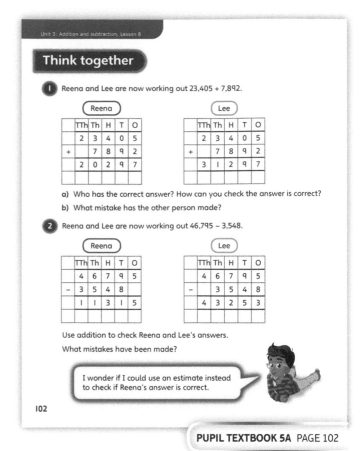

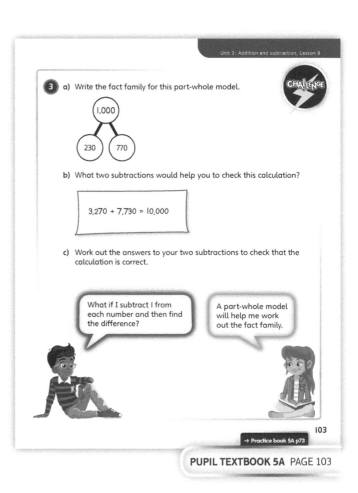

Practice

WAYS OF WORKING   Independent thinking

IN FOCUS   Question ❶ consolidates children's understanding of addition and subtraction as inverse operations, and the importance of using an inverse calculation to check answers. Question ❷ consolidates children's understanding of the number sentences that can be written from a part-whole model and encourages them to use addition to check if a calculation is correct. Questions ❸ and ❹ require children to find the mistake that each child has made. Encourage children to use the inverse operation and not to just redo the calculation when checking it. Question ❺ asks children to find the original calculation when given the inverse calculation. Encourage them to come up with more than one solution for this.

STRENGTHEN   Give children blank part-whole models and encourage children to fill them in with the numbers from each question, so that they can see if they need to add or subtract in order to check the inverse calculation.

DEEPEN   Question ❺ can be explored further by giving children other inverse calculations and asking them to find the original calculation. Give children missing number problems, such as ⬚ + 346 = 742, ask them to work out the missing number and then to use the inverse operation to check they have got the correct answer.

ASSESSMENT CHECKPOINT   Can children use the inverse operation to check the answers to calculations and to find and correct mistakes with confidence?

ANSWERS   Answers for the **Practice** part of the lesson can be found in the *Power Maths* online subscription.

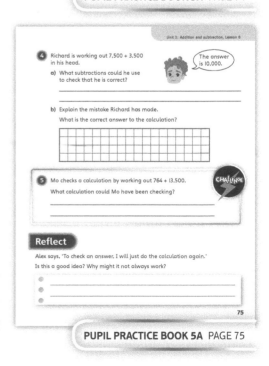

PUPIL PRACTICE BOOK 5A PAGE 73

PUPIL PRACTICE BOOK 5A PAGE 74

Reflect

WAYS OF WORKING   Independent thinking

IN FOCUS   This **Reflect** activity checks that children are able to explain why using the inverse operation to check a calculation is more useful than just redoing the calculation, which may be more prone to mistakes.

ASSESSMENT CHECKPOINT   Can children explain why using the inverse operation is an important and useful way to check and correct answers?

ANSWERS   Answers for the **Reflect** part of the lesson can be found in the *Power Maths* online subscription.

After the lesson ⏸

- Do children understand the importance of checking their answers?
- Can children identify the inverse operation for addition and subtraction?
- Which children need to use part-whole models to help work out what the inverse operation is?

PUPIL PRACTICE BOOK 5A PAGE 75

Multi-step addition and subtraction problems ❶

Learning focus

In this lesson, children will learn what strategies to use to solve problems that involve adding and subtracting whole numbers with more than four digits.

Before you teach

- Do children know how to lay out column addition and subtraction and make multiple exchanges?
- Do children understand key vocabulary, such as 'more than' and 'less than'?

NATIONAL CURRICULUM LINKS

Year 5 Number – addition and subtraction

Solve addition and subtraction multi-step problems in contexts, deciding which operations and methods to use and why.

ASSESSING MASTERY

Children can solve problems that involve a combination of adding and subtracting whole numbers with more than four digits and make multiple exchanges.

COMMON MISCONCEPTIONS

Children may not notice that the vocabulary used in the problem can be used to work out what calculation is needed. Ask:
- *Are there any key words that will give you a clue? What does 'total' or 'difference' mean?*

Children may not carry out both steps of a two-step problem. Ask:
- *What are you trying to find out? What kind of answer are you expecting to get? Have you used all of the information from the question?*

STRENGTHENING UNDERSTANDING

Children should first focus on solving problems with just addition or subtraction, before combining them. Encourage children to use a bar model or a number line to aid their understanding.

GOING DEEPER

Give children problems that are based on information represented in a list or table where they need to identify the correct information first before solving the problem.

KEY LANGUAGE

In lesson: more than, altogether, difference, total, method, addition, subtraction, combined, comparison

STRUCTURES AND REPRESENTATIONS

Bar model, number line, column addition, column subtraction

RESOURCES

Place value grid, counters

 In the eTextbook of this lesson, you will find interactive links to a selection of teaching tools.

Quick recap ⟳

Recap column addition and subtraction covered in this unit. Write on the board several fluency-based questions. How many questions can children answer in 5 minutes?

Discover

WAYS OF WORKING Pair work

ASK

- Question **1** a): *What is the price of the used sports car? What is the price of the new sports car? What does 'difference' mean?*
- Question **1** b): *How much money does Jen have? What do you know about how much Holly has?*

IN FOCUS Question **1** a) requires children to identify amounts of money and to find the difference between them, using subtraction. Question **1** b) involves working out the total amount of money that Jen and Holly have between them, which is a two-step addition problem.

PRACTICAL TIPS Show an example of a real car sales website or magazine and discuss why it might be useful to compare amounts of money in this context.

ANSWERS

Question **1** a): £16,725 – £7,560 = £9,165. The difference in price between the cost of the new sports car and the used one is £9,165.

Question **1** b): Jen and Holly have £6,650 altogether.

Share

WAYS OF WORKING Whole class teacher led

ASK

- Question **1** a): *What calculation do you need to do if you are finding the difference? What method could you use to subtract?*
- Question **1** b): *If Holly has £1,450 more than Jen, how can you work out how much money Holly has? How can you work out how much money Jen and Holly have altogether?*

IN FOCUS For question **1** a), show children the bar model and ask how they know that finding the difference leads us to carrying out a subtraction. Ensure children are confident in the correct layout for a 5-digit subtract 4-digit calculation. Take the opportunity to reinforce the place value of each digit when carrying out the calculation and to address the need for an exchange. For question **1** b), discuss how 'more than' indicates the need for an addition, but ensure that children do not just add the two numbers together. Show the bar model, which should help children to see why this will not work. Discuss how we could work out how much money Holly has and then ask children if this is our answer to the whole question. Discuss how we need to work out how much money Jen and Holly have altogether and so need to do another addition.

Think together

Think together

WAYS OF WORKING Whole class teacher led (I do, We do, You do)

ASK

- Question ❶: *How much money do Jen and Holly have altogether? How much does the new sports car cost? What calculation do you need to do?*
- Question ❷: *How much does each car cost? What method can you use to find how much they cost in total?*
- Question ❸: *What does 'combined' mean? How can you work out the combined cost of the SUV and the electric car? How much does the family car cost? How will you find how much more this is?*

IN FOCUS In question ❶, some children may not understand that the word 'more' is now being associated with a subtraction. Use the bar model to show why we need to subtract. Encourage children to use column subtraction. Question ❷ requires children to problem solve with more than two whole numbers. Show the bar model and discuss how we need to carry out an addition calculation. Discuss the different methods that could be used, adding all three numbers at once in a column addition or adding two numbers in a column addition then adding the other number to this answer. Question ❸ encourages children to solve a problem using both addition and subtraction. Encourage children to find the combined cost of the SUV and the electric car to begin with. Discuss how we then need to do a subtraction in order to find the difference in price.

STRENGTHEN To support understanding, children should represent all of the problems on a bar model, writing in any information they know onto the bar model and including a question mark to show what they are trying to find out.

DEEPEN Question ❸ can be explored further by asking children to find other combined prices to compare to the price of another car by finding the difference. Give a price difference and see if children can work out which pair or group of cars you are comparing.

ASSESSMENT CHECKPOINT Can children solve problems that involve addition, subtraction and a combination of the two, with whole numbers that have four or more digits?

ANSWERS

Question ❶: 16,725 – 6,650 = £10,075
Jen and Holly need £10,075 more to buy the new sports car.

Question ❷: 19,579 + 28,370 + 16,725 = £64,674
The three cars cost £64,674 in total.

Question ❸: 19,579 + 8,298 = £27,877
The SUV and the electric car cost £27,877.
28,370 – 27,877 = £493
The family car costs £493 more than this.

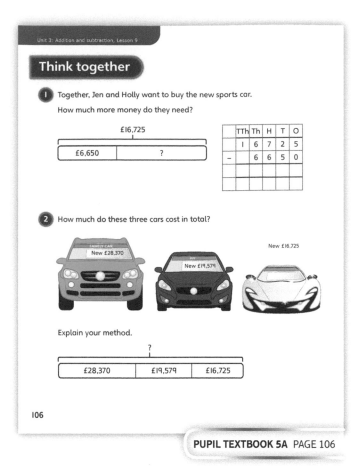

PUPIL TEXTBOOK 5A PAGE 106

PUPIL TEXTBOOK 5A PAGE 107

Practice

WAYS OF WORKING Independent thinking

IN FOCUS For question ❶, ask children to focus on the calculations that arise from each bar model and to show their method for each. Encourage children to discuss the different ways they could solve question ❶ c). Question ❷ provides an addition problem that requires children to interpret a number line in order to work out what calculation they need to do. Question ❸ aims to consolidate children's understanding of problem solving within a context. Encourage children to use the bar model and to add in any information as they complete the calculation. Question ❹ involves a combination of addition and subtraction. Ensure children lay out the column subtraction correctly. Make sure they realise that 8,000 comes last in the question but needs to be at the top in the column subtraction.

STRENGTHEN For question ❺, support children in interpreting the table. Start by asking simple questions about the table, for example, ask: *How many eggs were sold on Saturday? How many eggs were sold on Sunday? How many eggs were sold at the weekend in total?* Lead on from this to identifying what they need to do to find the answer to the actual question.

DEEPEN Question ❷ can be explored further by asking children to draw a bar model from the information shown on the number line.

ASSESSMENT CHECKPOINT Can children confidently add and subtract whole numbers with four or more digits and solve problems that involve combinations of these calculations?

ANSWERS Answers for the **Practice** part of the lesson can be found in the *Power Maths* online subscription.

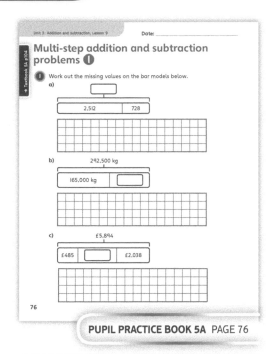

PUPIL PRACTICE BOOK 5A PAGE 76

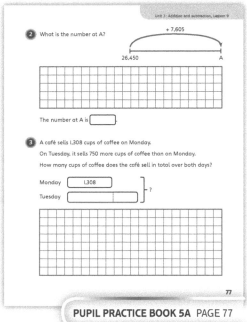

PUPIL PRACTICE BOOK 5A PAGE 77

Reflect

WAYS OF WORKING Independent thinking

IN FOCUS This **Reflect** activity checks children's understanding of using addition and subtraction calculations in a problem-solving context. Encourage children to write a problem where the numbers have four or more digits and encourage them to use key vocabulary, such as 'more', 'total', 'difference'. Ask children to also solve their problem once they have written it.

ASSESSMENT CHECKPOINT Can children use appropriate language to write a two-step problem that requires using both addition and subtraction to solve it?

ANSWERS Answers for the **Reflect** part of the lesson can be found in the *Power Maths* online subscription.

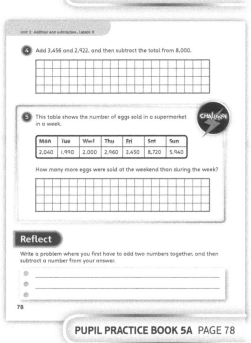

PUPIL PRACTICE BOOK 5A PAGE 78

After the lesson

- Can children apply what they know about addition and subtraction of whole numbers with four or more digits in problem-solving contexts?
- Would a class display of key vocabulary from word problems be useful?
- Which children need support to identify each step that is required to solve a two-step problem?

143

Multi-step addition and subtraction problems ②

Learning focus

In this lesson, children will learn how to solve more complex addition and subtraction multi-step problems that involve interpreting and identifying the information in order to solve the problem.

Before you teach

- Do children know how to lay out column addition and subtraction?
- Are children confident in drawing bar models including comparison bar models?
- Do children understand key vocabulary, such as more and fewer?

NATIONAL CURRICULUM LINKS

Year 5 Number – addition and subtraction

Solve addition and subtraction multi-step problems in contexts, deciding which operations and methods to use and why.

ASSESSING MASTERY

Children can solve more complex multi-step problems that involve adding and subtracting whole numbers, where the information is represented in tables or needs to be extracted from sentences.

COMMON MISCONCEPTIONS

Children may not understand what calculations a problem is asking them to do and so will just add all of the numbers given, particularly when the information is presented in a table. Ask:
- *What does the table tell you? What information will you use from the table? What model could you draw to help you?*

STRENGTHENING UNDERSTANDING

Encourage children to draw a bar model or a number line to aid their understanding of the information provided in each problem and to draw out what they already know and what calculations they will need to do to find the answer.

GOING DEEPER

Ask children to solve problems based on other mathematical contexts that still involve addition and subtraction. For example, working out the perimeter of a field when given the length and the width.

KEY LANGUAGE

In lesson: how much, more, fewer, total, add, left, two-way table

STRUCTURES AND REPRESENTATIONS

Bar model, number line, column addition, column subtraction

RESOURCES

Optional: water, measuring jug, blank comparison bar models

 In the eTextbook of this lesson, you will find interactive links to a selection of teaching tools.

Quick recap 🔍

Ask a child to pick a 2-digit number.

Ask another child to pick a different 2-digit number.

Ask them to add their numbers together and find the difference between their numbers. Tell the rest of the class. Ask the class if they can work out the two starting numbers.

Discover

WAYS OF WORKING Pair work

ASK

- Question **1** a): *What model does the fuel gauge look like? Can you read it like a number line?*
- Question **1** a) *How much fuel is left? How much fuel was there to begin with?*
- Question **1** b): *How much fuel is left? How much fuel is used each hour? How much fuel is used in two hours?*

IN FOCUS Question **1** a) requires children to identify an amount from a scale and to use this to find the amount of fuel that has been used. In question **1** b), children will need to use more than one calculation to work out how much fuel is left after two hours, which is a two-step problem.

PRACTICAL TIPS Use water in a measuring jug to reinforce the concept of capacity measured in litres. Gradually pour the water out and demonstrate how the scale of the jug is used to work out how much has been used and how much is left. Link this to the fuel gauge in the picture.

ANSWERS

Question **1** a): The plane has used 32,000 litres of fuel so far.

Question **1** b): There will be 37,840 litres of fuel left after two more hours of flying.

Multi-step addition and subtraction problems ❷

Discover

1 a) How much fuel has the plane used so far?

b) Each hour the plane uses 13,580 litres of fuel.

How much fuel will be left after two more hours of flying?

108

PUPIL TEXTBOOK 5A PAGE 108

Share

WAYS OF WORKING Whole class teacher led

ASK

- Question **1** a): *What is the fuel gauge going up in? What number lies exactly half-way between 60,000 and 70,000?*
- Question **1** b): *How could you work out how much fuel is used in two hours? What calculation do you need to do to work out how much fuel is left after two hours?*

IN FOCUS For question **1** a), show the fuel gauge and the number line. Can children explain how they know they are going up in increments of 5,000 litres? Ask children how many litres of fuel there were to begin with and how we could work out how much fuel is left. Discuss the use of the number line and how we can think about the next 10,000 after 65,000 and count up from there to get our answer. Discuss the alternative method of using column subtraction and link this to the lesson about checking answers.

For question **1** b), show the bar model and ask children what calculation we need to do to work out how much fuel is used in two hours. Reinforce that this is not the final answer to the question. Show children the next bar model to help them understand why we now need to subtract. Some children may spot the alternative method of subtracting 13,580 twice from 65,000 which is also suitable and could be used as a check.

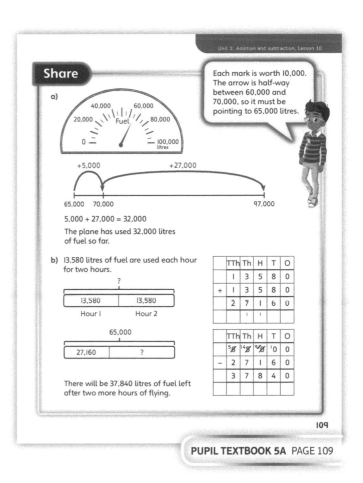

PUPIL TEXTBOOK 5A PAGE 109

Think together

WAYS OF WORKING Whole class teacher led (I do, We do, You do)

ASK

- Question **1**: *Which numbers are before 2 pm? Which are after 2 pm? How will you work out how many passengers there are before 2 pm and after 2 pm?*
- Question **2**: *How will you work out how many passengers the small plane holds? What calculation could you do to work out the total number of passengers on both planes?*
- Question **3**: *Which numbers in the table are not affected by length of flight? Which numbers in the table tell you how much fuel is needed per hour? How will you calculate how much fuel is needed for four hours?*

IN FOCUS The information in question **3** is presented in a table, and several numbers and calculations are involved. Encourage children to think about which numbers in the table will be affected by the length of the flight and which will not. Ask children to identify the amount of flight fuel and spare fuel needed for every hour of flying and question how we could work out how much would be needed for four hours. Some children may try to add four times. Encourage them to think of a more efficient way. Discuss the need to add the other numbers from the table to get the total amount of fuel needed for a four hour flight.

STRENGTHEN Write out question **2** and ask children to highlight the key numbers (416, 280) and words (fewer, how many, total), each in a different colour. Provide a blank comparison bar model and ask children to fill in each piece of information that they know, using the same colours. Reinforce the importance of making the link between the information in the word problem and how it can be represented in a bar model.

DEEPEN Question **3** can be explored further by asking children to find the amount of fuel needed for other lengths of flight. Take the opportunity for children to practise other key topics, such as multiplying by 10 (for a ten hour flight) or finding a half (for a half hour flight).

ASSESSMENT CHECKPOINT Can children use addition and subtraction of whole numbers to solve multi-step problems that involve interpreting tables and information?

ANSWERS

Question **1**: 14,569 + 11,118 = 25,687
25,687 passengers passed through before 2 pm.
23,277 + 5,946 = 29,223
29,223 passengers passed through after 2 pm.
More passengers passed through the airport after 2 pm.

Question **2**: 416 − 280 = 136
416 + 136 = 552
The two planes can carry 552 passengers in total.

Question **3**: 12,500 + 2,500 = 15,000
15,000 × 4 = 60,000
60,000 + 5,600 + 5,150 = 70,750
The pilot will need 70,750 litres of fuel for a four hour flight.

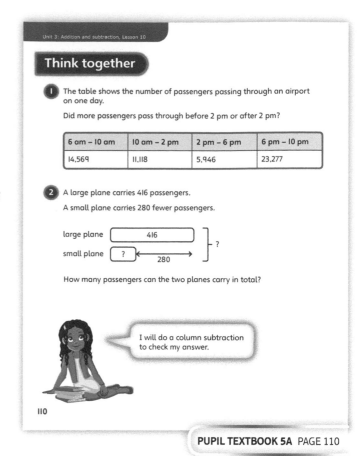

PUPIL TEXTBOOK 5A PAGE 110

PUPIL TEXTBOOK 5A PAGE 111

Practice

WAYS OF WORKING Independent thinking

IN FOCUS Question ❶ focuses on adding then subtracting to solve a problem. Encourage children to show their method for each part of their calculation. In question ❷ a), encourage children to think carefully about how they know Tex makes more toys (both numbers in Tex's column are greater than the numbers in the same rows in Karl's column). For question ❷ b), ensure children do not just work out the total number of toys that Tex makes. Question ❸ is a missing number problem with three numbers. Encourage children to draw a bar model to represent the problem and discuss the different methods that are possible to solve it, such as adding the two numbers together and subtracting from 30,000, or subtracting both numbers from 30,000. Question ❹ aims to consolidate children's understanding of problem solving within a context. Encourage children to use a comparison bar model, asking questions to help them do this correctly, for example, ask: *Which bar will be longest? How do you know? Which bar will be shortest?* Encourage children to keep re-reading the question to ensure they work out the total number of apples.

STRENGTHEN Encourage children to draw a bar model to help them see what calculations are needed, or provide blank bar models so they can fill the numbers in themselves.

DEEPEN In question ❺, the information is presented on a number line. This can be explored further by giving children a different starting number on the number line and a different jump size. This could also involve counting back on the number line and asking how many jumps are needed to get past 0.

ASSESSMENT CHECKPOINT Can children identify the relevant information needed to confidently solve multi-step problems involving adding and subtracting whole numbers, for example, in question ❹?

ANSWERS Answers for the **Practice** part of the lesson can be found in the *Power Maths* online subscription.

Reflect

WAYS OF WORKING Independent thinking

IN FOCUS This **Reflect** activity checks children's understanding of addition and subtraction calculations and comparison of large numbers. Encourage children to explain which method they would use to work out the answers to both calculations and how they would then compare their answers to find which is bigger.

ASSESSMENT CHECKPOINT Can children add, subtract and compare 5- and 6-digit whole numbers?

ANSWERS Answers for the **Reflect** part of the lesson can be found in the *Power Maths* online subscription.

After the lesson

- Do children understand how addition and subtraction skills can be applied to solving problems?
- What mistakes are being most commonly made?

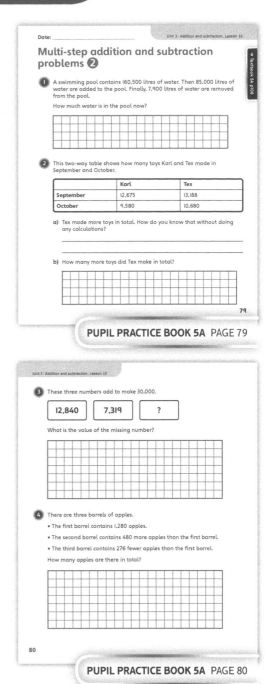

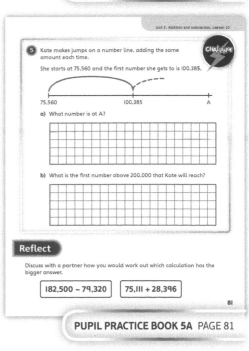

PUPIL PRACTICE BOOK 5A PAGE 79

PUPIL PRACTICE BOOK 5A PAGE 80

PUPIL PRACTICE BOOK 5A PAGE 81

Solve missing number problems

Learning focus

In this lesson, children will solve missing number problems involving addition and subtraction. A variety of contexts will be used and children should use the most appropriate and efficient strategies learned in the previous lessons.

Before you teach

- Are children secure with mental strategies for addition and subtraction?
- Do children have a preferred method? If so, how will you ensure they get appropriate practise of other methods?
- Which children will need a blank number line to help them work out their answers?

NATIONAL CURRICULUM LINKS

Year 5 Number – addition and subtraction

Solve addition and subtraction multi-step problems in contexts, deciding which operations and methods to use and why.

ASSESSING MASTERY

Children can work out missing number problems using appropriate mental or written methods. They can show their thinking using number lines and part-whole models.

COMMON MISCONCEPTIONS

Children may get mixed up about whether they need to add or subtract to find the missing number. Ask:
- *What calculation are you being asked to do? Can you draw a part-whole model or bar model to help you see what operation to use?*

They may also try to fall back on their preferred method, even if it is not the most efficient one. For example, if children are always using a formal written method, you could ask:
- *Is that the most efficient method to use? Is there another method you could try?*

STRENGTHENING UNDERSTANDING

Provide children with number lines and blank part-whole models to support their thinking. It may also be useful to provide calculators so that children can quickly sense check their own answers.

GOING DEEPER

Ask children to create some of their own problems for their partner to solve. Ask: *Can you create a problem that gives an answer of 17? 3? 54?*

KEY LANGUAGE

In lesson: solve, method, addition, subtraction, partition

STRUCTURES AND REPRESENTATIONS

Number lines, part-whole models

RESOURCES

Mandatory: number lines, blank part-whole models

Optional: number cards, shape cards, calculators

 In the eTextbook of this lesson, you will find interactive links to a selection of teaching tools.

Quick recap

Ask missing number questions such as
$170 + \square = 200$ or $300 - 180 = \square$.
Discuss methods children would use to work out the answers.

Discover

WAYS OF WORKING Pair work

ASK

- Question ❶: *Which strategy should we use? Will the same strategy work for both parts a) and b)?*
- Question ❶: *What numbers should we mark on a blank number line to help us?*
- Question ❶: *Does it matter how spaced out the jumps are?*

IN FOCUS In questions ❶ a) and ❶ b), children work out missing numbers in a puzzle context. They should be able to work out the answers mentally through counting on or counting back.

PRACTICAL TIPS Use shape cards to recreate the puzzle, with the answers written on the back of each card. The answers can be revealed once the children have had a go at it. Children could then create their own newspaper puzzles.

ANSWERS

Question ❶ a): The value of the triangle is 27.

Question ❶ b): The value of the circle is 37.

PUPIL TEXTBOOK 5A PAGE 112

Share

WAYS OF WORKING Whole class teacher led

ASK

- Question ❶ a): *Why has Dexter added 7 first?*
- Question ❶ b): *Compare counting back and Flo's method of counting on. Why do they both work? Which method do you prefer?*

IN FOCUS In question ❶, children are exposed to different methods of working out the missing numbers. Both parts of the question have a visual representation of a number line and counting on or counting back. Children should recognise that they are finding the difference between two numbers each time.

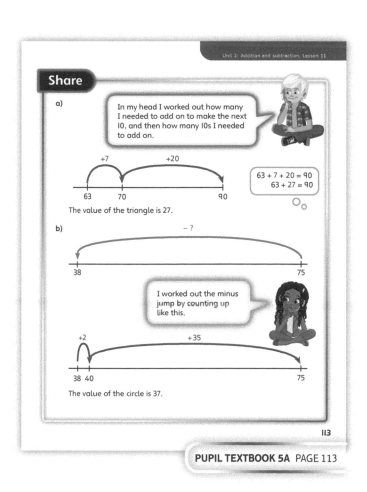

PUPIL TEXTBOOK 5A PAGE 113

Think together

WAYS OF WORKING Whole class teacher led (I do, We do, You do)

ASK

- Question **1**: *Will you count on or back to work out the missing numbers?*
- Question **2**: *Have you found all three methods? Are there any other methods you could use?*
- Question **3**: *What method is each of the children using? Which is your preferred method and why?*

IN FOCUS In questions **1**, **2** and **3** children find missing numbers by counting on or back. They have the opportunity to compare different methods and decide which their preferred method is.

STRENGTHEN Provide children with blank number lines so that they can show their jumps on or back. This will also help them keep track of where they are in their thought process.

DEEPEN Ask children to compare questions they think would be more appropriate for a mental method and when it is more efficient or accurate to use a formal written method.

ASSESSMENT CHECKPOINT Children can use a variety of methods to find missing numbers in problems. They can discuss which method is most efficient and why.

ANSWERS

Question **1**: The value of the red triangle is 62.
The value of the yellow triangle is 33.
The value of the blue triangle is 2,600.

Question **2**: The value of the yellow square is 440.

Question **3** a): Both methods involve finding the difference between two numbers so the answer will be the same whether they use counting on or counting back.

Question **3** b): This will be each child's personal choice.

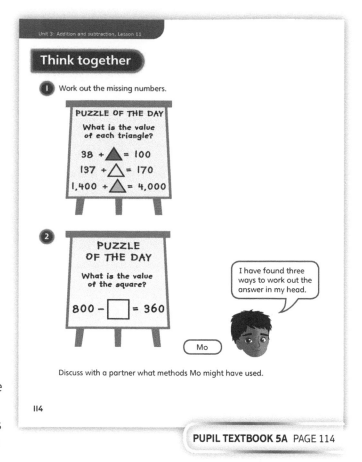

PUPIL TEXTBOOK 5A PAGE 114

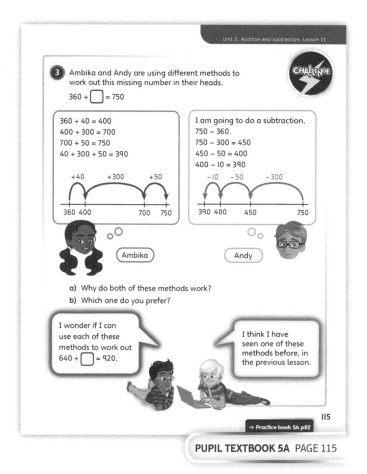

PUPIL TEXTBOOK 5A PAGE 115

Practice

WAYS OF WORKING Independent thinking

IN FOCUS Children practise working out missing numbers. They work with numbers with up to four digits with a focus on using mental strategies. In question ①, the thinking process is modelled and structured for children so that they can then use this strategy in other questions. Question ② provides number lines for children to use to help them model their strategy. Question ③ highlights a common misconception where children may incorrectly try to use their number bond knowledge. Questions ④ and ⑤ provide further practise of abstract missing number calculations. Question ⑥ provides number lines for children to model their strategy and Question ⑦ provides more of a challenge, with three numbers involved in the number sentence.

STRENGTHEN Provide children with blank number lines so that they can sketch their thinking process.

DEEPEN Ask children to create their own missing number problems for their partner to answer. Can they write them in order of difficulty? Ask: *Why are some questions easier than others?*

ASSESSMENT CHECKPOINT Children can work out a variety of missing number problems with numbers with up to four digits. They use an appropriate and efficient strategy and can explain their thinking.

ANSWERS Answers for the **Practice** part of the lesson can be found in the *Power Maths* online subscription.

Reflect

WAYS OF WORKING Pair work

IN FOCUS Children discuss which methods they can use to work out missing number problems. Although there are multiple methods, children should also discuss which they think are more efficient and why.

ASSESSMENT CHECKPOINT Children talk about the positives and draw backs of using particular methods for working out missing numbers.

ANSWERS Answers for the **Reflect** part of the lesson can be found in the *Power Maths* online subscription.

After the lesson ⏸

- Are children confident working mentally?
- Do children still need to use a number line or are they able to visualise a number line in their head?
- Do children keep referring back to one strategy or are they able to be flexible with their approach depending on the question?

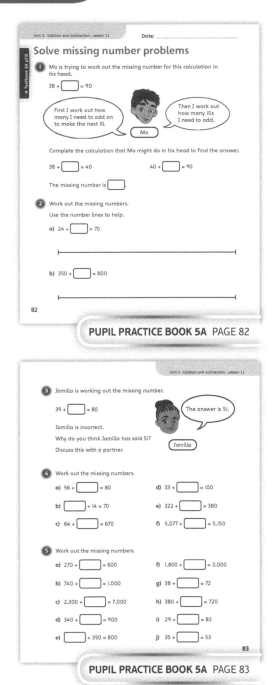

PUPIL PRACTICE BOOK 5A PAGE 82

PUPIL PRACTICE BOOK 5A PAGE 83

PUPIL PRACTICE BOOK 5A PAGE 84

Solve comparison problems

Learning focus

In this lesson, children will solve problems involving comparison. They should see links between numbers to help them work out missing numbers rather than rely on formal written methods for calculating.

Before you teach

- Are children confident with addition and subtraction of numbers with up to four digits?
- Do children have a secure understanding of place value?
- Have children been exposed to bar models before?

NATIONAL CURRICULUM LINKS

Year 5 Number – addition and subtraction

Solve addition and subtraction multi-step problems in contexts, deciding which operations and methods to use and why.

ASSESSING MASTERY

Children can use the relationship between addition and subtraction to work out missing numbers. They have a deep understanding of equivalence and use this to justify answers.

COMMON MISCONCEPTIONS

Children may have the misconception that if one number increases the other number should also increase. Consider the example $134 + 471 = 144 +$ ☐. Children may think that because 134 on the left increases by 10 to give 144 on the right, then the missing number should be $471 + 10 = 481$. In fact, they should decrease 471 by 10 to compensate for the increase to 134. The missing answer is 461. Ask:

- *What will happen to the answer if both numbers on the left increase to give the numbers on the right? Will the value of the left-hand side be the same, or different to, the value of the right-hand side?*

STRENGTHENING UNDERSTANDING

Bar models are a great visual representation for these types of problem. They not only help children find the correct answer but also reinforce the meaning of equality.

GOING DEEPER

Give children a calculation then ask them to find five more calculations that are equivalent to it. What strategies did they use to work them out?

KEY LANGUAGE

In lesson: equivalent, increase, decrease

STRUCTURES AND REPRESENTATIONS

Bar model

RESOURCES

Optional: bar model, number line

 In the eTextbook of this lesson, you will find interactive links to a selection of teaching tools.

Quick recap 🔎

Ask children to write an addition that gives the same answer as 7 + 5. Discuss their methods. Can they write a subtraction that gives the same answer? Can they write a calculation that gives the same answer as 77 + 55, without working out the answer?

Discover

Unit 3: Addition and subtraction, Lesson 12

Solve comparison problems

Discover

WAYS OF WORKING Pair work

ASK

- Question ❶: *Can you see any relationships between the numbers?*
- Question ❶: *Which number is different from the rest?*
- Question ❶: *What is the difference between the numbers?*

IN FOCUS Question ❶ gives children the opportunity to start visualising what a comparison problem might look like. They should realise that 123, 133 and 120 are all closely linked.

PRACTICAL TIPS Create bar models that children can hold up at the front of the class. Stick them on the wall to show they are all equivalent.

ANSWERS

Question ❶ a): 404 is on the card with the red circle.

Question ❶ b): 417 is on the card with the blue triangle.

❶ a) What number is on the card with the red circle?

b) What number is on the card with the blue triangle?

116

PUPIL TEXTBOOK 5A PAGE 116

Share

WAYS OF WORKING Whole class teacher led

ASK

- Question ❶ a): *What is the difference between 123 and 133?*
- Question ❶ a): *Do you think we should add 10 or subtract 10 from 414? Why?*
- Question ❶ b): *What is the difference between 123 and 120?*
- Question ❶ b): *Do you think we should add 3 or subtract 3 from 414 to make them equivalent? Why?*

IN FOCUS Children begin to see relationships between numbers. They should see that if one side of an equation is increased, the other side needs to decrease and vice versa.

Unit 3: Addition and subtraction, Lesson 12

Share

a) 133 is **10 more** than 123.

This means that the number on the card with the red circle must be 10 less than 414.

I used a number fact instead of doing the addition.

| 123 | 414 |
| 133 | ● |

404

404 is the number on the card with the red circle.

b) 120 is **3 less** than 123.

This means that the number on the card with the blue triangle must be 3 more than 414.

I also saw a number fact I could use instead of adding.

| 123 | 414 |
| 120 | ▲ |

417

417 is the number on the card with the blue triangle.

117

PUPIL TEXTBOOK 5A PAGE 117

Think together

WAYS OF WORKING Whole class teacher led (I do, We do, You do)

ASK
- Questions ❶ and ❷: *What has stayed the same? What has changed?*
- Questions ❶ and ❷: *Will each missing number increase or decrease from the original number?*
- Question ❸: *Do the same rules apply when using subtraction?*

IN FOCUS In questions ❶ and ❷, children compare what has changed and what has stayed the same within calculations. They then need to identify the change to decide how to calculate the missing numbers.

STRENGTHEN Provide bar model templates for children to complete or partly completed bar models for any children who need an extra level of support.

DEEPEN Ask children to create their own problems that are similar to the examples used in questions ❷ and ❸.

ASSESSMENT CHECKPOINT Children can find missing numbers and explain how they have found them. They should be able to come up with a rule for how they have adjusted their numbers.

ANSWERS

Question ❶: 215 + 136 = 214 + **137**
215 + 136 = 213 + **138**
215 + 136 = 225 + **126**

Question ❷: 3,000 + 2,750 = 2,000 + **3,750**
3,000 + 2,750 = 2,000 + **3,750**

Question ❸: For the addition, Max can increase 280 by 10 to compensate for 540 decreasing by 10.
For the subtraction, Max can decrease 280 by 10, so the difference remains the same on both sides of the equals sign.

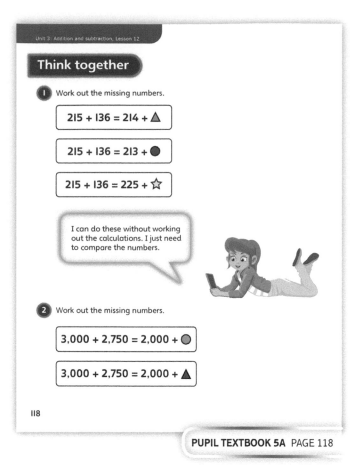

PUPIL TEXTBOOK 5A PAGE 118

PUPIL TEXTBOOK 5A PAGE 119

Practice

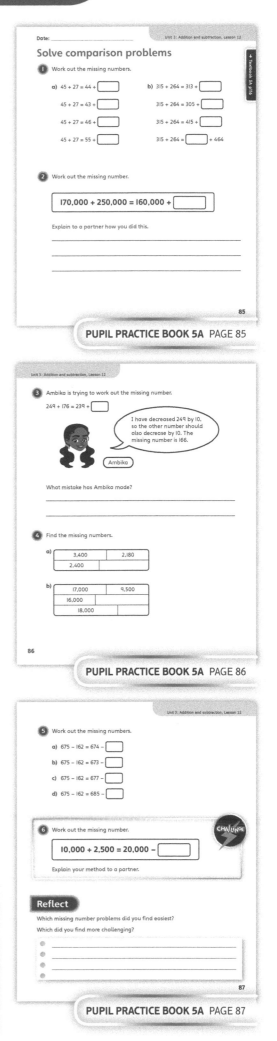

WAYS OF WORKING Pair work

IN FOCUS Question ① a) begins with simple examples of comparison problems where there is only a small change. Children should be able to justify their answers fairly easily. They can then use this to progress onto more difficult examples in question ① b). Although question ② uses 6-digit numbers, children should recognise that this does not make the question harder. Question ③ requires children to explain the mistake in the method used. In Question ④ b), children are exposed to an example that has two missing values to find. Can they write this out as a calculation? Question ⑤ provides multiple linked calculations for children to solve, and question ⑥ involves both addition and subtraction in the same number sentence, for an added challenge.

STRENGTHEN Provide children with blank or partially filled in bar models to help them.

DEEPEN Ask children to create some of their own problems. They could write some just using bar models like in question ④, as well as some written as calculations. Ask: *Which do you think are easier to solve? Why?*

ASSESSMENT CHECKPOINT Children can confidently solve comparison problems and explain why they are increasing or decreasing a number.

ANSWERS Answers for the **Practice** part of the lesson can be found in the *Power Maths* online subscription.

Reflect

WAYS OF WORKING Pair work

IN FOCUS Children discuss which questions they found easier or harder. This reflection will help them when exposed to similar examples in the future.

ASSESSMENT CHECKPOINT Children will recognise that some questions are harder than others. It is likely that children find the subtraction examples trickier than the addition examples.

ANSWERS Answers for the **Reflect** part of the lesson can be found in the *Power Maths* online subscription.

After the lesson ⏸

- Did children find the bar models useful to show equivalence? Did they find the questions written as calculations or those with bar models easier?
- Are children able to justify why they increase or decrease a number rather than just trying to remember a rule?
- Did children make any basic arithmetic mistakes that need addressing?

PUPIL PRACTICE BOOK 5A PAGE 85

PUPIL PRACTICE BOOK 5A PAGE 86

PUPIL PRACTICE BOOK 5A PAGE 87

End of unit check

Don't forget the unit assessment grid in your *Power Maths* online subscription.

WAYS OF WORKING Group work teacher led

IN FOCUS These questions cover the whole unit and are designed to draw out misconceptions and misunderstandings.

Check that children can use the formal written method for addition and subtraction and can correctly align the numbers.

Children should realise that to find a missing number they will need to use an inverse operation, and be able to do this mentally using knowledge of bonds to 100.

When children are solving word problems including those with several steps, encourage them to represent each problem using a bar model to help them work out if they need to add or subtract.

ANSWERS AND COMMENTARY

Children who have mastered the concepts in this unit understand the place value of each digit in 5-digit numbers and can look for key language to see if they are requested to add or subtract in order to answer the calculation. They will have confidence in selecting efficient mental methods to solve problems and know that rounding can help them check their answer.

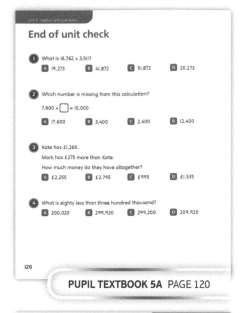

PUPIL TEXTBOOK 5A PAGE 120

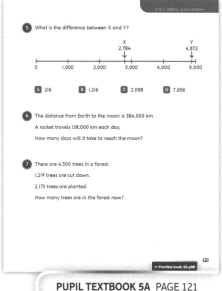

PUPIL TEXTBOOK 5A PAGE 121

Q	A	WRONG ANSWERS AND MISCONCEPTIONS	STRENGTHENING UNDERSTANDING
1	D	A suggests children have forgotten to exchange. B suggests they have lined up the numbers incorrectly and done an incorrect exchange. C suggests they have lined up the numbers incorrectly.	Use place value grids and counters to support column addition and subtraction. Work through step by step showing any counter exchanges, linking the place value grid to the column method.
2	C	For A, children have added when they needed to subtract. B is an incorrect bond to 100. D is a bond to 20,000, not 10,000.	
3	B	A is an incorrect addition. For C, children have subtracted instead of added. For D, they have just added the two numbers.	For question 2, help children understand the part-whole method by drawing a simple bar model or part-whole model to show that they should subtract instead of add together.
4	B	For A, C or D, children have made a mistake with the place value of 300,000 or 80.	
5	C	For A, children have counted up to 3,000 only. B shows they have found the difference up to 4,000. D suggests they have added the numbers together.	Use a bar model to represent each situation and help children decide if they need to add or subtract.
6	3 days	Children should explore both addition and subtraction methods.	
7	5,456 trees	Do children realise they need to do both a subtraction and an addition?	

My journal

WAYS OF WORKING Independent thinking

ANSWERS AND COMMENTARY 100,000 – 39,480 = 60,520 (Tuesday)
60,520 – 39,480 = 21,040 (difference between Monday and Tuesday).
Encourage children to work out the missing number first, by using the
addition and subtraction methods that they have encountered in this unit.
Look at different ways that children find the answer. Do they subtract
39,480 from 100,000 or subtract 1 from both numbers and then subtract
39,479 from 99,999?

Children's stories should use the words Monday and Tuesday and associate
the number 39,480 with Monday. The number 100,000 should be associated
with the sum of Monday and Tuesday. They should use language such as:
*How many more on Tuesday than Monday? How many fewer on Monday? How
much less on Monday? What is the difference between Monday and Tuesday?*
Ensure that the numbers are not unrealistic (for example: 'Max cycles
39,480 km on Monday').

Power check

WAYS OF WORKING Independent thinking

ASK

• *Are you happy to decide when and why it is more efficient to add or subtract
two numbers mentally?*
• *Do you know why it is useful to work out an estimate to an addition or
subtraction by rounding?*
• *Do you feel confident solving word problems by working out what
information you have and what you need to find?*

Power puzzle

WAYS OF WORKING Pair work

IN FOCUS In this **Power puzzle**, children have to work out the missing
values using a variety of addition and subtraction methods. In the first
example, children will need to start by working out numbers where there is
one value missing from a row or column. Once they have worked out these
missing values, they should see that they can now work out the rest.

In the second example, children are asked to make up their own answers
to make the grid correct. Encourage children to double check with addition
that all the row and column totals are correct.

ANSWERS AND COMMENTARY

a) The missing values are highlighted in the table below.

	20,000	50,000	40,000
60,000	3,722	32,932	23,346
30,000	15,441	11,102	3,457
20,000	837	5,966	13,197

b) One possible solution is:

	2,600	2,000	1,400
3,000	1,650	1,150	200
2,000	800	700	500
1,000	150	150	700

PUPIL PRACTICE BOOK 5A PAGE 88

PUPIL PRACTICE BOOK 5A PAGE 89

After the unit ⏸

• How many children can confidently add and subtract using formal
written methods?
• Can children choose an appropriate mental or written method to
add or subtract numbers?

Strengthen and **Deepen** activities
for this unit can be found in the
Power Maths online subscription.

Unit 4
Multiplication and division ①

Don't forget to watch the Unit 4 video!

Mastery Expert tip! 'I made sure I gave children enough time to use the concrete and pictorial representations, alongside their abstract learning. This really helped secure their understanding of the concepts and it also meant they were engaged throughout!'

WHY THIS UNIT IS IMPORTANT

This unit will develop children's multiplicative reasoning. Children will begin by developing their understanding of multiples, common multiples, factors and common factors, recognising what they are and how they are found. These concepts will be closely linked to familiar and new concrete and pictorial representations to secure their understanding. Following this, children will learn about prime numbers and how they are different to composite numbers. Children will then investigate square and cube numbers, linked to their concrete understanding of the shape namesakes.

Having learnt about properties of numbers, children will learn how to multiply and divide by 10, 100 and 1,000 and use this knowledge to multiply and divide by multiples of 10, 100 and 1,000.

WHERE THIS UNIT FITS

→ Unit 3: Addition and subtraction
→ **Unit 4: Multiplication and division (1)**
→ Unit 5: Fractions (1)

In this unit, children develop their understanding of the multiplicative properties of numbers.

Before they start this unit, it is expected that children:
- know their times-tables to 12 fluently and can count in 10s, 100s and 1,000s
- understand and use the operations of multiplication and division confidently
- recognise the place value of 4-digit numbers.

ASSESSING MASTERY

Children will be able to reliably find multiples and factors of given numbers. They will be able to explain the unique properties of prime, square and cube numbers and represent them in concrete, pictorial and abstract ways, including using the notations $(^2)$ and $(^3)$. They will be able to reliably multiply and divide whole numbers by 10, 100 and 1,000, recognising patterns and explaining how these affect a number's place value. Finally, they will be able to confidently use their understanding of these concepts to multiply and divide whole numbers by multiples of 10, 100 and 1,000. They will show their ability to solve problems using their new learning.

COMMON MISCONCEPTIONS	STRENGTHENING UNDERSTANDING	GOING DEEPER
Children may confuse factors and multiples.	To secure understanding of factors and multiples, use arrays to demonstrate the different factors a given number has.	Encourage children to investigate which types of numbers have the most factors. Can they make generalities about the numbers with the most or least factors?
Children may recognise multiplying and dividing by 10, 100 and 1,000 as just 'adding or taking off a 0' from a number.	Encourage children to make or draw the calculations they are trying to solve. Discuss what their representation shows. Does the system of adding a 0 always work? Why not?	Encourage children to create their own video tutorial or poster, explaining why the 'just add a 0' procedural method does not work consistently and what happens to a number's place value.

UNIT STARTER PAGES

Use these pages to introduce the focus to children. You can use the characters to explore different ways of working too.

STRUCTURES AND REPRESENTATIONS

Array: Arrays are a visual representation of multiplication and division. They are an excellent tool for showing equal groups within a number.

Bar model: The bar model enables children to more easily represent a problem. In the context of this unit, it is used to show different types of calculations.

70	70	70	70	70	70	70

Multiplication square: Multiplication squares are used in this unit to demonstrate and investigate the patterns found in different types of numbers.

×	1	2	3	4	5	6	7	8	9	10	11	12
1	1	2	3	4	5	6	7	8	9	10	11	12
2	2	4	6	8	10	12	14	16	18	20	22	24
3	3	6	9	12	15	18	21	24	27	30	33	36

KEY LANGUAGE

Here is some key language that children will need to know as part of the learning in this unit:

→ multiple
→ factor
→ prime number
→ composite number
→ square number, square (x^2)
→ cube number, cube (x^3)
→ multiply, multiplication, times
→ divide, division
→ estimate
→ place value
→ ones, tens, hundreds, thousands, tens of thousands
→ common factor, common multiple, lowest common multiple

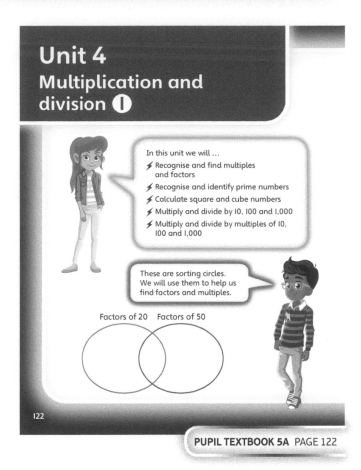

Unit 4
Multiplication and division ❶

In this unit we will …
⚡ Recognise and find multiples and factors
⚡ Recognise and identify prime numbers
⚡ Calculate square and cube numbers
⚡ Multiply and divide by 10, 100 and 1,000
⚡ Multiply and divide by multiples of 10, 100 and 1,000

These are sorting circles. We will use them to help us find factors and multiples.

Factors of 20   Factors of 50

122

PUPIL TEXTBOOK 5A PAGE 122

We will need some maths words. Look for the words you do not already know. What might they mean?

prime number   composite number
square number   cube number   square (x^2)
cube (x^3)   lowest common multiple   multiply
divide   multiple   factor

We will use multiplication squares too! They will help us spot patterns in the numbers we learn about!

123

PUPIL TEXTBOOK 5A PAGE 123

Multiples

Learning focus

In this lesson, children will learn the meaning of the mathematical term 'multiple'. They will spot patterns in multiples of numbers and use these to make generalisations and predictions.

Before you teach

- How confident are children with their multiplication tables?
- Could this limit their understanding or progress in this lesson?
- How will you mitigate this risk?

NATIONAL CURRICULUM LINKS

Year 5 Number – multiplication and division

Identify multiples and factors, including finding all factor pairs of a number, and common factors of two numbers.

ASSESSING MASTERY

Children can understand and explain the term 'multiple' and can reliably find multiples of numbers, explaining where and why there are patterns. They can link this with their understanding of times-tables.

COMMON MISCONCEPTIONS

Children may assume that any number ending in *x* is a multiple of *x* (for example, 6, 16 and 26 are all multiples of 6). While this may be the case with some multiples (multiples of 2 and 5), it is not always the case. Ask:
- *Can you count in x on this 100 square? Colour in all of the numbers that are a multiple of x.*
- *Are all numbers ending x coloured? Why or why not?*

STRENGTHENING UNDERSTANDING

For children whose fluency with times-tables or understanding of multiplication as an operation may be weaker, use arrays to help make their understanding concrete. Ask: *Can you show me a group of four? What would two groups of four look like as an array? How about three? What do you notice about the array? Can you record the multiplication as a calculation? Can you record the linked division?*

GOING DEEPER

To deepen understanding of multiples, children can be given clues about a number that they need to identify. For example, ask: *My number is a multiple of 5, 6 and 2 but not of 7. What is my number?* Also, children could be encouraged to create their own 'guess my number' challenge for a partner.

KEY LANGUAGE

In lesson: multiple, remainder

Other language to be used by the teacher: multiply, times, multiplication

STRUCTURES AND REPRESENTATIONS

100 square, arrays, number lines

RESOURCES

Mandatory: 100 square, whiteboards

Optional: blank counters, sorting circles, cubes

 In the eTextbook of this lesson, you will find interactive links to a selection of teaching tools.

Quick recap

Practise rapid recall of multiplication facts. Give children a whiteboard and pen. Read out a times-table question and give them a set amount of time to write their answer, then all show their boards at the same time.

Discover

WAYS OF WORKING Pair work

ASK

- Question ① a): *What are the multiples of 4?*
- Question ① a): *Could you use the multiples you have found so far to predict higher multiples?*
- Question ① b): *Is Luis correct? How can you prove your ideas?*

IN FOCUS Question ① b) will help to approach the potential misconception that all numbers ending in *x* are a multiple of *x*. Children could be encouraged to find other similar examples. Ask: *Why is 14 not a multiple of 4 even though it has a 4 in it? How can you prove 14 is not a multiple of 4?*

PRACTICAL TIPS Children should have access to a 100 square and follow the process shown in the **Discover** scenario. If possible, this could be done outside in the playground as in the picture, or, if not, then with counters and printed 100 squares.

ANSWERS

Question ① a):

1	2	3	4	5	6	7	8	9	10
11	12	13	14	15	16	17	18	19	20
21	22	23	24	25	26	27	28	29	30
31	32	33	34	35	36	37	38	39	40
41	42	43	44	45	46	47	48	49	50
51	52	53	54	55	56	57	58	59	60
61	62	63	64	65	66	67	68	69	70
71	72	73	74	75	76	77	78	79	80
81	82	83	84	85	86	87	88	89	90
91	92	93	94	95	96	97	98	99	100

Question ① b): Luis is incorrect. 74 is not a multiple of 4.

Share

WAYS OF WORKING Whole class teacher led

ASK

- Question ① a): *How can you find the multiples of 4?*
- Question ① a): *How do the counters show the multiples clearly?*
- Question ① a): *Did you find all of the multiples shown?*
- Question ① a): *Did you notice any patterns in the multiples? How did they help you to continue finding more multiples?*

IN FOCUS It will be important to focus on the pictorial representations of the multiples of 4 shown in question ① a). This will help children link their understanding of multiplication and arrays with the new vocabulary they are learning. It may help to give children the opportunity to build the arrays shown.

PUPIL TEXTBOOK 5A PAGE 124

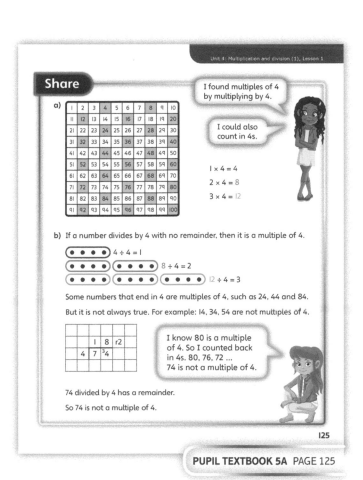

PUPIL TEXTBOOK 5A PAGE 125

Think together

WAYS OF WORKING Whole class teacher led (I do, We do, You do)

ASK

- Questions ❶ and ❷: *Can you see a pattern in the multiples already? How will this continue? What can you say is the same about all of the multiples of x?*
- Question ❷: *How can you find multiples after 100?*

IN FOCUS Question ❶ scaffolds children's ability to find multiples by allowing them to continue the already started pattern on the 100 square. Question ❷ develops from this, asking children to find all the multiples of 5 using a blank 100 square.

STRENGTHEN Strengthen understanding of question ❸ a) by providing children with a 100 square. Ask: *How does the 100 square help? Can you use it to predict a 3-digit multiple of 6? From the numbers you have found, can you see whether the statement in ❸ b) is always, sometimes or never true?*

DEEPEN While solving question ❸ c), develop children's reasoning by asking: *What numbers have you found? What is special about the numbers in the middle segment of the sorting circles? How can you use this to help you find multiples of 4 more easily in the future? Why are there no numbers in the multiple of 4 only segment?*

ASSESSMENT CHECKPOINT Questions ❷ and ❸ will assess children's ability to find multiples of numbers and record them in different ways. Question ❸ will also be useful to help assess children's recognition that not all multiples ending in x are a multiple of x.

ANSWERS

Question ❶: Multiples of 2 have 0, 2, 4, 6 or 8 in the 1s digit. Numbers that are multiples of 2 are all even. Numbers that are **not** multiples of 2 are all odd.

Question ❷: 5, 10, 15, 20, 25, 30, 35, 40, 45, 50, 55, 60, 65, 70, 75, 80, 85, 90, 95, 100
Multiples of 5 have 0 or 5 in the 1s digit.
Even multiples of 5 all end in 0.
Odd multiples of 5 all end in 5.

Question ❸ a): For example:

	Multiple of 6	Not a multiple of 6
Ends in a 6	6, 36, 66, 96	16, 26, 46, 56, 76
Does not end in a 6	12, 18, 24, 30, 42	1, 2, 3, 4, 5, 7, 8, 9, 10, 11, 13, 14, 15

Question ❸ b): The statement is sometimes true, as evidenced in the chart in question ❸ a).

Question ❸ c): For example:

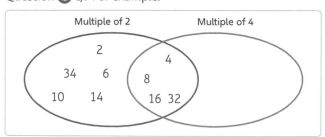

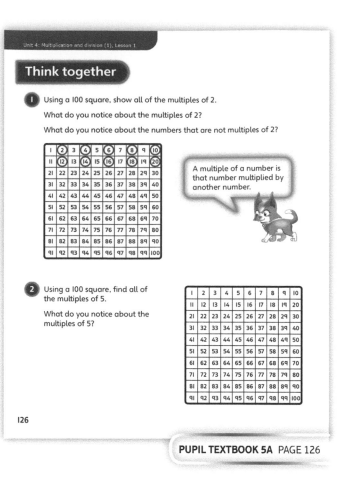

Think together

❶ Using a 100 square, show all of the multiples of 2.
What do you notice about the multiples of 2?
What do you notice about the numbers that are not multiples of 2?

A multiple of a number is that number multiplied by another number.

❷ Using a 100 square, find all of the multiples of 5.
What do you notice about the multiples of 5?

126

PUPIL TEXTBOOK 5A PAGE 126

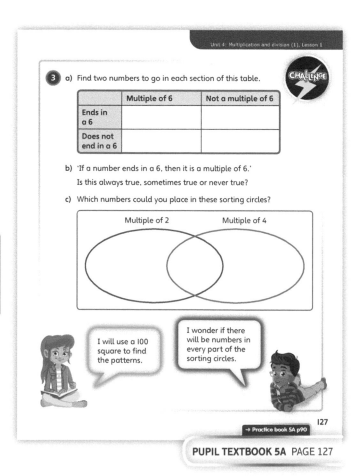

❸ a) Find two numbers to go in each section of this table.

CHALLENGE

	Multiple of 6	Not a multiple of 6
Ends in a 6		
Does not end in a 6		

b) 'If a number ends in a 6, then it is a multiple of 6.'
Is this always true, sometimes true or never true?

c) Which numbers could you place in these sorting circles?

Multiple of 2   Multiple of 4

I will use a 100 square to find the patterns.

I wonder if there will be numbers in every part of the sorting circles.

127

→ Practice book 5A p90

PUPIL TEXTBOOK 5A PAGE 127

Practice

Independent thinking

IN FOCUS Question ❶ helps scaffold children's independent understanding of multiples by offering them different pictorial representations of them. It may help children to build their own versions using cubes or counters and then draw what they have made.

Question ❻ encourages children to recognise that a multiple of any even number is also a multiple of 2.

STRENGTHEN When working with greater numbers, such as in question ❺, encourage children to write down, for example, 25, 30, 35, as well as 125, 130 and 135 and recognise what is the same and what is different.

DEEPEN Children could deepen their understanding of the generalisations that can be made around multiples by investigating question ❻ in more detail. Ask: *What happens if you change the question from multiples of 2 and 6 to multiples of 3 and 5? What is different and what stays the same? Is it the same section that remains empty? Why or why not? Is it possible to find combinations where all boxes are filled? Explain your reasoning.*

ASSESSMENT CHECKPOINT Use question ❸ to assess whether children can work systematically to list multiples. Use question ❹ to assess whether children can recognise multiples of given numbers and use question ❺ to assess whether they can continue sequences of multiples that are outside of their multiplication facts.

ANSWERS Answers for the **Practice** part of the lesson can be found in the *Power Maths* online subscription.

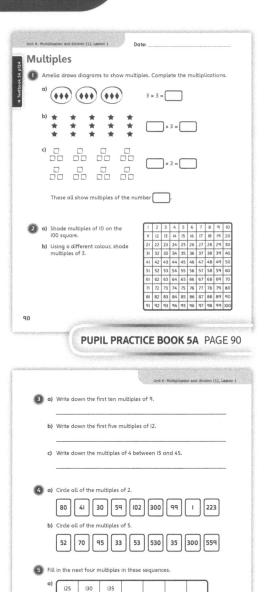

PUPIL PRACTICE BOOK 5A PAGE 90

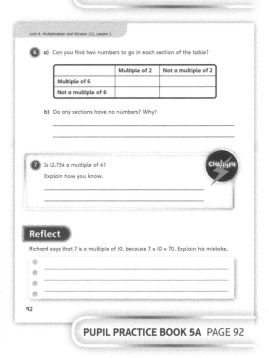

PUPIL PRACTICE BOOK 5A PAGE 91

Reflect

Independent thinking

IN FOCUS This question will demonstrate whether children have a secure understanding of multiples and how to find them, and are able to use this to diagnose another child's error.

ASSESSMENT CHECKPOINT It is important to look out for children's ability to understand that 70 *is* a multiple of 10, but 7 is the number of 10s needed to make 70.

ANSWERS Answers for the **Reflect** part of the lesson can be found in the *Power Maths* online subscription.

After the lesson

- Having concluded the lesson, are there any times-tables that children will benefit from more practice in than others?
- Do children understand the patterns between the digit in the 1s position and the multiple?

PUPIL PRACTICE BOOK 5A PAGE 92

Common multiples

Learning focus

In this lesson, children will build on their learning from the previous lesson on multiples as they begin to explore common multiples.

Before you teach ⏸

- How confident are children with their times-tables?
- Can children find multiples of a given number?
- Can children recognise multiples of numbers?

NATIONAL CURRICULUM LINKS

Year 5 Number – multiplication and division

Identify multiples and factors, including finding all factor pairs of a number, and common factors of two numbers.

ASSESSING MASTERY

Children can understand and explain the term 'multiple' and can reliably find multiples of numbers. They use this knowledge to find common multiples of pairs of numbers.

COMMON MISCONCEPTIONS

Children may assume that the only common multiple of any pair of numbers is the product of the two numbers. While this method will always find a common multiple, children should recognise that there could be common multiples lower than this. Ask:

- *Write out the first ten multiples of both numbers. Are there any other common multiples?*

STRENGTHENING UNDERSTANDING

For children whose fluency with times-tables or understanding of multiplication as an operation may be weaker, use arrays to help them find multiples and aid their understanding of multiplication.

GOING DEEPER

Ask children to find a pair of numbers that have another given number as a common multiple. Ask children to find a pair of numbers whose first common multiple is their product, and another pair where the first common multiple is less than their product.

KEY LANGUAGE

In lesson: multiple, common multiple, **lowest common multiple**

Other language to be used by the teacher: multiply, times, multiplication

STRUCTURES AND REPRESENTATIONS

100 square, arrays, number lines

RESOURCES

Mandatory: 100 square

Optional: blank counters, sorting circles

 In the eTextbook of this lesson, you will find interactive links to a selection of teaching tools.

Quick recap 🔁

Choose a number and ask children to say the first five multiples of that number aloud as a class. Repeat for other numbers.

Discover

WAYS OF WORKING Pair work

ASK

• Question **1** a): *What number did this boy (point to the 2nd child) have? What number did this boy (point to the 4th child) have? Why did they both say 'Fizz'? What number is both of their numbers a multiple of?*
• Question **1** a): *What number did this girl (point to the third child) have? Why do you think she said 'Buzz'?*
• Question **1** a): *Why did this girl (point to the 6th child) say 'Fizz Buzz'?*
• Question **1** a): *What other numbers will be 'Fizz Buzz'?*

IN FOCUS Question **1** a) provides children with a practical introduction into common multiples. They should recognise what numbers are represented by 'Fizz' and 'Buzz', and what they have in common.

PRACTICAL TIPS Play the game as a class. Ask children to explain any patterns that they notice in their answer.

ANSWERS

Question **1** a): When the number is a multiple of 2, the child says 'Fizz'. When the number is a multiple of 3, the child says 'Buzz'.

Question **1** b): The first three common multiples of 4 and 6 are 12, 24 and 36.

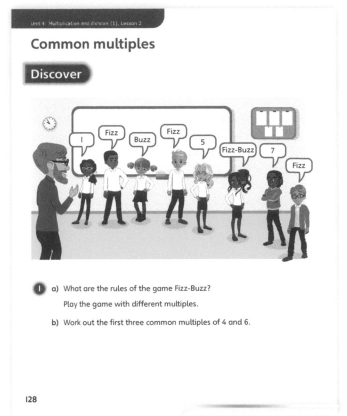

PUPIL TEXTBOOK 5A PAGE 128

Share

WAYS OF WORKING Whole class teacher led

ASK

• Question **1** a): *What numbers are replaced with 'Fizz'? What do you notice?*
• Question **1** a): *What numbers are replaced with 'Buzz'? What do you notice?*
• Question **1** a): *Why is 'Fizz Buzz' written above the number 6?*
• Question **1** b): *What do you notice about the common multiples of 4 and 6?*

IN FOCUS In question **1** b), it is important to highlight the strategic approach that is being used to list the multiples and identify common multiples.

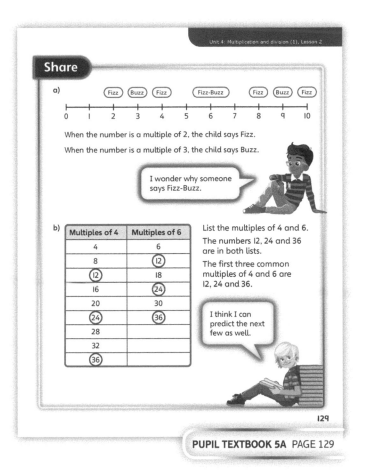

PUPIL TEXTBOOK 5A PAGE 129

Think together

Whole class teacher led (I do, We do, You do)

ASK

- Question ①: *How can you find the first ten multiples of each number? How does this help you to find the common multiples? Is this all the common multiples or are there more?*
- Question ①: *What do you notice about the common multiples?*
- Question ②: *What do you notice about the lowest common multiple of each pair of numbers?*

IN FOCUS Question ① encourages children to use a strategic approach to identifying common multiples. In question ②, children should notice that the lowest common multiple can be either one of the numbers, or it can be the product of the two numbers, or a smaller multiple of both numbers, so there is not one quick rule to identifying it.

STRENGTHEN Encourage children to continue to list the multiples starting from 1 × the number to ensure they do not omit any multiples.

DEEPEN In question ③, children explore the potential misconception that the lowest common multiple is always the product of the numbers. They could be encouraged to find pairs of numbers where this does and does not work.

ASSESSMENT CHECKPOINT Use question ① to assess whether children can find common multiples. Use question ② to assess whether they can find the lowest common multiple of a pair of numbers.

ANSWERS

Question ①: Multiples of 6: 6, 12, 18, 24, 30, 36, 42, 48, 54, 60
Multiples of 9: 9, 18, 27, 36, 45, 54, 63, 72, 81, 90
The common multiples of 9 and 6 are 18, 36 and 54.

Question ② a): 12

Question ② b): 10

Question ② c): 24

Question ③ a): Alex knows that the product of two numbers is a common multiple of both numbers.

Question ③ b): Multiples of 10: 10, 20, 30, …
Multiples of 15: 15, 30, 45, …
The lowest common multiple of 10 and 15 is 30.

Question ③ c): All multiples of the lowest common multiple of two numbers will be common multiples of those numbers.

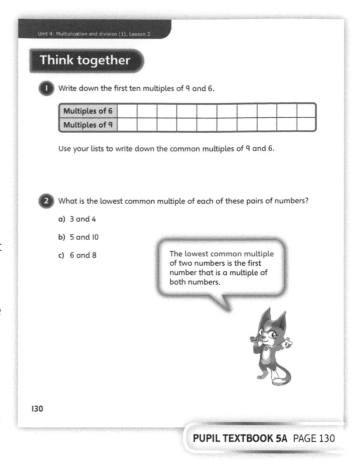

PUPIL TEXTBOOK 5A PAGE 130

PUPIL TEXTBOOK 5A PAGE 131

Practice

WAYS OF WORKING Independent thinking

IN FOCUS Questions ❶, ❷ and ❸ encourage children to use a strategic approach to identifying common multiples and lowest common multiples. They should then apply this in question ❹.

STRENGTHEN Encourage children to continue to list the multiples, starting from 1 × the number to ensure they do not omit any multiples.

DEEPEN In question ❺, children should explain their reasoning and could be encouraged to explore other pairs of numbers and explain any patterns that they notice.

ASSESSMENT CHECKPOINT Use questions ❶ and ❷ to assess whether children can find common multiples of pairs of numbers. Use questions ❸ and ❹ to assess whether children can find the lowest common multiple of pairs of numbers.

ANSWERS Answers for the **Practice** part of the lesson can be found in the *Power Maths* online subscription.

Reflect

WAYS OF WORKING Pair work

IN FOCUS Children work in pairs to explain methods for identifying lowest common multiples.

ASSESSMENT CHECKPOINT Children should explain clearly how to find the lowest common multiple of any pair of numbers. Ensure they do not think that the lowest common multiple is always the product of the two numbers.

ANSWERS Answers for the **Reflect** part of the lesson can be found in the *Power Maths* online subscription.

After the lesson

- Do children understand how to list or find multiples of any number?
- Do children understand how to find common multiples of any pair of numbers?
- Can children identify the lowest common multiple of any pair of numbers?

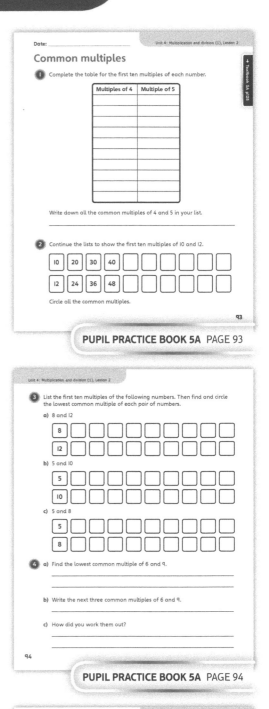

PUPIL PRACTICE BOOK 5A PAGE 93

PUPIL PRACTICE BOOK 5A PAGE 94

PUPIL PRACTICE BOOK 5A PAGE 95

Factors

Learning focus

In this lesson, children will learn the meaning of the mathematical term 'factor' and use multiplication and division to find factors. They will spot patterns in factors of numbers and use these to make generalisations and predictions.

Before you teach

- How can you develop the connection between multiples and factors in this lesson?
- What real-life contexts can you draw on to help children secure their understanding in this lesson?

NATIONAL CURRICULUM LINKS

Year 5 Number – multiplication and division

Identify multiples and factors, including finding all factor pairs of a number, and common factors of two numbers.

ASSESSING MASTERY

Children can understand and explain the term 'factor', reliably find factors of numbers and explain where and why there are patterns in factors. They can link their understanding of times-tables with their new understanding of factors and the concept of dividing numbers into equal groups.

COMMON MISCONCEPTIONS

Children may confuse the term 'factor' with the term 'multiple'. Ask:
- *What do you need to do to find a multiple? How is that different to what you are doing today to find a factor? How would you explain the difference to someone who was confused?*

STRENGTHENING UNDERSTANDING

To help children more fluently find factors of numbers, it would be useful to identify times-tables they are less confident with and practise these with them. This could be achieved through a variety of methods including using times-tables rhymes, songs and chants, playing games in pairs – matching calculations with numbers and creating arrays, and other concrete and pictorial representations of a times-table.

GOING DEEPER

Children could be encouraged to investigate whether a number can be both a factor and a multiple of itself. Ask: *Can it be true that 3 is both a factor and a multiple of 3? Explain your ideas.*

KEY LANGUAGE

In lesson: factors, remainder, division, multiplication

Other language to be used by the teacher: multiple, divide, multiply

STRUCTURES AND REPRESENTATIONS

Number lines, arrays

RESOURCES

Mandatory: squared paper, counters

Optional: multilink cubes

 In the eTextbook of this lesson, you will find interactive links to a selection of teaching tools.

Quick recap

In preparation for finding factors, ask children to build arrays for the number 24 and see how many different arrays they can find. What are the dimensions of each array? Why can they not build an array that has exactly five counters in each row?

Discover

WAYS OF WORKING Pair work

ASK

- Question ❶ a): *How can you show what equal groups can be made with 24 chairs?*
- Question ❶ a): *What mathematical representation could show your thinking best?*
- Question ❶ b): *How could you investigate what equal groups are in 25?*

IN FOCUS Question ❶ a) will allow children to experiment with factors. This will offer them a good opportunity to begin linking factors with their times-table knowledge.

Question ❶ b) is important as it enables children to begin making conjectures about factors, looking for patterns and making generalisations about how factors change depending on the given number.

PRACTICAL TIPS This activity could be linked to setting the school hall out for an assembly or for lunch. Children could be encouraged to rearrange the chairs in their classroom to accommodate a number of people, with all rows of chairs being equal.

ANSWERS

Question ❶ a): $1 \times 24 = 24$; 1 and 24 are factors of 24.
$2 \times 12 = 24$; 2 and 12 are factors of 24.
$3 \times 8 = 24$; 3 and 8 are factors of 24.
$4 \times 6 = 24$; 4 and 6 are factors of 24.

Question ❶ b): There are fewer arrangements as the factors of 25 are 1, 5 and 25.

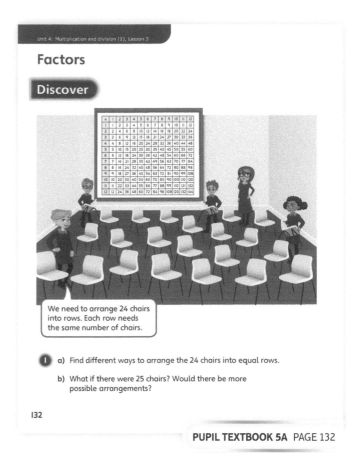

PUPIL TEXTBOOK 5A PAGE 132

Share

WAYS OF WORKING Whole class teacher led

ASK

- Question ❶ a): *How do the arrays show the factors clearly?*
- Question ❶ a): *Is there another calculation for each set of factors? Explain your ideas.*
- Question ❶ b): *How many factors of 25 did you find? How did you show them?*
- Question ❶ b): *Why does 25 have fewer factors than 24?*

IN FOCUS For questions ❶ a) and ❶ b), it is important to link the factors of the numbers given to children's understanding of arrays. This will ensure children make the link between their concrete understanding and the concept of dividing numbers into equal groups to give factors. Children should be encouraged to build or draw the appropriate arrays for any given numbers they are investigating. Children need to understand the link between factors and commutativity. Discuss how knowing $3 \times 8 = 24$ also implies $8 \times 3 = 24$, so we know both 3 and 8 are factors of 24.

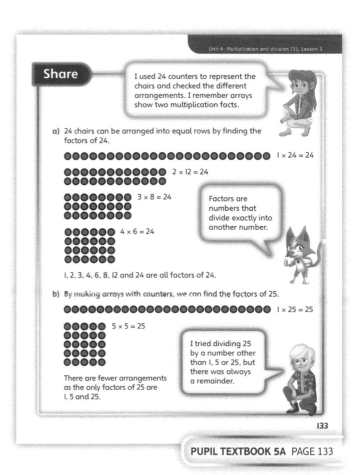

PUPIL TEXTBOOK 5A PAGE 133

169

Think together

WAYS OF WORKING Whole class teacher led (I do, We do, You do)

ASK

- Question **1**: *What factors does the array show? Explain how you know. How will you prove if there are any more factors of 16?*
- Question **2** a): *How will the number line help us find if 5 is a factor of 16? How does using multiplication help us find factors?*
- Question **2** b): *How will the number line help us find if 4 is a factor of 22? How does using division help us find factors?*
- Question **3**: *How will the lists be similar for 30 and 40? How will they be different for 30 and 40?*

IN FOCUS Question **1** is important as it links children's understanding of arrays to their new understanding of factors. It also demonstrates multiple ways of recording factors of a number.

Question **2** a) helps children recognise how factors can be found using increasingly formal methods of recording and calculating. It will also demonstrate clearly the link between multiples and factors and how, when counting in 5s, 16 is not one of the multiples, thereby proving that 5 is not a factor of 16.

STRENGTHEN To strengthen understanding and to help draw the correct arrays on squared paper for question **1**, offer children counters or multilink cubes to allow them to build the arrays first. Ask: *Can you use the squares on the paper to draw the same array accurately?*

DEEPEN When solving question **3** b), deepen children's reasoning by encouraging them to generalise about the factors they have been investigating. Ask: *Is there a way to always reliably know when you have found all of the factors of a number? Is there any way of predicting how many factors a number will have?*

ASSESSMENT CHECKPOINT Question **3** will assess children's ability to reliably find the factors of a given number using both multiplication and division. Look for children's recorded evidence of their thinking as this will demonstrate their recognition of the link between factors and multiplicative reasoning.

ANSWERS

Question **1**: The factors of 16 are 1, 2, 4, 8, 16.

Question **2** a): 5 is not a factor of 16 because 16 is not in the 5 times-table. The nearest multiple of 5 is 15 and the following multiple is 20.

Question **2** b): 4 is not a factor of 22 because 22 is not divisible by 4 without a remainder.

Question **3** a): The factors of 30 are 1, 2, 3, 5, 6, 10, 15 and 30.

Question **3** b): The factors of 40 are 1, 2, 4, 5, 8, 10, 20 and 40.
They will be able to stop their lists when they find their calculations contain the same numbers, but with the factors the other way around. In this case it will be at 5×6 or $30 \div 5$.

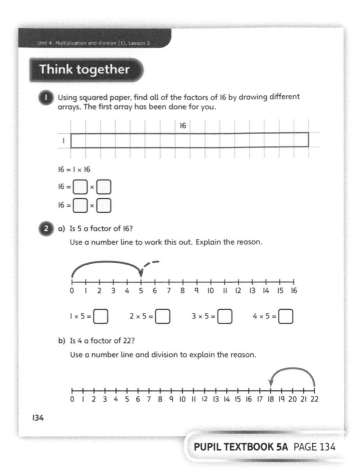

PUPIL TEXTBOOK 5A PAGE 134

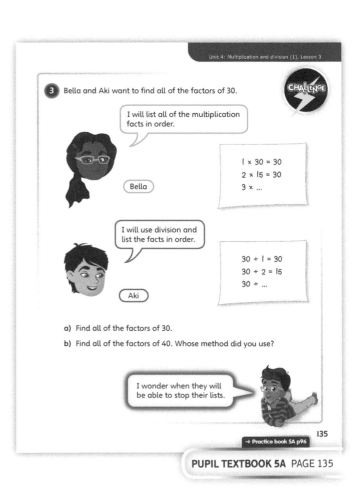

PUPIL TEXTBOOK 5A PAGE 135

Practice

WAYS OF WORKING Independent thinking

IN FOCUS Questions ❶ and ❷ link children's independent understanding of the concrete and pictorial representations of multiplication with their understanding of factors. Question ❷ develops this by requiring children to draw an array as well as listing the factors shown by it.

Question ❸ is important as it encourages children to provide their reasoning as to why a number either is or is not a factor of another number. It will be important to carefully observe children's responses to ensure their full understanding.

Question ❼ is important as it makes the link between factors and multiples clear and overt.

STRENGTHEN If children are not sure whether they have found all of the factors of a number in questions ❹, ❺ and ❻, ask: *How could you organise your thinking so that you can prove you have found all of the factors? How will you know when you have tested all of the possible factors?*

DEEPEN While children solve question ❼, deepen their understanding of using factors and multiples in context by asking: *Which type of number is more useful, factors or multiples? Why? Is there a time where finding multiples would be more useful than factors and vice versa?*

ASSESSMENT CHECKPOINT Question ❹ will assess children's ability to work systematically to find all of the factors of a number using multiplication.

ANSWERS Answers for the **Practice** part of the lesson can be found in the *Power Maths* online subscription.

Reflect

WAYS OF WORKING Independent thinking

IN FOCUS This question will give children the opportunity to show their understanding of what a factor is: that it is the number being multiplied and does not have a bearing on whether a number is odd or even.

ASSESSMENT CHECKPOINT Children can reliably explain what a factor is and how to work out what the factors of 70 are.

ANSWERS Answers for the **Reflect** part of the lesson can be found in the *Power Maths* online subscription.

After the lesson ⏸

- How confident were children with the concept of factors?
- Did the lesson make the link between factors and multiples explicit? If not, how will you ensure this link is clear in future lessons?
- How could you include the concepts in this lesson in other areas of the curriculum?

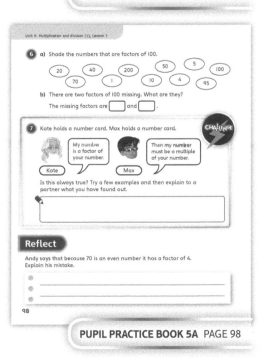

PUPIL PRACTICE BOOK 5A PAGE 96

PUPIL PRACTICE BOOK 5A PAGE 97

PUPIL PRACTICE BOOK 5A PAGE 98

Common factors

Learning focus

In this lesson, children build on their learning on factors from the previous lesson to identify common factors.

Before you teach

Before you teach

- How confident are children with their times-tables?
- Can children find the factors of a given number?
- How confident are children with division?

NATIONAL CURRICULUM LINKS

Year 5 Number – multiplication and division

Identify multiples and factors, including finding all factor pairs of a number, and common factors of two numbers.

ASSESSING MASTERY

Children can understand and explain the term 'factor' and can reliably find factors of numbers. They use this knowledge to find common factors of pairs of numbers.

COMMON MISCONCEPTIONS

Children may forget to include 1 and the number itself when listing factors. If they forget to include 1, children may think that some pairs of numbers do not have any common factors. Ask:

- *Have you found all the factors of the numbers? Can you use your times-tables to check? What do you multiply 1 by to make each of the numbers?*

STRENGTHENING UNDERSTANDING

For children whose fluency with times-tables or understanding of multiplication and division may be weaker, use arrays to help make their understanding concrete and find factors.

GOING DEEPER

Ask children to find a pair of numbers that have another given number as a common factor. Ask them to find a pair of numbers whose highest common factor is 1 and another pair of numbers where the highest common factor is one of the numbers.

KEY LANGUAGE

In lesson: factor, **common factor**

Other language to be used by the teacher: multiply, times, multiplication, divide, division, remainder

STRUCTURES AND REPRESENTATIONS

Arrays, number lines

RESOURCES

Mandatory: counters, whiteboards, digit cards

 In the eTextbook of this lesson, you will find interactive links to a selection of teaching tools.

Quick recap 🔾

Choose a number and ask children to list the factors of that number on a whiteboard. Repeat for other numbers, ensuring they do not miss any out.

Discover

WAYS OF WORKING Pair work

ASK

• Question ① a): *What are the factors of 18? How do you know that you have found them all?*
• Question ① b): *What are the factors of 12? How do you know that you have found them all?*
• Question ① b): *Which of the numbers are common factors of 18 and 12? How do you know?*

IN FOCUS Children explore common factors using sorting circles. In question ① b), they should recognise that the numbers in the middle of the sorting circles are factors of both numbers, and use their understanding of 'common' in this context from their previous learning on common multiples.

PRACTICAL TIPS Provide children with enlarged sorting circles and digit cards to recreate the activity practically. Give them different digit cards and different sorting circle headings to provide further practice.

ANSWERS

Question ① a): The factors of 18 are 1, 2, 3, 6, 9 and 18.

Question ① b):

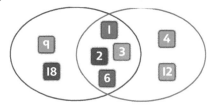

Factors of 18   Factors of 12

The numbers 1, 2, 3 and 6 are in the part where the two circles overlap. They are called common factors. They are factors of both 12 and 18.

Share

WAYS OF WORKING Whole class teacher led

ASK

• Question ① a) *How do the arrays help to identify factors? Are there any other arrays that can be made using exactly 18 counters?*
• Question ① b): *What is special about the numbers in the overlapping part of the sorting circles?*

IN FOCUS Ensure children understand the link between the arrays and the factors of the numbers. In question ① b), they should use their understanding of the word 'common' from the lesson on common multiples to understand why the numbers in the overlapping section are common factors.

Common factors

Discover

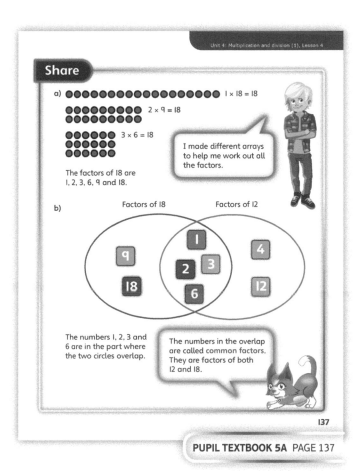

① a) Which of the numbers are factors of 18?

   b) Where should all the number cards go?
   What numbers are in the part where the two circles overlap?

PUPIL TEXTBOOK 5A PAGE 136

Share

a) ●●●●●●●●●●●●●●●●●● $1 \times 18 = 18$
   ●●●●●●●●● $2 \times 9 = 18$
   ●●●●●●●●●
   ●●●●●● $3 \times 6 = 18$
   ●●●●●●
   ●●●●●●

   The factors of 18 are 1, 2, 3, 6, 9 and 18.

   I made different arrays to help me work out all the factors.

b)   Factors of 18   Factors of 12

   The numbers 1, 2, 3 and 6 are in the part where the two circles overlap.

   The numbers in the overlap are called common factors. They are factors of both 12 and 18.

137

PUPIL TEXTBOOK 5A PAGE 137

Think together

WAYS OF WORKING Whole class teacher led (I do, We do, You do)

ASK

- Question **1**: *How do you know if the number is a factor of 24 or 30? What does it mean if it is a factor of both?*
- Question **2**: *Why should you list the factors first? How do you know you have not missed any?*

IN FOCUS In question **1**, children should use their understanding of division to identify whether each number is a factor of the given numbers, and then recognise that, if there is a tick in both boxes of any given column, then that number is a common factor of 24 and 30.

STRENGTHEN Provide children with arrays to support them in finding factors and ensuring they can identify common factors.

DEEPEN In question **3**, encourage children to explain their reasoning for each part. Can they find another pair of numbers like 18 and 36? Can they find a pair of numbers where the only common factor is 1?

ASSESSMENT CHECKPOINT Use questions **1** and **2** to assess whether children can identify common factors of numbers. Use question **3** to assess the depth of children's understanding of common factors.

ANSWERS

Question **1**:

Number	I	2	3	4	5	6
Multiples of 24	✓	✓	✓	✓		✓
Multiples of 30	✓	✓	✓		✓	✓

1, 2, 3 and 6 are common factors of 24 and 30.

Question **2**:

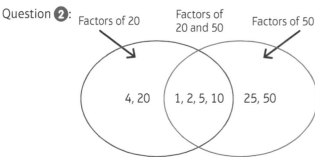

Question **3**:

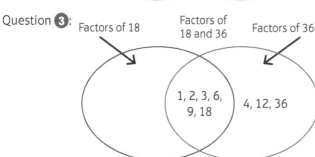

Ambika is correct, 1 is a common factor of any two numbers.
Andy is correct, all factors of 18 are also factors of 36, because 18 itself is a factor of 36.

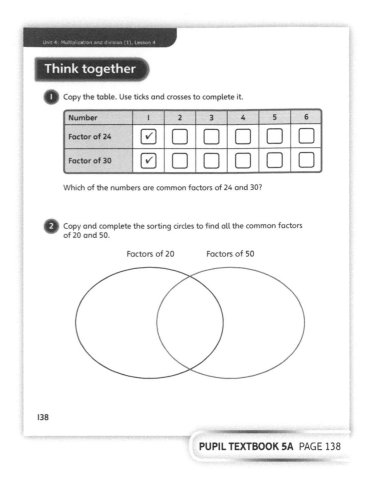

Think together

1 Copy the table. Use ticks and crosses to complete it.

Number	I	2	3	4	5	6
Factor of 24	✓	☐	☐	☐	☐	☐
Factor of 30	✓	☐	☐	☐	☐	☐

Which of the numbers are common factors of 24 and 30?

2 Copy and complete the sorting circles to find all the common factors of 20 and 50.

Factors of 20   Factors of 50

138

PUPIL TEXTBOOK 5A PAGE 138

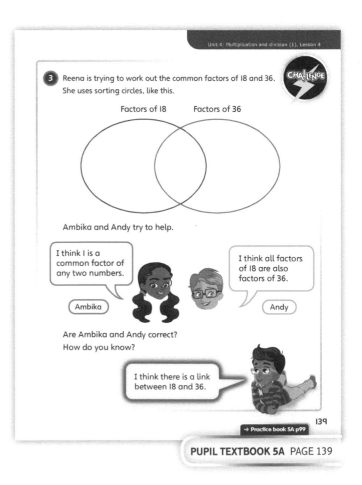

3 Reena is trying to work out the common factors of 18 and 36. She uses sorting circles, like this.

Factors of 18   Factors of 36

Ambika and Andy try to help.

I think I is a common factor of any two numbers.
Ambika

I think all factors of 18 are also factors of 36.
Andy

Are Ambika and Andy correct? How do you know?

I think there is a link between 18 and 36.

139

→ Practice book 5A p99

PUPIL TEXTBOOK 5A PAGE 139

Practice

WAYS OF WORKING Independent thinking

IN FOCUS These questions provide the opportunity for children to practise finding common factors. In question **2** c), encourage them to recognise the importance of listing all of the factors first.

STRENGTHEN Encourage children to make or draw arrays to support them in finding factors. Ask them to circle any numbers that appear in both lists.

DEEPEN In question **6**, children should explain their method and reasoning behind their thinking. You could encourage children to find two numbers that have another number, which you stipulate, as a common factor.

THINK DIFFERENTLY In question **5**, children could choose pairs of even numbers to start with and explore the similarities and differences in the lists of common factors before generalising.

ASSESSMENT CHECKPOINT Use questions **1** to **4** to assess whether children can find common factors of given pairs of numbers. Use questions **5** and **6** to assess children's depth of understanding of common factors.

ANSWERS Answers for the **Practice** part of the lesson can be found in the *Power Maths* online subscription.

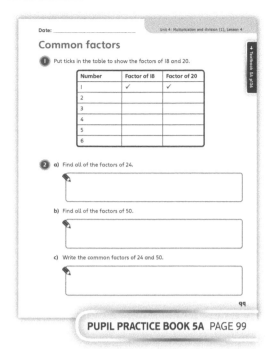

PUPIL PRACTICE BOOK 5A PAGE 99

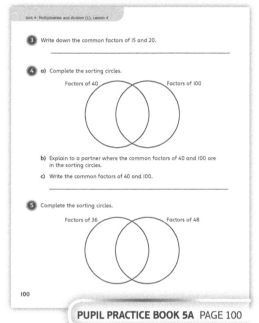

PUPIL PRACTICE BOOK 5A PAGE 100

Reflect

WAYS OF WORKING Independent thinking

IN FOCUS This question is designed to ensure children recognise that every number has at least two factors: the number itself and 1.

ASSESSMENT CHECKPOINT Children should recognise the statement is true and be able to explain why in their own words.

ANSWERS Answers for the **Reflect** part of the lesson can be found in the Power Maths online subscription.

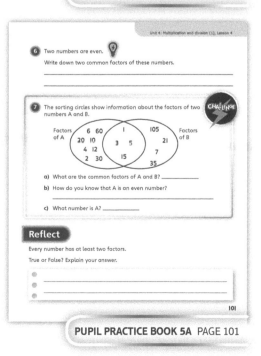

PUPIL PRACTICE BOOK 5A PAGE 101

After the lesson ⏸

- Can children list the factors of a given number?
- Can children find common factors of a pair of numbers?

Prime numbers

Learning focus

In this lesson, children will learn about prime numbers and how they are different to other numbers. They will learn the correct vocabulary of prime and composite numbers and how to differentiate between them.

Before you teach

- Are children confident with finding factors?

NATIONAL CURRICULUM LINKS

Year 5 Number – multiplication and division

Know and use the vocabulary of prime numbers, prime factors and composite (non-prime) numbers.

ASSESSING MASTERY

Children can reliably identify prime and composite numbers, recognising and explaining that prime numbers have only two factors, 1 and themselves. They can identify all prime numbers up to 100 and can use this understanding to solve problems.

COMMON MISCONCEPTIONS

Children may recognise that almost every even number is composite. Having recognised this pattern, they may mistakenly assume that 2 is composite because it is even. Ask:
- *Show me how many arrays you can make with two counters. How many factors does 2 have?*

Children may think that 1 is a prime number as its factor is 1 and is also itself. Ask:
- *How many different factors does 1 have? How many different factors does a prime number have?*

STRENGTHENING UNDERSTANDING

To strengthen understanding of the concept of prime numbers, give children as many practical opportunities as possible to investigate and play with prime numbers. For example, setting up teams, sharing toys or food or setting out table places. When sharing in these different contexts, children should be encouraged to use the vocabulary of 'prime' if they find a number that has only two factors: 1 and itself.

GOING DEEPER

Children can be encouraged to investigate whether the generalisations and patterns they have made and found in the lesson continue in numbers over 100. Ask: *What patterns or rules did you find when finding prime numbers? Do you think these patterns will carry on after 100? Why? Try to prove it.*

KEY LANGUAGE

In lesson: prime numbers, composite numbers, factors, remainder, division

Other language to be used by the teacher: divide, even, multiples, arrays

STRUCTURES AND REPRESENTATIONS

Arrays, 100 square

RESOURCES

Mandatory: counters

 In the eTextbook of this lesson, you will find interactive links to a selection of teaching tools.

Quick recap

Give children 12 counters each. How many different arrays can they make using all twelve counters? Now take one counter from each child. How many different arrays can they make using all eleven counters? What do they notice?

Discover

WAYS OF WORKING Pair work

ASK

- Question ❶ a): *How many rugby players are there? How many ways can you share 13 equally?*
- Questions ❶ a) and b): *How can you prove that you have found all of the factors?*
- Question ❶ b): *How is 9 different from 7? How are the factors different?*

IN FOCUS Questions ❶ a) and b) will provide children with opportunities to discuss and experiment with the factors in prime numbers. It will be important for them to begin noticing patterns in prime numbers, so they can begin generalising about prime number properties.

PRACTICAL TIPS This concept could be introduced in PE while setting up team games. Alternatively, for a more classroom-based approach, children could be asked to share stickers out equally between them. Give them a challenge that allows them to keep the stickers only if they can share them out equally without splitting any into smaller pieces.

ANSWERS

Question ❶ a): The players can only be in 1 group of 13 players or 13 groups of 1 player.

Question ❶ b): The team of 9 tennis players can split into equal groups in more ways than the team of 7 basketball players.

Share

WAYS OF WORKING Whole class teacher led

ASK

- Question ❶ a): *What factors have you found for 13?*
- Question ❶ a): *How can you prove that you have found all of the factors?*
- Question ❶ b): *How do you know that 7 is a prime number?*

IN FOCUS Question ❶ a) introduces children to the vocabulary of 'prime numbers', with Sparks providing a definition. It will be important for children to practise using this information in their explanations.

Think together

WAYS OF WORKING Whole class teacher led (I do, We do, You do)

WAYS OF WORKING Whole class teacher led (I do, We do, You do)

ASK

- Question **1** a): *Can you use counters to help you? How can you prove that you have found all of the factors?*
- Question **1** b): *What was the most efficient method of finding out if the number was prime or composite?*
- Question **2**: *Which numbers do you predict will be prime?*
- Question **3**: *Can you see a pattern emerging?*

IN FOCUS Question **1** will help to tackle the assumption that all odd numbers are prime. It is also important as it gives children a larger number, closer to 100, to investigate. Discuss with children how they went about finding out if the number was prime or composite and which was the most efficient method.

STRENGTHEN To strengthen understanding in question **3**, encourage children to use arrays to help find the factors of each number. Encourage them to look for patterns that may make identifying primes and composites easier. Ask: *Can you show me the arrays that x can make? How will finding patterns help you to find the next prime number?*

DEEPEN In question **3**, use children's findings to draw out generalisations. Ask: *What do you notice about the numbers you have circled? Can you see any patterns that might help you identify prime numbers in the future?*

ASSESSMENT CHECKPOINT Question **3** will assess children's ability to find prime numbers up to 100. When explaining their findings, they should show recognition that all the circled numbers have only two factors.

ANSWERS

Question **1** a): Factors of 63 are 1, 3, 7, 9, 21, 63.

Question **1** b): 63 is a composite number.

Question **2**:

Number	Factors	How many factors?	Is it a prime or composite number?
12	1, 2, 3, 4, 6, 12	6	Composite
11	1, 11	2	Prime
10	1, 2, 5, 10	4	Composite
9	1, 3, 9	3	Composite
8	1, 2, 4, 8	4	Composite
7	1, 7	2	Prime
6	1, 2, 3, 6	4	Composite
5	1, 5	2	Prime
4	1, 2, 4	3	Composite
3	1, 3	2	Prime
2	1, 2	2	Prime

Question **3**: The prime numbers between 0 and 100 are 2, 3, 5, 7, 11, 13, 17, 19, 23, 29, 31, 37, 41, 43, 47, 53, 59, 61, 67, 71, 73, 79, 83, 89, 97. There are 25 prime numbers between 0 and 100.

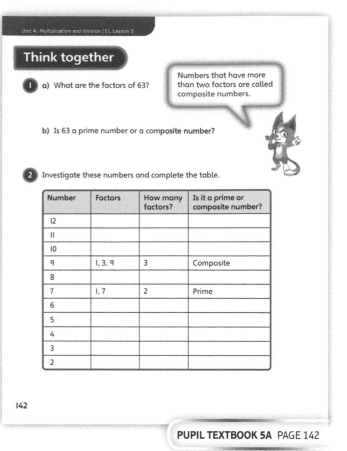

PUPIL TEXTBOOK 5A PAGE 142

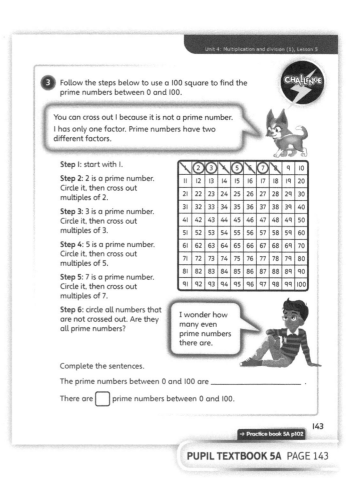

PUPIL TEXTBOOK 5A PAGE 143

Practice

WAYS OF WORKING Independent thinking

IN FOCUS Question ❶ shows a prime number as an array and asks children to explain why this number is prime. This helps to strengthen the link between the pictorial representation of the array and the abstract understanding of prime numbers.

Question ❸ is important as it helps to tackle the misconception that all even numbers are composite numbers. It also strengthens the understanding that, after 2, all even numbers are composite.

Questions ❹ and ❺ offer a valuable opportunity to continue developing generalisations about prime numbers.

STRENGTHEN When solving question ❻, children may be put off by the new and unfamiliar arrangement of the number grid. Ask: *What do you notice about the number grid? How is it similar and different to number grids you have used before? What might change in the patterns of prime numbers because of the new grid?*

DEEPEN When considering their ideas for question ❺, deepen children's reasoning by asking: *What conjecture could you make about these numbers? How could you prove your ideas? Can you prove your thinking using resources or a picture?*

ASSESSMENT CHECKPOINT Questions ❹ and ❺ give a good opportunity to assess children's fluency with prime numbers. They should be able to demonstrate understanding of the generalisations that can be made about prime numbers and prove them using the resources in the lesson.

ANSWERS Answers for the **Practice** part of the lesson can be found in the *Power Maths* online subscription.

Reflect

WAYS OF WORKING Pair work

IN FOCUS This question is important as it will give a quick assessment of children's ability to find prime numbers and how confident they are with the link between prime numbers and their knowledge of factors and multiplication. Children could be encouraged to share their diagram with their partner and discuss how and why the pictures are the same and different.

ASSESSMENT CHECKPOINT Children should be able to use arrays to demonstrate whether a number is prime or composite and explain what the diagram proves clearly and concisely.

ANSWERS Answers for the **Reflect** part of the lesson can be found in the *Power Maths* online subscription.

After the lesson ⏸

- Were children able to confidently find prime numbers all the way to 100?
- How was children's independence and resilience developed in this lesson?
- Were children able to sufficiently demonstrate their problem-solving abilities?

PUPIL PRACTICE BOOK 5A PAGE 102

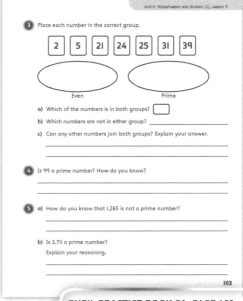

PUPIL PRACTICE BOOK 5A PAGE 103

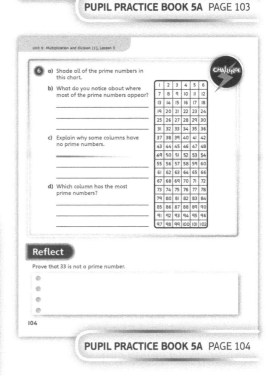

PUPIL PRACTICE BOOK 5A PAGE 104

Square numbers

Learning focus

In this lesson, children will learn about square numbers and how to recognise and represent square numbers pictorially before linking this to using notation, including squared (^{2}). Children will find square numbers in the multiplication grid and use them to solve calculations and problems.

Before you teach ⏸

- How will you make sure that the concrete resources you use make the link between shape and number explicit?
- Are children sufficiently confident with factors and multiples?

NATIONAL CURRICULUM LINKS

Year 5 Number – multiplication and division

Recognise and use square numbers and cube numbers, and the notation for squared (^{2}) and cubed (^{3}).

ASSESSING MASTERY

Children can understand the term 'square' and how it relates to number as well as shape, and use the correct notation when making calculations using squares. They can reliably find square numbers and can explain how the number links to the properties of the shape.

COMMON MISCONCEPTIONS

As finding a square number requires multiplying a number by itself, children may interpret this as x multiplied by 2, instead of x multiplied by x. Show children the appropriate square and ask:
- *If you look at this square as an array, what multiplication does it represent? How is that different to $x \times 2$?*

STRENGTHENING UNDERSTANDING

Children could be encouraged to draw different squares using squared paper. Discuss the properties of squares and how each side is the same length. These properties can then be linked to arrays. Ask: *What squares have you made? What array does your square look like? Explain your answer.*

GOING DEEPER

Children can investigate relationships between prime numbers and square numbers. Ask: *Is it always true, sometimes true or never true that the sum of two primes is a square number?*

KEY LANGUAGE

In lesson: square number, squares, x^2, factors

Other language to be used by the teacher: multiply, multiplied, times, dimensions

STRUCTURES AND REPRESENTATIONS

Arrays, multiplication grid

RESOURCES

Optional: multilink cubes, counters, squared paper, chessboard, pencils, rulers

 In the eTextbook of this lesson, you will find interactive links to a selection of teaching tools.

Quick recap 🔁

Give children squared paper, a pencil and a ruler. Ask them to draw a square of side length 3 cm. How many squares are inside it? What about a side length of 4 cm?

Discover

Square numbers

Discover

WAYS OF WORKING Pair work

ASK

* Question ① a): *What would be the best way to count the squares?*
* Question ① a): *How many squares can you see?*
* Question ① b): *How will you know if you have found all of the squares?*

IN FOCUS Question ① a) is important as it gives children their first opportunity to count the squares within a square. Discuss how children went about counting them. Did they use an efficient way?

Question ① b) is a good opportunity to discuss other squares and begin children's pattern recognition in square numbers.

PRACTICAL TIPS Children could be encouraged to make different squares using different resources. Discuss the dimensions of the squares and how this relates to the area inside. To make this point especially clear, use multilink cubes or squared paper to allow children to count the squares inside the shape.

ANSWERS

Question ① a): There are 64 small squares on the chessboard altogether.

Question ① b): Possible squares:
1×1
2×2
3×3
4×4
5×5
6×6
7×7
8×8

① a) How many small squares are there on the chessboard altogether?

b) What other size squares can you find on the chessboard?

144

PUPIL TEXTBOOK 5A PAGE 144

Share

WAYS OF WORKING Whole class teacher led

ASK

* Question ① a): *How did you count the squares?*
* Question ① a): *Have you seen the square sign before? Where have you seen it used?*
* Question ① b): *Why do you think these numbers are called 'square' numbers?*
* Question ① b): *Can you draw the larger square numbers on squared paper?*

IN FOCUS Question ① a) is important as it introduces children to the vocabulary and mathematical notation of square numbers. It will be important for children to have plenty of opportunity to practise using both of these.

Question ① b) provides a key opportunity to link children's understanding of the square shapes with their new understanding of how square numbers work.

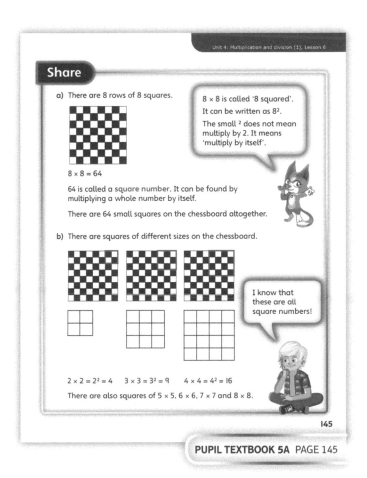

Share

a) There are 8 rows of 8 squares.

$8 \times 8 = 64$

8×8 is called '8 squared'. It can be written as 8^2.

The small ^{2} does not mean multiply by 2. It means 'multiply by itself'.

64 is called a square number. It can be found by multiplying a whole number by itself.

There are 64 small squares on the chessboard altogether.

b) There are squares of different sizes on the chessboard.

I know that these are all square numbers!

$2 \times 2 = 2^2 = 4$ $3 \times 3 = 3^2 = 9$ $4 \times 4 = 4^2 = 16$

There are also squares of 5×5, 6×6, 7×7 and 8×8.

145

PUPIL TEXTBOOK 5A PAGE 145

Think together

Whole class teacher led (I do, We do, You do)

ASK

- Question **1**: *How are the arrays similar and different to a square? What is the most efficient way of finding how many counters there are?*
- Question **2**: *Is there a pattern in the square numbers you have found? Can you explain it?*
- Question **3**: *What is wrong with Jamilla's square? What is the clearest way of representing a square number? Why?*

IN FOCUS Question **2** allows children to observe the patterns in square numbers. It provides an excellent opportunity for children to begin predicting the sequence of square numbers beyond 12 × 12.

Question **3** helps address the possible misconception that the squares around the outside of a square are all that need counting to find square numbers, instead of the inside as well. Question **3** b) aims to make children second guess whether 16 is a square number or not, as the counters are arranged as two groups of 8, potentially causing a cognitive conflict where they will assume a mistake has been made.

STRENGTHEN To strengthen understanding in question **3**, ask: *Can you make a complete square with 12 cubes? Explain why not. How has Jamilla managed to make a square? Where has she gone wrong?*

DEEPEN Question **2** can be deepened by asking children to spot and explain the pattern that emerges in the squared numbers. Ask: *What pattern can you spot? Is there a pattern in how the numbers increase? How might this pattern help you predict the next square numbers in the sequence?*

ASSESSMENT CHECKPOINT Question **3** will offer the opportunity to assess whether children can reliably identify a square number. Look for children demonstrating their understanding of the concrete representations of square numbers and applying this to their explanations.

ANSWERS

Question **1**: $5^2 = 5 \times 5 = 25$
　　　　　　10 squared is 100.

Question **2**: 1, 4, 9, 16, 25, 36, 49, 64, 81, 100, 121, 144

Question **3** a): Jamilla is incorrect as she has not made a complete, solid square.

Question **3** b): 16 is a square number. Look for children making or drawing a 4 by 4 square.

PUPIL TEXTBOOK 5A PAGE 146

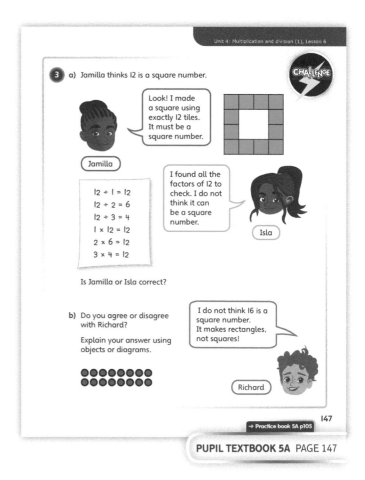

PUPIL TEXTBOOK 5A PAGE 147

Practice

WAYS OF WORKING Independent thinking

IN FOCUS Question ❶ scaffolds children's independent thinking about square numbers. Offering the pictures alongside the written representation will enable children to visualise the square numbers in later questions.

Question ❷ offers an opportunity for children to develop and demonstrate their understanding of the squared (^{2}) notation. Be sure that children understand that 5^2 means 5×5, and not 5×2.

Question ❸ allows children to create a pictorial representation of a square number. Be sure that children link this to the number sentence below so that they become fluent with the more efficient abstract way of recording.

STRENGTHEN To support children with understanding square numbers, provide them with counters or cubes so they can create corresponding arrays.

DEEPEN While solving question ❼, ask: *Do you agree with Isla? Can you find evidence that supports your ideas and proves them?*

ASSESSMENT CHECKPOINT Question ❻ will offer an opportunity to assess children's ability to recognise and identify square numbers with no scaffolding. Look for children using their understanding of arrays, number patterns and times-tables when identifying which of the numbers are squares.

ANSWERS Answers for the **Practice** part of the lesson can be found in the *Power Maths* online subscription.

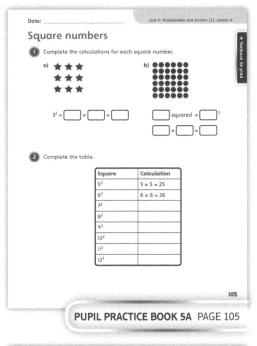

PUPIL PRACTICE BOOK 5A PAGE 105

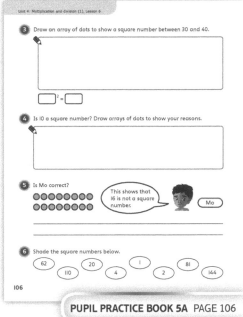

PUPIL PRACTICE BOOK 5A PAGE 106

Reflect

WAYS OF WORKING Independent thinking

IN FOCUS This question will offer an opportunity to assess children's fluency with square numbers. Children should be able to record some squares from memory by this point and know efficient methods of finding ones they have not committed to memory.

ASSESSMENT CHECKPOINT Look for which square numbers children have committed to memory. Use this opportunity to assess which children still rely on drawing square arrays and which children have moved into more abstract recording.

ANSWERS Answers for the **Reflect** part of the lesson can be found in the *Power Maths* online subscription.

After the lesson ⏸

- Have children moved beyond the concrete and pictorial representations of squared numbers and showed fluency with the abstract concepts, particularly the mathematical notation of squared (^{2})?
- Were children confidently able to identify square numbers in the multiplication grid and recognise the sequence of square numbers in the grid?
- How will you support those children who still need help with fully understanding the abstract concepts and using the mathematical notation?

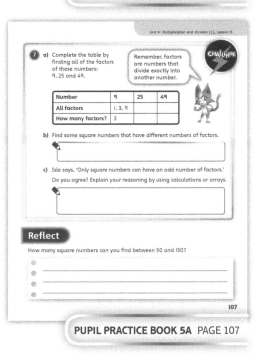

PUPIL PRACTICE BOOK 5A PAGE 107

Cube numbers

Learning focus

In this lesson, children will learn how to recognise and represent cube numbers pictorially before linking this to using notation, including cubed (^{3}). They will learn how to find cube numbers and use them to solve calculations and problems.

Before you teach

- Do children understand the mathematical notation for square numbers?
- For children who found the abstract notation tricky, how will you offer support in this lesson?

NATIONAL CURRICULUM LINKS

Year 5 Number – multiplication and division

Recognise and use square numbers and cube numbers, and the notation for squared (^{2}) and cubed (^{3}).

ASSESSING MASTERY

Children can understand the term 'cube' and how it relates to number as well as shape. They can reliably find cube numbers and use the notation for cube numbers (^{3}), and can explain how the number links to the properties of the shape, particularly in regards to dimensions.

COMMON MISCONCEPTIONS

As finding a cube number requires multiplying a number by itself three times, children may interpret this as x multiplied by 3, instead of x multiplied by x multiplied by x. Show children the appropriate cube and ask:
- *How is a cube different to a square? How many dimensions does a cube have? How many times will you need to multiply this number? What does the cube sign mean?*

STRENGTHENING UNDERSTANDING

Children should be given the opportunity to create cubes of different dimensions, prior to the lesson. Discuss the properties of the cube and how it was made. Ask: *How many smaller cubes were needed to make your big cube? How could you describe the dimensions of your cube?*

GOING DEEPER

Ask: *Is it always true, sometimes true or never true that the sum of four cube numbers is a cube number?*

KEY LANGUAGE

In lesson: cube number, cubes, multiplied

Other language to be used by the teacher: multiply, dimension

RESOURCES

Mandatory: multilink cubes

Optional: base 10 equipment, puzzle cube

 In the eTextbook of this lesson, you will find interactive links to a selection of teaching tools.

Quick recap

Give children some multilink cubes and explore making larger cubes with them. How many small cubes do they use each time to make larger cubes?

Discover

Cube numbers

WAYS OF WORKING Pair work

ASK

- Question ❶ a): *How could you test how many small cubes were used?*
- Question ❶ a): *How many small cubes would be needed to make the next biggest cube?*
- Question ❶ b): *Why does Isla think 2 × 2 × 2 = 6? What is her mistake?*
- Question ❶ b): *Does what you learnt about the square sign apply in a similar way to the cube sign?*

IN FOCUS Question ❶ b) is important as it approaches the misconception that the cube notation means 'multiply by 3'. This will be a key piece of learning during the lesson, so it is important to focus on this point during the class discussion.

PRACTICAL TIPS Children should be encouraged to make the cubes they see in the picture using multilink cubes. This will help secure their understanding of why a number is cubed in later discussions. If children investigate different-sized cubes, then a record of these, and their related multiplication facts, could be displayed in the classroom for future use.

ANSWERS

Question ❶ a): 8 small cubes make up the puzzle cube.

Question ❶ b): Isla saw 2 multiplied 3 times, so did 2 × 3 = 6 (2 + 2 + 2 rather than 2 × 2 × 2).

Discover

a) How many small cubes make up the puzzle cube?

b) Explain Isla's mistake.

148

PUPIL TEXTBOOK 5A PAGE 148

Share

WAYS OF WORKING Whole class teacher led

ASK

- Question ❶ a): *How did you visualise the cube?*
- Question ❶ a): *Did it make a difference to the answer how the cube was visualised? Why?*
- Question ❶ a): *How many dimensions does a cube have?*
- Question ❶ a): *How is the cube sign similar and different to the square sign?*
- Question ❶ a): *How does the cube sign relate to the dimensions of the cube?*
- Question ❶ b): *What would you say to Isla to help her correct her mistake?*

IN FOCUS Question ❶ a) gives an excellent opportunity to link the concrete representation of a cube with the abstract concept of cubing numbers. If they haven't already, children should be encouraged to practically recreate what is shown in the picture to help secure their understanding.

Share

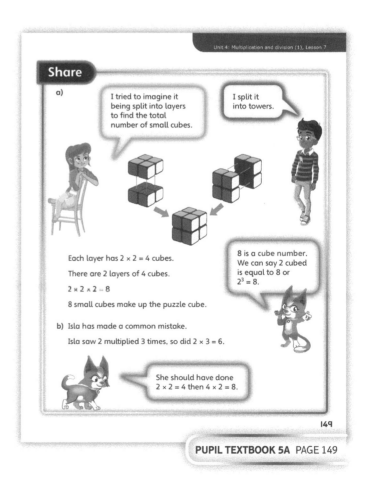

Think together

Think together

WAYS OF WORKING Whole class teacher led (I do, We do, You do)

ASK

- Question **1**: *How many cubes do you need?*
- Question **2** b): *Why does Zac think 2 is a cube number? Could it be one?*
- Questions **3** a) and b): *How will you investigate these calculations? What resources or methods will you use?*

IN FOCUS Question **1** allows children to link a concrete representation to the abstract notion of cube numbers.

Question **2** a) gives children the opportunity to investigate what happens when you cube 1. This can be compared to what happens when 10 is cubed, providing an interesting opportunity for children to discuss patterns and make generalisations about place value and cubing.

Question **2** b) deals with the misconception that a cube number is the starting number rather than the product.

STRENGTHEN To help children visualise 10^3 in question **2** a), provide them with base 10 equipment. Ask: *Can you use the base 10 equipment to show 10^3? Can you explain how the base 10 equipment has shown 10^3?*

DEEPEN Question **3** will deepen children's understanding by helping them to link their previous learning about the distributive properties of multiplication to their new understanding of cube numbers. Ask: *How can partitioning the numbers help make the multiplication easier?*

ASSESSMENT CHECKPOINT Question **3** a) and b) will assess not only children's ability to cube numbers reliably, but also their ability to use their fluency in number and calculations to help them cube bigger numbers.

ANSWERS

Question **1** a): 27 cubes are needed.

Question **1** b): $3^3 = 27$

Question **2** a): $1^3 = 1 \times 1 \times 1 = 1$
$10^3 = 10 \times 10 \times 10 = 1,000$

Question **2** b): Zac is incorrect. The cube number is 8 (the product), not 2.

Question **3** a): $5^3 = 125$

Question **3** b): $6^3 = 216$

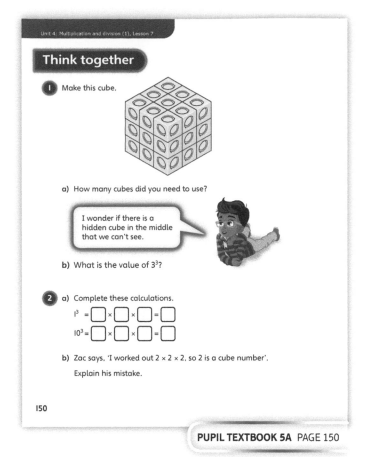

PUPIL TEXTBOOK 5A PAGE 150

PUPIL TEXTBOOK 5A PAGE 151

Practice

WAYS OF WORKING Independent thinking

IN FOCUS Question ❶ will help children to differentiate between correctly cubing and the potential misconceptions that can arise when trying to cube a number. It also shows pictorial and concrete representations of the concept of cubing and how it is different to multiplying by 3 or adding the same number three times.

Question ❻ helps children link the distributive law to their understanding of cubing numbers. This is important as it will help them more successfully cube larger numbers in the future.

STRENGTHEN To strengthen children's understanding of how to approach each calculation in question ❷, ask: *Can you build what is shown in the picture? Explain what you are building as you do so. How would you record each step?*

DEEPEN Expand question ❻ by providing children with cubes to work out 4³. Encourage them to split the cubes in similar ways to Luis to break down the calculation and write out their workings in full.

THINK DIFFERENTLY Question ❹ allows children to demonstrate their secure understanding by recognising and correcting the mistakes. Each mistake covers a different misconception, so it will be important to observe if any examples are found to be trickier across the class. This will show if an area of teaching needs revisiting.

ASSESSMENT CHECKPOINT Question ❷ will allow for the assessment of whether children can reliably calculate cube numbers. Look for their ability to link the pictorial representation with the abstract calculation.

Question ❺ offers a good opportunity to assess children's fluency with numbers when finding cube numbers.

ANSWERS Answers for the **Practice** part of the lesson can be found in the *Power Maths* online subscription.

Reflect

WAYS OF WORKING Independent thinking

IN FOCUS This question allows for an overall assessment of children's progress. Children should be able to confidently find the first five cube numbers, explaining how they did so. Comparing children's methods in a class discussion will highlight the different ways the cube numbers can be reached.

ASSESSMENT CHECKPOINT Look for children confidently cubing numbers using the concrete, pictorial and abstract representations from the lesson. Children should be able to use the distributive law to help them with larger numbers.

ANSWERS Answers for the **Reflect** part of the lesson can be found in the *Power Maths* online subscription.

After the lesson ⏸

- Did children recognise how the distributive law is useful when cubing larger numbers?
- Could you have made this more explicit? If so, how?
- How confident are children with the abstract mathematical notation? Could they link it to their learning about squares?

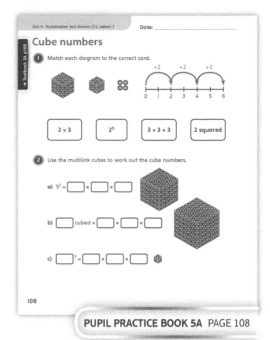

PUPIL PRACTICE BOOK 5A PAGE 108

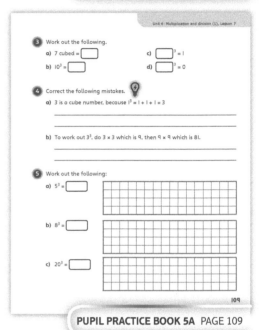

PUPIL PRACTICE BOOK 5A PAGE 109

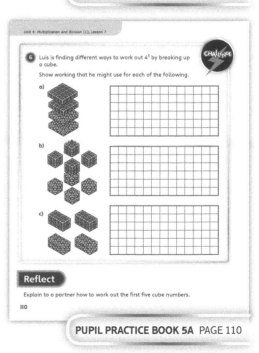

PUPIL PRACTICE BOOK 5A PAGE 110

Multiply by 10, 100 and 1,000

Learning focus

In this lesson, children will use their understanding of place value to develop their ability to fluently multiply whole numbers by 10, 100 and 1,000.

Before you teach

- How confident are children with place value and using base 10 equipment?
- How confident are children already when multiplying whole numbers by 10, 100 and 1,000?

NATIONAL CURRICULUM LINKS

Year 5 Number – multiplication and division

Multiply and divide whole numbers and those involving decimals by 10, 100 and 1,000.

ASSESSING MASTERY

Children can reliably multiply whole numbers by 10, 100 and 1,000, link their understanding of place value to their calculations and confidently represent their thinking using concrete, pictorial and abstract representations.

COMMON MISCONCEPTIONS

A common misconception when multiplying by 10, 100 and 1,000 is to just 'add a 0 on to the end'. This is a damaging habit as this will cause errors when multiplying by 100 and 1,000, and later on, decimals. Ask:

- *Show me what you mean by 'add a 0'. What does 'add a 0' actually mean? Make your original number and your final number using base 10 equipment. How are they similar and different?*

STRENGTHENING UNDERSTANDING

Children could be encouraged to count in 10s, 100s and 1,000s before this lesson to secure their fluency with the patterns in those number sequences. This could also be approached through chants, rhymes and songs.

GOING DEEPER

Encourage children to create contextual word problems to multiply by 10, 100 or 1,000. Challenge children to make them as real to life as possible so that they are required to consider the real-life application of this skill.

KEY LANGUAGE

In lesson: multiply, multiplication, ten, one hundred, one thousand, 10, 100, 1,000

Other language to be used by the teacher: place value

STRUCTURES AND REPRESENTATIONS

Place value grid, number lines

RESOURCES

Mandatory: base 10 equipment, place value counters

Optional: printed place value grids, toy car parts

 In the eTextbook of this lesson, you will find interactive links to a selection of teaching tools.

Quick recap ↻

Show children, for example, 30 made from base 10 equipment and ask: *How many 10s and how many 1s are there? How do you know?* Ensure children recall that there are ten 1s in 10 and ten 10s in 100. Repeat with other multiples of 10 made from base 10 equipment.

Discover

Pair work

ASK

- Question ① a): *How many wheels are on each car?*
- Question ① a): *How can you test how many wheels would be needed for 10 cars?*
- Questions ① a) and b): *What patterns can you spot in the numbers?*

IN FOCUS Questions ① a) and b) will give children their first opportunity to multiply 1-digit numbers by 10 and 100, allowing an initial assessment of children's confidence before starting the main part of the lesson.

PRACTICAL TIPS This introduction would lend itself well to junk modelling. Children could be given the parts to make one junk model car. The questions posed in the **Discover** section could then be posed about their car parts.

ANSWERS

Question ① a): 40 wheels are needed for 10 cars.

400 wheels are needed for 100 cars.

Question ① b): 20 lamps are needed for 10 cars.

200 lamps are needed for 100 cars.

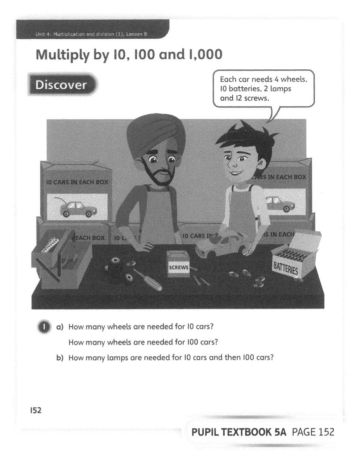

PUPIL TEXTBOOK 5A PAGE 152

Share

Whole class teacher led

ASK

- Question ① a): *How does the array and base 10 equipment make the multiplication clear?*
- Question ① a): *What would be different if the car needed 6 wheels?*
- Question ① a): *Which method is more efficient, counting in 4s or 10s? Why?*
- Question ① b): *How do the numbers change and stay the same as you multiply by 10 and 100?*
- Question ① b): *Can you use the base 10 equipment to explain how the numbers change?*

IN FOCUS When focusing on these two questions, it is important that children recognise the patterns in the numbers. Question ① a) also reinforces the proportionality of each multiple of 4. This should be supported with the use of concrete resources, so children can manipulate and experience the difference between one 4, ten 4s and one hundred 4s.

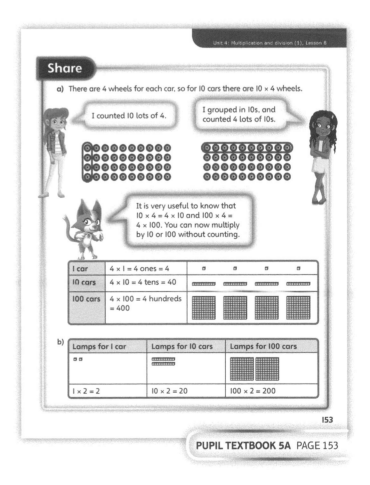

PUPIL TEXTBOOK 5A PAGE 153

Think together

Think together

WAYS OF WORKING Whole class teacher led (I do, We do, You do)

ASK

- Question ❶ a): *How does using the base 10 equipment make the multiplication clear? What is similar and different about each calculation?*
- Question ❶ b): *How does the number line show the multiplication clearly?*
- Question ❷: *How would you explain Aki's mistake to him?*
- Question ❸ a): *What has happened to the digit '3'? How has the exchange altered the value of the digit?*
- Question ❸ c): *What happens when you repeatedly multiply by 10?*

IN FOCUS Questions ❷ and ❸ are important as they tackle the misconception of 'add a 0'. Question ❸ a) uses exchanging to show that a number moved from the ones column to the tens column is being multiplied by 10, rather than 'just adding a 0'.

STRENGTHEN If children are finding it difficult to diagnose Aki's misconception in question ❷, they could build the calculations using concrete resources.

DEEPEN Question ❸ deepens children's understanding of how multiplying by 10 is linked to place value. Ask: *How can you use multiplying by 10 to multiply by 100 and 1,000? How could you use this to multiply by greater numbers?*

ASSESSMENT CHECKPOINT Questions ❶ a) and b) will assess children's ability to more fluently multiply whole numbers by 10, 100 and 1,000. Question ❸ will assess children's fluency with how multiplying by 100 and 1,000 is linked to multiplying by 10. It will also highlight misconceptions about place value and 'just adding a 0' when multiplying by 10.

ANSWERS

Question ❶ a): 7 × 10 = 70
7 × 100 = 700
7 × 1,000 = 7,000

Question ❶ b): 12 × 10 = 120
12 × 100 = 1,200
12 × 1,000 = 12,000

Question ❷: Aki has solved the first calculation correctly, 23 × 100 = 2,300.
Aki has solved the second calculation incorrectly by multiplying 20 by 10, instead of 100.

Question ❸ a): Each set of ten 1s counters are exchanged for one 10 place value counter, giving three 10s counters in the tens column.
10 × 3 = 30

Question ❸ b): 3 × 10 = 30
17 × 10 = 170
The numbers move to the next column in the place value grid when multiplying by 10.

Question ❸ c): 3 × 10 × 10 = 300
17 × 10 × 10 = 1,700
Multiplying by 10 and then 10 again is the same as multiplying by 100.

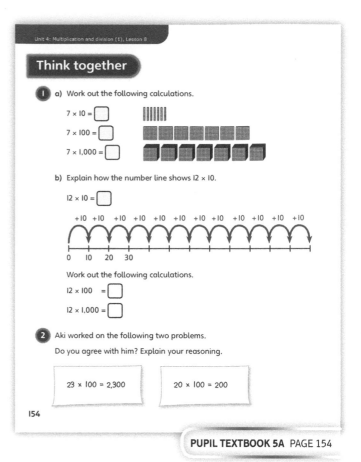

PUPIL TEXTBOOK 5A PAGE 154

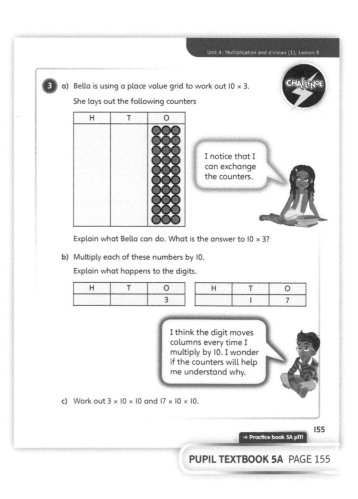

PUPIL TEXTBOOK 5A PAGE 155

Practice

Independent thinking

IN FOCUS These questions provide the opportunity for children to practise multiplying whole numbers by 10, 100 and 1,000. Visual representations are used to support understanding in question ❶. Question ❸ encourages children to spot patterns and make links to place value.

STRENGTHEN Provide children with base 10 equipment or place value counters and a place value grid to represent calculations and support them in finding the answers.

DEEPEN In question ❻, children make connections between multiplying by 100 and 1,000, and multiplying by 10 multiple times. Children could be given a target number and tasked with finding as many different questions as possible that involve multiplying by 10, 100 and 1,000 that give that answer.

THINK DIFFERENTLY Question ❷ explores common mistakes that children might make when multiplying whole numbers by 10, 100 and 1,000.

ASSESSMENT CHECKPOINT Question ❶ assesses whether children can multiply 1-digit numbers by 10, 100 and 1,000. Use question ❷ to assess whether children can spot and explain common mistakes. Use questions ❸ to ❺ to assess whether children can multiply by 10, 100 and 1,000.

ANSWERS Answers for the **Practice** part of the lesson can be found in the *Power Maths* online subscription.

PUPIL PRACTICE BOOK 5A PAGE 111

PUPIL PRACTICE BOOK 5A PAGE 112

Reflect

WAYS OF WORKING Independent thinking

IN FOCUS This question provides the opportunity for children to explain their learning about multiplying whole numbers by 10, 100 and 1,000 in their own words.

ASSESSMENT CHECKPOINT Children should be able to accurately explain how to multiply by 10, 100 and 1,000 and link this to place value.

ANSWERS Answers for the **Reflect** part of the lesson can be found in the *Power Maths* online subscription.

After the lesson ⏸

- Did children recognise the usefulness of distributive properties when multiplying by 100 and 1,000?
- What concrete manipulatives were most effective in this lesson? Why?

PUPIL PRACTICE BOOK 5A PAGE 113

Divide by 10, 100 and 1,000

Learning focus

In this lesson, children will use their understanding of place value to develop their ability to fluently divide whole numbers by 10, 100 and 1,000.

Before you teach

- Are children confident multiplying by 10, 100 and 1,000?
- How will you strengthen understanding of inverse operations?

NATIONAL CURRICULUM LINKS

Year 5 Number – multiplication and division

Multiply and divide whole numbers and those involving decimals by 10, 100 and 1,000.

ASSESSING MASTERY

Children can reliably divide whole numbers by 10, 100 and 1,000, link their understanding of place value to their calculations and represent their thinking using concrete, pictorial and abstract representations.

COMMON MISCONCEPTIONS

Children may recognise dividing by 10, 100 and 1,000 as just 'taking away a 0'. Ask:
- *What does it mean to 'take away a 0'? How are dividing by 10 and taking away 0 similar/different?*

STRENGTHENING UNDERSTANDING

Recap division and factors with children before this lesson, if necessary. Give children opportunities to experience this concept by sharing objects into groups and linking this to the division calculation. Make arrays and note the related division and multiplication facts.

GOING DEEPER

Give children this statement to investigate: *Is it always true, sometimes true or never true that when you divide a 3-digit whole number by 10, 100 or 1,000 it will result in a number greater than 1?*

KEY LANGUAGE

In lesson: divide, division, ten (10), one hundred (100), one thousand (1,000)

Other language to be used by the teacher: place value, ones, tens, hundreds, thousands, tens of thousands, share

STRUCTURES AND REPRESENTATIONS

Bar models, place value grids, number lines

RESOURCES

Mandatory: place value counters, base 10 equipment

Optional: printed place value grids

 In the eTextbook of this lesson, you will find interactive links to a selection of teaching tools.

Quick recap

Play 'Multiplying by 10, 100 and 1,000' bingo. Give children a set of numbers from which they choose six. Now present them with calculations that involve multiplying by 10, 100 and 1,000 to give those numbers as answers. The first child to cross out all their numbers wins. You could also include missing number style questions to give a brief insight into dividing by 10, 100 and 1,000.

Discover

WAYS OF WORKING Pair work

ASK

- Question ➊: *Is it better to come first, second or third in this competition? Explain why.*
- Question ➊: *How much do you win for coming first, second or third?*

IN FOCUS In questions ➊ a) and b), children may assume that the people who come second win more as the shared prize pot is far greater. The conversation from these questions will give children their first opportunity to understand the effect of dividing a whole number by 10, 100 and 1,000.

PRACTICAL TIPS To help engage children in this area of learning, you could create a small competition or race where the first ten children to succeed are able to share a prize pot. This activity, while difficult to achieve for 100 and 1,000 participants, will give children hands-on, contextual experience with the main concept in the lesson. It may also give an opportunity to recap that division is not commutative and the order of operations matters.

ANSWERS

Question ➊ a): Each 1st prize winner will receive £38.

Question ➊ b): Each 2nd prize winner will receive £12.

PUPIL TEXTBOOK 5A PAGE 156

Share

WAYS OF WORKING Whole class teacher led

ASK

- Question ➊ a): *What calculation does the bar model show? Explain how.*
- Question ➊ a): *What would you need to multiply 10 by to have a product of 380?*
- Question ➊ b): *How much would each person have won if the prize pot for 2nd place was £15,000?*
- Question ➊ b): *What would the total prize pot for 2nd place have to be if second prize winners were winning more than first prize winners? Explain how you know.*

IN FOCUS Both questions ➊ a) and ➊ b) are important to focus on as they show how children may solve the calculations in multiple ways. Making this clear will help develop children's fluency when solving similar calculations in the future.

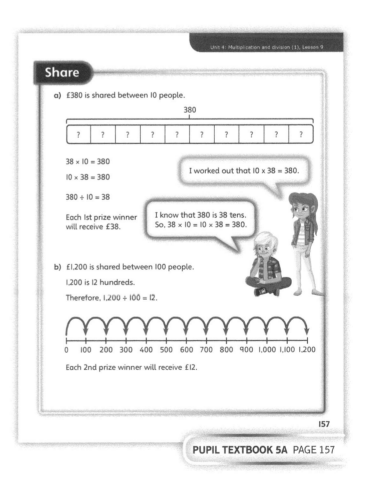

PUPIL TEXTBOOK 5A PAGE 157

Think together

Whole class teacher led (I do, We do, You do)

ASK

- Question **1**: *What patterns can you spot in these calculations? What can you say about what happens every time you divide by 10, 100 or 1,000?*
- Question **3**: *How does your knowledge of base 10 equipment help you with this question?*

IN FOCUS Question **1** gives children the opportunity to begin generalising about dividing by 10, 100 and 1,000. It will be important to discuss the patterns they see and link what they know of place value and dividing to the quotients found in the division calculations.

Question **3** b) is important as it requires children to use place value when dividing by 10, 100 and 1,000, to complete multi-step calculations. It will be important to discuss how, similarly to multiplying by 10 and 10 again, dividing by 10 and then by 10 again is equivalent to dividing by 100 (as an example).

STRENGTHEN If children need support with question **3**, encourage them to use base 10 equipment to develop the concept of exchange in place value, helping them visualise the problem more clearly.

DEEPEN When solving question **1**, deepen children's understanding by encouraging them to generalise. Ask: *What patterns do you notice? What is similar and different about the calculations? How can this help you solve similar calculations more quickly in the future?*

ASSESSMENT CHECKPOINT Question **3** will provide good evidence of children's understanding of the concepts in this lesson, and their ability to link them to their learning from the previous lesson and about place value. Look for children's recognition that, for example, dividing by 10, then by 10 again, is equivalent to dividing by 100.

ANSWERS

Question **1** a): $30 \div 10 = 3$, $300 \div 100 = 3$, $3,000 \div 1,000 = 3$

Question **1** b): $310 \div 10 = 31$, $3,100 \div 100 = 31$, $31,000 \div 1,000 = 31$

Question **1** c): $300 \div 10 = 30$, $3,000 \div 100 = 30$, $30,000 \div 1,000 = 30$
Each set of calculations gives the same answer.

Question **2**: 4,000 is four 1,000s.
$4 \times 1,000 = 4,000$
$4,000 \div 1,000 = 4$
Each third prize winner receives £4.

Question **3** a): $4,000 \div 10 = 400$; $4,000 \div 100 = 40$
$3,200 \div 10 = 320$; $3,200 \div 100 = 32$
When you divide the numbers by 10 or 100, the digits stay the same but move right in the place value grid.

Question **3** b): Max is correct, it is the same as dividing by 1,000.

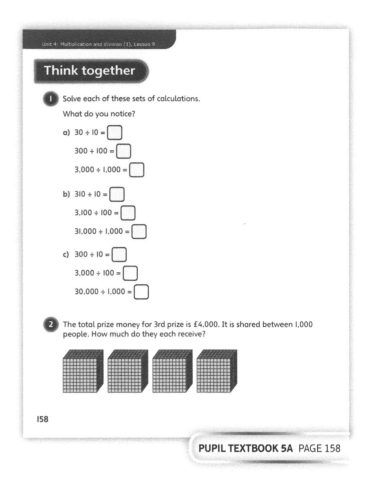

PUPIL TEXTBOOK 5A PAGE 158

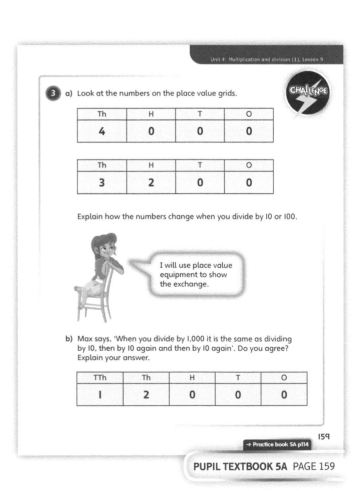

PUPIL TEXTBOOK 5A PAGE 159

Practice

WAYS OF WORKING Independent thinking

IN FOCUS Question ③ begins to show children how their learning on dividing by 10, 100 and 1,000 can be used in real-life contexts.

When solving question ④, remind children of the patterns they noticed and the generalisations they made earlier in the lesson. How can their thinking and ideas help them?

Question ⑥ is important as it encourages children to think algebraically, considering what they know about each side of the equality and how knowing one number will affect the other unknown number.

STRENGTHEN In question ⑥, ask children to use counters on a place value grid to help them. Ask: *How can you use the place value grid to begin finding out what the triangle will be worth? What number do you need to multiply or divide by 10?*

DEEPEN Question ⑤ deepens children's fluency, reasoning and problem-solving by putting their learning into a context. It also tries to catch children out by requiring them to divide and multiply by 10. This is not initially clear in the question and will only become clear once children have considered the problem carefully. Ask: *How did you find out how many marbles were in each jar? What did you do to find out how many jars were needed?*

ASSESSMENT CHECKPOINT Question ⑥ will assess children's ability to use their understanding of multiplying and dividing by 10, 100 and 1,000 to solve a mathematical problem. Look for their ability to apply the patterns and solutions they have found to other examples when finding more solutions with the same relationship.

ANSWERS Answers for the **Practice** part of the lesson can be found in the *Power Maths* online subscription.

Reflect

WAYS OF WORKING Independent thinking

IN FOCUS This question will help assess whether children have fully understood the effect that dividing by 10, 100 and 1,000 has on the place value of digits in a number.

ASSESSMENT CHECKPOINT Look for fluent explanations of which calculation is incorrect and why, and accurate use of place value. Children should use multiplication and division and demonstrate their understanding of their inverse relationship to check their calculations.

ANSWERS Answers for the **Reflect** part of the lesson can be found in the *Power Maths* online subscription.

After the lesson ⏸

- Are children equally confident with dividing by 10, 100 and 1,000 as they were with multiplying?
- Are children able to explain dividing by 10, 100 and 1,000 in terms of place value?
- Can children confidently use place value grids, concrete objects and pictorial representations to aid their thinking?

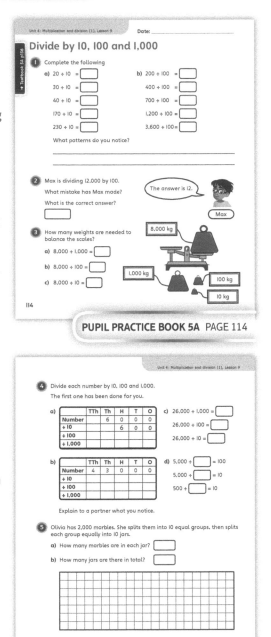

PUPIL PRACTICE BOOK 5A PAGE 114

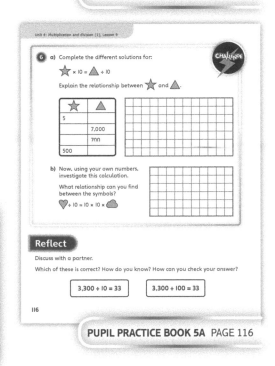

PUPIL PRACTICE BOOK 5A PAGE 115

PUPIL PRACTICE BOOK 5A PAGE 116

Multiples of 10, 100 and 1,000

Learning focus

In this lesson, children will use their knowledge and understanding of multiplying and dividing by 10, 100 and 1,000 to reliably multiply numbers by multiples of 10, 100 and 1,000 using known multiplication facts.

Before you teach

- Are there any misconceptions from the previous two lessons on multiplying and dividing by 10, 100 and 1,000 that will inhibit learning in this lesson?

NATIONAL CURRICULUM LINKS

Year 5 Number – multiplication and division

Multiply and divide whole numbers and those involving decimals by 10, 100 and 1,000.

ASSESSING MASTERY

Children can use their understanding of multiplying and dividing by 10, 100 and 1,000, and their understanding of multiplication facts, to reliably multiply and divide given numbers by multiples of 10, 100 and 1,000. They can use this ability to solve mathematical problems, explaining their reasoning and giving clear evidence of their thinking.

COMMON MISCONCEPTIONS

Children may solve a calculation such as 180 ÷ 6 by solving 18 ÷ 6, recording the answer as 3 in error. Ask:
- *Can you show me both of those calculations using a concrete or pictorial representation?*

Children may assume that 20 × 50 = 100 (an equal number of 0s on either side of the equals sign). Ask:
- *What is 2 × 5? What is 20 × 5? Can you represent 20 × 5 and 20 × 50 using resources or a picture?*

STRENGTHENING UNDERSTANDING

Develop children's fluency with multiples of 10 by counting through a times-table and linking it to the same times-table multiplied by 10, 100 or 1,000. Ask: *Can you count up in the 4 times-table? What will be different and what will stay the same if you count in the 40 times-table? 400? 4,000? What patterns can you find?*

GOING DEEPER

Encourage children to create their own lesson or tutorial on how to multiply and divide with multiples of 10, 100 and 1,000. Ask: *How will you make sure your explanation is clear to someone who does not yet understand?* Children could then teach each other and discuss how and where they would improve the lessons.

KEY LANGUAGE

In lesson: multiple, multiply, multiplication, divide, division, ten (10), hundred (100), thousand (1,000), metres (m), grams (g)

Other language to be used by the teacher: place value, ones, tens, hundreds, thousands, tens of thousands

STRUCTURES AND REPRESENTATIONS

Bar models, number lines

RESOURCES

Mandatory: base 10 equipment, place value counters

Optional: multiplication squares, calendar

 In the eTextbook of this lesson, you will find interactive links to a selection of teaching tools.

Quick recap

Choose a number that is a multiple of 10, 100 or 1,000, for example 500. Ask children if the number is a multiple of 10, 100 or 1,000. Repeat for a selection of numbers. Children could be given cards that say 'multiple of 10', 'multiple of 100' and 'multiple of 1,000' and asked to hold the correct card(s) up for each number.

Discover

ASK

- Question ① a): *How many days are there in April?*
- Question ① a): *What type of calculation do you need to do here?*
- Question ① b): *How could knowing how many words Emma will learn in ten days help you solve this question?*

IN FOCUS Question ① a) offers children their first opportunity to begin calculating with multiples of 10.

Question ① b) is important as it encourages children's fluency and flexibility when calculating with the types of numbers used in the lesson.

PRACTICAL TIPS Children could be encouraged to set their own target for a 30-day month. They could be encouraged to set three levels of challenge for themselves: a low, middle and high number of words to enable them to investigate the effect of multiplying different numbers by a multiple of 10.

ANSWERS

Question ① a): Emma plans to learn 150 words in April.

Question ① b): Ebo knows that $10 \times 30 = 300$. So, he knows 5×30 must be half of 300, which is 150.

Multiples of 10, 100 and 1,000

Discover

I plan to learn five words of French every day in April. How many will that be?

I know how many words it would be if you learnt ten each day. I will use that to find the answer a different way.

Emma

Ebo

① a) How many words does Emma plan to learn in April?

b) What do you think Ebo's method could be?

160

PUPIL TEXTBOOK 5A PAGE 160

Share

ASK

- Question ① a): *How does the base 10 equipment make the multiplication clear?*
- Question ① a): *How are the base 10 equipment representation and the number line similar? How are they different?*
- Question ① a): *How is 5 × 3 similar to 5 × 30? How can you use multiplication facts to help you solve a trickier calculation like this?*
- Question ① b): *How does the bar model make Ebo's way of calculating clear?*
- Question ① b): *Did you find another way of solving this?*

IN FOCUS Questions ① a) and ① b) both offer the opportunity to scaffold children's understanding of the concepts covered in the questions by helping them make links between the different representations and methods of the same calculation.

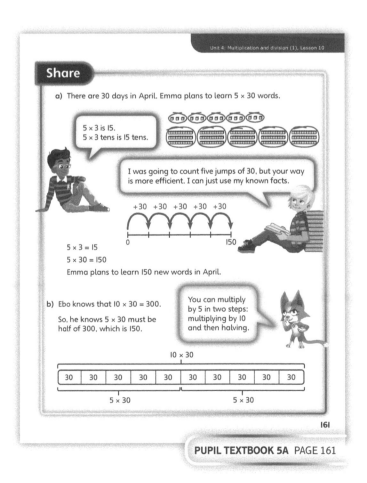

Share

a) There are 30 days in April. Emma plans to learn 5 × 30 words.

5 × 3 is 15.
5 × 3 tens is 15 tens.

I was going to count five jumps of 30, but your way is more efficient. I can just use my known facts.

+30 +30 +30 +30 +30

0   150

5 × 3 = 15
5 × 30 = 150

Emma plans to learn 150 new words in April.

b) Ebo knows that 10 × 30 = 300.

So, he knows 5 × 30 must be half of 300, which is 150.

You can multiply by 5 in two steps: multiplying by 10 and then halving.

10 × 30

| 30 | 30 | 30 | 30 | 30 | 30 | 30 | 30 | 30 | 30 |

5 × 30   5 × 30

161

PUPIL TEXTBOOK 5A PAGE 161

Think together

WAYS OF WORKING Whole class teacher led (I do, We do, You do)

ASK

- Question ❶: *How will you represent the problem?*
- Question ❷: *What is similar and different about the two arrays of place value counters?*
- Question ❸ a): *What number facts are useful to know when solving these calculations?*
- Question ❸ b): *Can you draw the rest of the bar model that would represent this problem?*

IN FOCUS Question ❶ uses base 10 equipment to show two possible methods of answering this question. Children should recognise that $180 \div 30$ can be more easily solved as $18 \div 3$.

Question ❷ reinforces the learning from question ❶ and represents it in the more abstract way of place value counters. Discussing what is similar and different about the two sets of place value counters will be key to ensuring children's understanding.

STRENGTHEN While solving questions ❶, ❷ and ❸, give children access to the concrete resources pictured in the questions. If children need help, ask: *Can you make the calculation using the resources you have used before? What does your concrete representation show?*

DEEPEN When solving question ❸ a), deepen children's ability to recognise generalisations by asking:
- *What do you notice about all of the calculations?*
- *How are the calculations similar and different?*
- *Can you come up with a rule that would help you calculate with any multiple of 10, 100 or 1,000?*

ASSESSMENT CHECKPOINT Questions ❶ and ❷ will assess children's ability to recognise the similarities between calculations, such as 4×300 and 4×3. Look for their recognition that, having identified and understood the pattern, they can use their simpler multiplication and division knowledge to solve trickier calculations.

ANSWERS

Question ❶: $180 \div 30 = 6$

Question ❷ a): $4 \times 3 = 12$; $4 \times 300 = 1,200$

Question ❷ b): $24 \div 6 = 4$; $2,400 \div 600 = 4$

Question ❸ a): $9 \times 3,000 = 27,000$
$400 \times 2,400 = 960,000$
$35,000 \div 7,000 = 5$
$2,400 \div 12 = 200$

Question ❸ b): $800 \times 6 = 400 \times 12$
Athlete B will have to train for 12 days before she has run as far as Athlete A. Children may reason that there are twice as many blocks of 400 as there are of 800 to fill the same bar model.

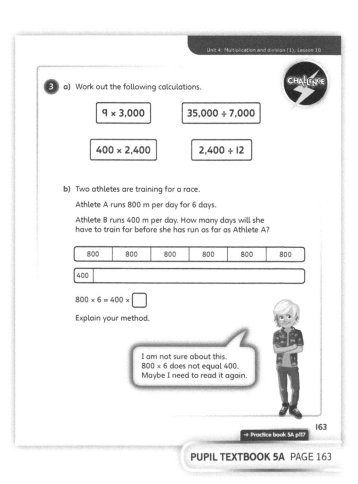

PUPIL TEXTBOOK 5A PAGE 162

PUPIL TEXTBOOK 5A PAGE 163

Practice

WAYS OF WORKING Independent thinking

IN FOCUS These questions provide the opportunity for children to practise working with multiples of 10, 100 and 1,000. Questions ① to ③ focus on multiplication and questions ④ and ⑤ focus on division, whilst later questions combine the two. Questions ①, ② and ④ use visual representations to support children's understanding. Questions ③ and ⑤ are designed to prompt children to spot patterns and make connections.

STRENGTHEN Provide children with base 10 equipment or place value counters to support them. Encourage them to unitise to deepen their understanding, for example 3 × 200 = 3 × two 100s = six 100s.

DEEPEN Children could be tasked to find equivalent calculations that involve multiples of 10, 100 and 1,000. Question ⑦ requires children to explore and correct a place value misconception when working with multiples of 10, 100 and 1,000.

THINK DIFFERENTLY For each part of question ⑥, encourage children to think about what other calculations would give the same answer.

ASSESSMENT CHECKPOINT Use questions ① to ③ to assess whether children can multiply by multiples of 10, 100 and 1,000. Use questions ④ to ⑥ to assess whether children can divide numbers that are multiples of 10, 100 and 1,000.

ANSWERS Answers for the **Practice** part of the lesson can be found in the *Power Maths* online subscription.

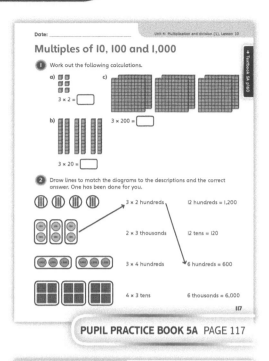

PUPIL PRACTICE BOOK 5A PAGE 117

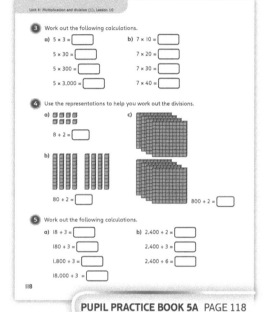

PUPIL PRACTICE BOOK 5A PAGE 118

Reflect

WAYS OF WORKING Independent thinking

IN FOCUS This activity provides children with the opportunity to explain their thinking around multiplying and dividing by multiples of 10, 100 and 1,000 in their own words.

ASSESSMENT CHECKPOINT Children should explain any connections that they notice about multiples of 10, 100 and 1,000.

ANSWERS Answers for the **Reflect** part of the lesson can be found in the *Power Maths* online subscription.

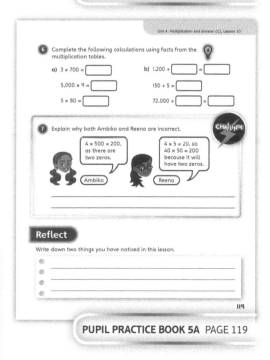

PUPIL PRACTICE BOOK 5A PAGE 119

After the lesson

- Were children able to apply their learning from the previous two lessons on multiplying and dividing by 10, 100 and 1,000 fluently to this lesson?

End of unit check

Don't forget the unit assessment grid in your *Power Maths* online subscription.

IN FOCUS

- Question **1** assesses children's ability to recognise and find factors and multiples.
- Question **2** assesses children's recognition and understanding of square numbers.
- Question **3** assesses whether children understand the difference between a factor and a multiple.
- Question **4** a) assesses children's ability to recognise multiplication facts they know and use them to solve more complex calculations.
- Question **4** b) assesses children's ability to work out prime numbers.
- Question **5** assesses children's fluency with multiplying and dividing whole numbers by multiples of 10, 100 and 1,000.
- Question **6** is a SATs-style question and assesses children's ability to find lowest common multiples and highest common factors.
- Question **7** is a problem-solving question working with a multiple of 10.

ANSWERS AND COMMENTARY Children who have mastered the concepts in this unit will reliably find multiples and factors of given numbers. They will be able to explain the unique properties of prime, square and cube numbers. They will be able to reliably multiply and divide whole numbers by 10, 100 and 1,000, and they will be able to confidently use their understanding of these concepts to multiply and divide whole numbers by multiples of 10, 100 and 1,000.

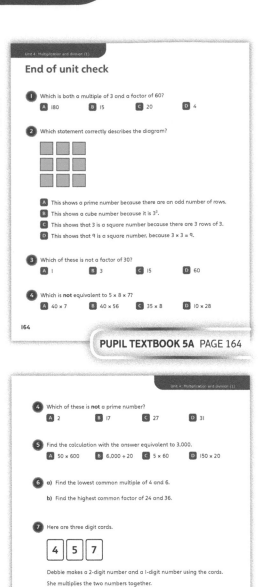

PUPIL TEXTBOOK 5A PAGE 164

PUPIL TEXTBOOK 5A PAGE 165

Q	A	WRONG ANSWERS AND MISCONCEPTIONS	STRENGTHENING UNDERSTANDING
1	B	A or C may show that children have muddled factors and multiples.	Encourage children to build or draw the numbers and calculations shown in the questions using the representations they are most confident with. Discuss what their representation shows them and how they can apply that to the question they are solving.
2	D	B suggests children think that $3 \times 3 = 3^3$.	
3	D	A, B or C suggests children have confused factors with multiples.	
4 a)	B	If children think that B is equivalent to $5 \times 8 \times 7$, they have incorrectly multiplied 5 by 8 and 8 by 7, whereas you should only multiply each number once.	
4 b)	C	A, B or D suggests children have incorrectly found other factors for these numbers.	
5	D	A, B or C suggests that children are finding inverses tricky, or that they are less confident with multiples of 10, 100 or 1,000.	
6		a) 12; b) 12	
7		75 and 4 or 74 and 5.	

My journal

WAYS OF WORKING Independent thinking

ANSWERS AND COMMENTARY

Children may record answers such as:
- I know 250 is not a square number because 15 squared is 225 and 16 squared is 256.
- 2,500 is a square number because 50 × 50 is 2,500.
- I know 2,500 is going to be square because 5 × 5 is 25. If I multiply both 5s by 10 then the answer must be multiplied by 100. 25 × 100 = 2,500.

If children are finding it difficult to unpick the reasoning in the question, ask:
- *What resources or representations have helped you find square numbers before?*
- *What calculation do you use to find a square number?*

Power check

WAYS OF WORKING Independent thinking

ASK
- *How do you feel about the new concepts you learnt in this unit?*
- *Could you teach someone about square and cube numbers?*
- *What is useful about multiplying and dividing by 10, 100 and 1,000?*

Power puzzle

WAYS OF WORKING Independent thinking

IN FOCUS Children need to solve problems by applying their understanding of the multiplicative properties of numbers. They should demonstrate their knowledge of multiples and factors and the notation for squared and cubed when trying different solutions to find each of the numbers; while also adhering to the requirement of using all four digits in their calculations each time.

ANSWERS AND COMMENTARY Children should be able to reliably identify factors of given numbers and describe how this will help them decide how to use each digit. If children are not sure how to begin finding factors, ask: *Can the number be divided by 2? What other numbers can it be divided by? Why would this be helpful?*

They should also recognise square and cube numbers and meaningfully apply the properties of square and cube numbers in their calculations. If children need extra support to recognise square or cube numbers, ask: *What is a square number? Can you show me what some square numbers look like using an array? What are the first cube numbers to 1,000? How might knowing this help you?*

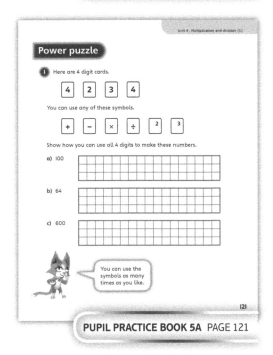

PUPIL PRACTICE BOOK 5A PAGE 120

PUPIL PRACTICE BOOK 5A PAGE 121

After the unit ⏸

- Are children confident explaining what a prime number is and finding factors?
- Do children understand how to multiply and divide by 10, 100 and 1,000 and use this knowledge to multiply and divide by multiples of 10, 100 and 1,000?

Strengthen and **Deepen** activities for this unit can be found in the *Power Maths* online subscription.

Unit 5
Fractions ❶

Mastery Expert tip! 'I make sure I give children as many opportunities to experience fractions in real-life contexts as possible, for example with weights and measures. It really helps to engage their interest and imagination and secure their conceptual understanding!'

Don't forget to watch the Unit 5 video!

WHY THIS UNIT IS IMPORTANT

Children will begin this unit by developing their understanding of how to find equivalent fractions by simplifying and expanding, and exploring how equivalent fractions represent the same fraction differently. Children will learn how to create a family of equivalent fractions by multiplying. These skills are vital for further work later on, when children are required to find equivalent fractions in order to add and subtract fractions with different denominators. Children will learn to convert between mixed numbers and improper fractions, and how to use these in real-life contexts, using pictorial representations to demonstrate their understanding. Finally, children will use their knowledge of equivalent fractions in order to compare or order fractions.

WHERE THIS UNIT FITS

→ Unit 4: Multiplication and division (1)

→ **Unit 5: Fractions (1)**

→ Unit 6: Fractions (2)

This unit builds on previous learning about equivalent fractions and converting between mixed numbers and improper fractions. In this unit, children will apply this learning and move on to comparing and ordering fractions, including those greater than 1.

Before they start this unit, it is expected that children:
- can recognise and identify a numerator and denominator in a fraction and explain what these represent
- understand how to find a missing numerator or denominator in simple fractions by multiplying or dividing
- know how to count in simple fractions from 0, saying answers as mixed numbers and improper fractions.

ASSESSING MASTERY

Children will demonstrate mastery by fluently finding equivalent fractions through simplifying and expanding. They will be able to convert between mixed numbers and improper fractions and explain in which contexts the two types of fraction are more appropriate to use. Children will be able to reliably compare and order different fractions, including those with different denominators.

COMMON MISCONCEPTIONS	STRENGTHENING UNDERSTANDING	GOING DEEPER
When converting a mixed number to an improper fraction, children may simply add the whole number to the numerator of the mixed number, rather than adding what the whole number is worth. For example, they may convert $4\frac{5}{6}$ into $\frac{9}{6}$, rather than $\frac{29}{6}$.	Give children opportunities to handle the fractions in a concrete way, i.e. give them paper plates to cut up into the given fraction. Ask: *What fraction does one whole plate represent? How do you know? If you had four whole plates, what fraction would they represent? How do you know?*	Children could be asked to create their own story problems that require the conversion of mixed numbers or improper fractions.

Children could be challenged to explain how $4\frac{8}{12}$ has been converted to the improper fraction $\frac{14}{3}$. This would require them to use their understanding of conversions and of simplifying fractions. |
| Children may assume that if the denominator of one fraction is bigger than another, then that fraction must be bigger, i.e. $\frac{3}{8}$ is larger than $\frac{3}{4}$ as 8 is bigger than 4. | Give children a fraction wall, and ask: *What do you notice about how $\frac{1}{4}$ and $\frac{1}{8}$ are similar and different to each other? Why are eighths smaller than quarters?* | |

Unit 5: Fractions ❶

UNIT STARTER PAGES

Introduce this unit using teacher-led discussion. Use these pages to demonstrate the focus of the unit to children. You can use the characters to explore different ways of working too!

STRUCTURES AND REPRESENTATIONS

Shape fractions: Shape fractions are used in this unit to show fractions pictorially. Different shapes are shared equally into different fractions to reinforce children's conceptual understanding.

Fraction strip: Fraction strips are used in this unit to show fractions in a more formal pictorial way.

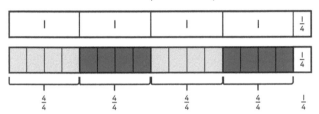

Number line: Number lines are used in this unit to represent number sequences.

Digit cards – fractions: Fraction cards are used to represent fractions in the problems and puzzles children will solve in this unit. They are also used to help children compare and order fractions.

KEY LANGUAGE

There is some key language that children will need to know as part of the learning in this unit.

→ equivalent
→ numerator, denominator
→ whole, fraction
→ improper fraction, mixed number
→ convert, sequence, order
→ greater than (>), less than (<), equal to (=)

PUPIL TEXTBOOK 5A PAGE 166

PUPIL TEXTBOOK 5A PAGE 167

Equivalent fractions

Learning focus

In this lesson, children will develop their understanding of equivalent fractions. They will learn to find and represent fractions equivalent to a unit fraction using manipulatives, pictures and abstract representations.

Before you teach ⏸

- How confident are children with the vocabulary related to fractions?
- Are children able to recognise simple fractions such as halves, quarters and fifths confidently?

NATIONAL CURRICULUM LINKS

Year 5 Number – fractions (including decimals and percentages)

Identify, name and write equivalent fractions of a given fraction, represented visually, including tenths and hundredths.

ASSESSING MASTERY

Children can recognise and find equivalent fractions for a given fraction. They can recognise and explain the links between equivalent fractions using the correct mathematical vocabulary and can confidently represent equivalent fractions in concrete, pictorial and abstract ways.

COMMON MISCONCEPTIONS

Children may add or subtract instead of multiplying or dividing when finding equivalent fractions; for example, stating that $\frac{1}{3}$ is equivalent to $\frac{3}{5}$ because $1 + 2 = 3$ and $3 + 2 = 5$. Ask:

- *Can you represent $\frac{1}{3}$ and $\frac{3}{5}$ on a fraction strip? Are the representations the same? Can the fractions be equivalent?*

STRENGTHENING UNDERSTANDING

Before beginning this lesson, and the unit, it may be beneficial for some children to recap their understanding of fractions. Offer opportunities to experience different representations of halves, quarters, thirds, fifths and tenths. Ask: *Are there any links between these fractions?*

GOING DEEPER

Children should explore the relationship between the numerator and denominator in each fraction, as well as the corresponding numerators and denominators in pairs of equivalent fractions.

KEY LANGUAGE

In lesson: equivalent fraction, numerator, denominator, fraction

Other language to be used by the teacher: multiply, divide, multiplication, division

STRUCTURES AND REPRESENTATIONS

Arrays, bar models, number lines

RESOURCES

Optional: blank paper (to fold), 2D shapes

 In the eTextbook of this lesson, you will find interactive links to a selection of teaching tools.

Quick recap 🔁

Show children a representation of a fraction on the board, for example, a bar model split into three equal parts with one part shaded. Ask them what fraction is shown. Ensure they understand what each number in the fraction represents. Repeat for other representations, then give children a unit fraction and ask them to represent it.

Discover

Equivalent fractions

WAYS OF WORKING Pair work

ASK

- Question ① a): *What number does the number line start from? What does it go up to?*
- Question ① a): *How many equal parts is the number line split into?*

IN FOCUS The focus of question ① a) is for children to understand that there are lots of ways of writing the fraction that the arrow is pointing to. Although one-third is the simplest, any equivalent fraction has the same value and so is in the same place.

PRACTICAL TIPS Provide children with their own copy of the number line that they can split up into more parts to support them in finding more equivalent fractions in question ① b).

ANSWERS

Question ① a): $\frac{1}{3}$

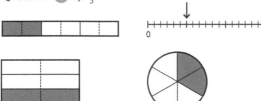

Question ① b): Several possible answers, e.g.
$$\frac{1}{3} = \frac{2}{6}, \frac{5}{15}, \frac{10}{30}, \frac{11}{33}.$$

Discover

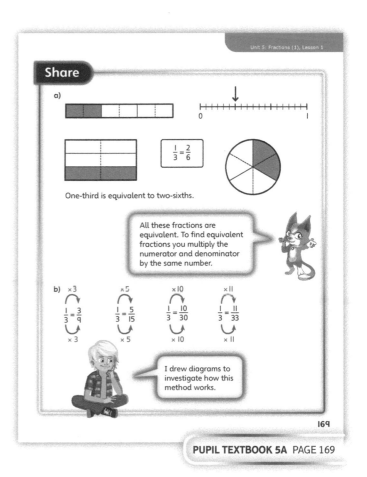

① a) Write one-third as a fraction.

Draw diagrams to show one-third in different ways.

Alter your diagrams to show that two-sixths is the same as one-third.

b) Write more fractions that are equivalent to one-third.

168

PUPIL TEXTBOOK 5A PAGE 168

Share

WAYS OF WORKING Whole class teacher led

ASK

- Question ① a): *What does the 1 represent? What does the 3 represent?*
- Question ① a): *What other diagrams can you draw? How can you split your diagram so it shows sixths, not thirds?*
- Question ① b): *How many different answers are there? How do you know?*

IN FOCUS The focus here is for children to understand different representations of one-third. They use these representations to find other equivalent fractions in question ① b).

Share

a)

One-third is equivalent to two-sixths.

$\frac{1}{3} = \frac{2}{6}$

All these fractions are equivalent. To find equivalent fractions you multiply the numerator and denominator by the same number.

b)
$$\overset{\times 3}{\frac{1}{3} = \frac{3}{9}} \quad \overset{\times 5}{\frac{1}{3} = \frac{5}{15}} \quad \overset{\times 10}{\frac{1}{3} = \frac{10}{30}} \quad \overset{\times 11}{\frac{1}{3} = \frac{11}{33}}$$
$$\underset{\times 3}{} \quad \underset{\times 5}{} \quad \underset{\times 10}{} \quad \underset{\times 11}{}$$

I drew diagrams to investigate how this method works.

169

PUPIL TEXTBOOK 5A PAGE 169

Think together

Whole class teacher led (I do, We do, You do)

ASK

• Question ❶: *What is the same about these fractions? What is different?*
• Question ❷: *What has the numerator/denominator been multiplied by? What do you need to multiply the denominator/numerator by?*

IN FOCUS The focus of question ❶ is for children to use diagrams and other representations to understand the abstract method for finding equivalent fractions. They should notice the multiplicative relationship in pairs of equivalent fractions.

STRENGTHEN Provide children with strips of paper that they can fold to represent fractions and then split further to support them in finding equivalent fractions.

DEEPEN In question ❸ a), ask children to prove, using diagrams or other representations, that Aki's method does not work. In question ❸ b), ask: *What do you notice about the relationship between the numerator and the denominator in each part? Does this always happen?*

ASSESSMENT CHECKPOINT Use question ❶ and ❷ to assess children's understanding of fractions equivalent to a unit fraction.

ANSWERS

Question ❶: $\frac{1}{2} = \frac{2}{4}$ $\frac{1}{2} = \frac{3}{6}$ $\frac{1}{2} = \frac{5}{10}$

Question ❷: $\frac{1}{5} = \frac{4}{20}$ $\frac{1}{5} = \frac{7}{35}$ $\frac{1}{5} = \frac{16}{80}$

Question ❸ a): No, as this method does not give equivalent fractions.

Question ❸ b): $\frac{1}{4} = \frac{2}{8} = \frac{3}{12} = \frac{4}{16} = \frac{5}{20} = \frac{6}{24}$

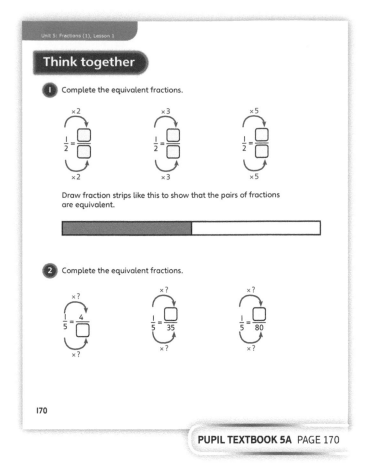

PUPIL TEXTBOOK 5A PAGE 170

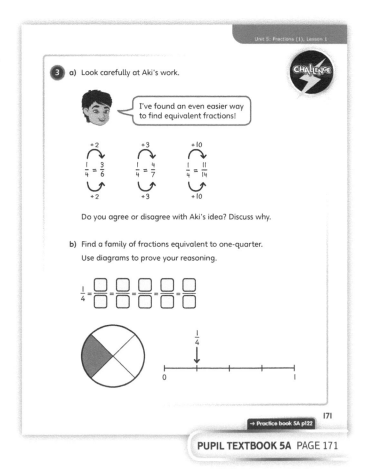

PUPIL TEXTBOOK 5A PAGE 171

Practice

WAYS OF WORKING Independent thinking

IN FOCUS The focus of this section is for children to develop their understanding of the abstract method for finding equivalent fractions using diagrams and other representations. Question ❶ uses pictorial representations before the section moves onto abstract representations in question ❷.

STRENGTHEN Provide children with strips of paper that they can fold to represent fractions and then alter to find equivalent fractions.

DEEPEN In question ❻, encourage children to explore the relationship between the numerator and the denominator in equivalent fractions, for example, noticing that in any fraction equivalent to one-hundredth, the denominator is 100 times the numerator. Ask them to investigate this for other fractions.

ASSESSMENT CHECKPOINT Use questions ❶ to ❹ to assess whether children can confidently find fractions equivalent to a unit fraction. Use question ❺ to assess whether children can spot and explain any common misconceptions.

ANSWERS Answers for the **Practice** part of the lesson can be found in the *Power Maths* online subscription.

Reflect

WAYS OF WORKING Independent thinking

IN FOCUS This activity is designed to assess children's understanding of how to find equivalent fractions.

ASSESSMENT CHECKPOINT Children should be able to describe a systematic approach to listing equivalent fractions for $\frac{1}{9}$.

ANSWERS Answers for the **Reflect** part of the lesson can be found in the *Power Maths* online subscription.

After the lesson ⏸

- Can children give an example of a fraction that is equivalent to a unit fraction?
- Can children find missing numerators in equivalent fractions?
- Can children find missing denominators in equivalent fractions?

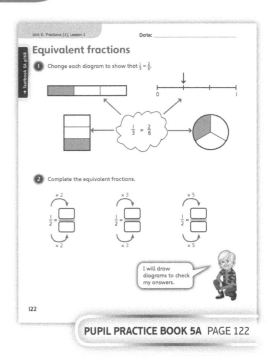

PUPIL PRACTICE BOOK 5A PAGE 122

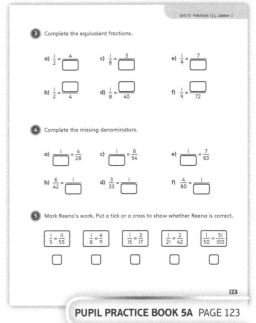

PUPIL PRACTICE BOOK 5A PAGE 123

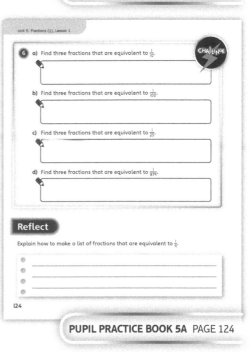

PUPIL PRACTICE BOOK 5A PAGE 124

Equivalent fractions – unit and non-unit fractions

Learning focus

In this lesson, children will further develop their understanding of equivalent fractions. They will learn to find and represent equivalent fractions using manipulatives, pictures and abstract representations.

Before you teach

- Can children represent and recognise simple fractions?
- Can children find a fraction equivalent to a unit fraction?
- Can children find missing numerators and denominators in pairs of equivalent fractions where one is a unit fraction?

NATIONAL CURRICULUM LINKS

Year 5 Number – fractions (including decimals and percentages)

Identify, name and write equivalent fractions of a given fraction, represented visually, including tenths and hundredths.

ASSESSING MASTERY

Children can recognise and find equivalent fractions for a given fraction. They can recognise and explain the links between equivalent fractions using the correct mathematical vocabulary and can confidently represent equivalent fractions in concrete, pictorial and abstract ways.

COMMON MISCONCEPTIONS

Children may add or subtract instead of multiplying or dividing when finding equivalent fractions; for example, stating that $\frac{1}{3}$ is equivalent to $\frac{3}{5}$ because $1 + 2 = 3$ and $3 + 2 = 5$. Ask:

- *Can you represent $\frac{1}{3}$ and $\frac{3}{5}$ on a fraction strip? Are the representations the same? Can the fractions be equivalent?*

STRENGTHENING UNDERSTANDING

If children find it difficult to find fractions equivalent to a non-unit fraction, link back to their learning on fractions equivalent to a unit fraction to support them. For example, if $\frac{1}{5}$ is equivalent to $\frac{2}{10}$ then $\frac{3}{5}$ must be equivalent to $\frac{6}{10}$.

GOING DEEPER

Children could investigate the statement: Is it always, sometimes or never true that any fraction can be written in lots of ways? Encourage children to create their own 'always, sometimes, never' statements about equivalent fractions to challenge a partner.

KEY LANGUAGE

In lesson: equivalent, numerator, denominator, fraction

Other language to be used by the teacher: multiply, divide, multiplication, division

STRUCTURES AND REPRESENTATIONS

Arrays, bar models, number lines

RESOURCES

Optional: blank paper (to fold), modelling clay, 2D shapes

 In the eTextbook of this lesson, you will find interactive links to a selection of teaching tools.

Quick recap 🔎

Show children a unit fraction on the board and ask them to show an equivalent fraction on their whiteboard. Ensure they can explain how they found the equivalent fraction using diagrams or other representations. Repeat for other unit fractions.

Discover

Unit 5: Fractions (1), Lesson 2

WAYS OF WORKING Pair work

ASK

- Question ❶ a): *What is the whole?*
- Question ❶ a): *What will $\frac{1}{3}$ look like?*
- Question ❶ b): *How can you use $\frac{1}{3}$ to find $\frac{2}{3}$?*
- Question ❶ b): *How else can you write $\frac{2}{3}$?*

IN FOCUS The focus of this activity is for children to use their understanding of finding fraction equivalents to unit fractions to find fraction equivalents to non-unit fractions.

PRACTICAL TIPS Provide children with strips of paper or modelling clay to represent the baguette, so that they can split it up to support their understanding.

ANSWERS

Question ❶ a):

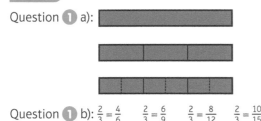

Question ❶ b): $\frac{2}{3} = \frac{4}{6}$ $\frac{2}{3} = \frac{6}{9}$ $\frac{2}{3} = \frac{8}{12}$ $\frac{2}{3} = \frac{10}{15}$

Equivalent fractions – unit and non-unit fractions

Discover

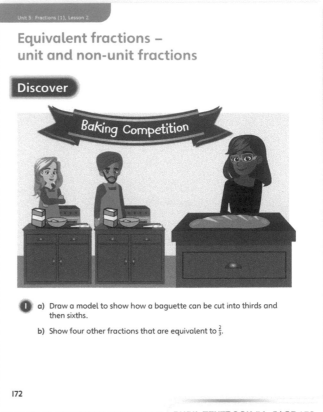

❶ a) Draw a model to show how a baguette can be cut into thirds and then sixths.

b) Show four other fractions that are equivalent to $\frac{2}{3}$.

172

Share

WAYS OF WORKING Whole class teacher led

ASK

- Question ❶ a): *If you cut one piece into two parts, what do you need to do to the other pieces so the parts are all equal?*
- Question ❶ b): *What else can you split each part into? Are there any other options?*

IN FOCUS Here children build on their understanding from the previous lesson to use bar models and other representations to find fractions equivalent to a non-unit fraction.

Share

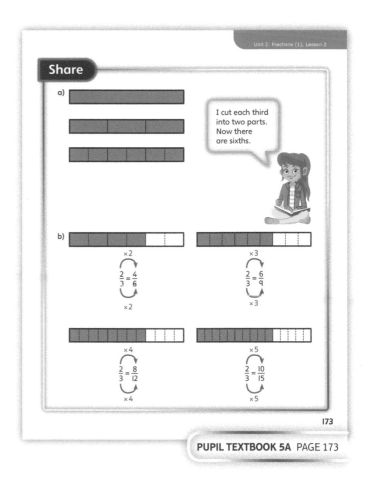

173

Think together

Unit 5: Fractions (1), Lesson 2

Think together

WAYS OF WORKING Whole class teacher led (I do, We do, You do)

ASK

- Question **1**: *How does the diagram show two-fifths? What has each fifth been split into? How many parts are there in total? How many are shaded?*
- Question **2**: *What can you split each interval into? What equivalent fractions does this tell you?*

IN FOCUS The focus of questions **1** and **2** is for children to use diagrams and other representations to find fractions equivalent to a non-unit fraction. They should build on their understanding from the previous lesson to support their understanding with this.

STRENGTHEN Provide children with strips of paper that they can fold to represent the fractions. This will support them in finding equivalent fractions.

DEEPEN In question **3**, ask children to explore Reena's method of finding equivalent fractions. Ask: *Does it work for all fractions? Or does it only work for a certain type of fraction?*

ASSESSMENT CHECKPOINT Use questions **1** and **2** to assess whether children can confidently find fractions equivalent to a non-unit fraction.

ANSWERS

Question **1** a): $\frac{2}{5} = \frac{6}{15}$

Question **1** b): $\frac{2}{5} = \frac{8}{20}$

Question **2**: Answers may vary, e.g.

$$\frac{2}{7} = \frac{4}{14} \qquad \frac{5}{7} = \frac{10}{14} \qquad \frac{6}{7} = \frac{12}{14}$$

Question **3**: Answers may vary, e.g.

$$\frac{1}{10} = \frac{2}{20} = \frac{3}{30} = \frac{7}{70}$$

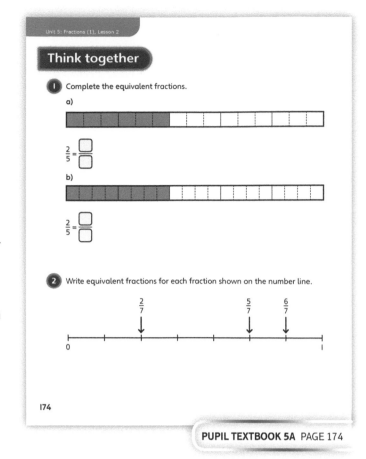

PUPIL TEXTBOOK 5A PAGE 174

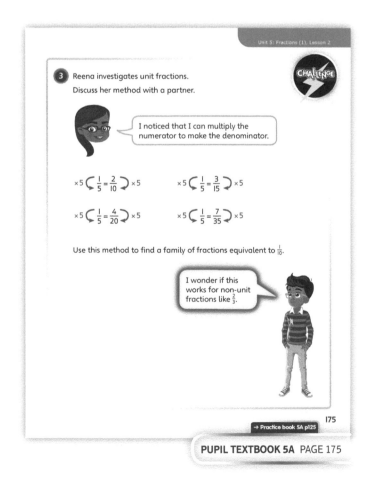

PUPIL TEXTBOOK 5A PAGE 175

Practice

WAYS OF WORKING Independent thinking

IN FOCUS These questions provide an opportunity for children to consolidate what they know about finding fractions equivalent to non-unit fractions. In questions ❶ and ❷, they use diagrams to support their understanding of the abstract method. Questions ❸ and ❹ move towards more abstract representations.

STRENGTHEN Provide children with strips of paper to fold and represent fractions that they can then use to find equivalent fractions.

DEEPEN In question ❺, encourage children to explore any patterns they notice in their answers. In question ❻, encourage children to explain their reasoning, including the use of inverse operations.

ASSESSMENT CHECKPOINT Use questions ❶ to ❹ to assess whether children can confidently find fractions equivalent to a non-unit fraction. Use question ❺ to assess whether children can find multiple equivalent fractions for a given fraction. Use question ❻ to assess children's understanding of the abstract method and whether they can use it to find missing numerators and denominators.

ANSWERS Answers for the **Practice** part of the lesson can be found in the *Power Maths* online subscription.

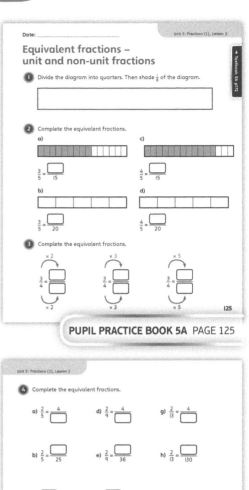

PUPIL PRACTICE BOOK 5A PAGE 125

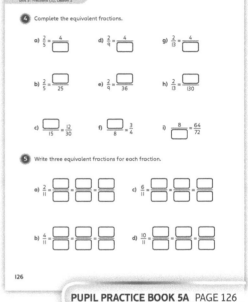

PUPIL PRACTICE BOOK 5A PAGE 126

Reflect

WAYS OF WORKING Independent thinking

IN FOCUS The focus of this activity is to assess children's understanding of the abstract method for finding equivalent fractions.

ASSESSMENT CHECKPOINT Children should use the multiplicative relationship between pairs of equivalent fractions to find further equivalent fractions. They should explain in their own words how they have done this.

ANSWERS Answers for the **Reflect** part of the lesson can be found in the *Power Maths* online subscription.

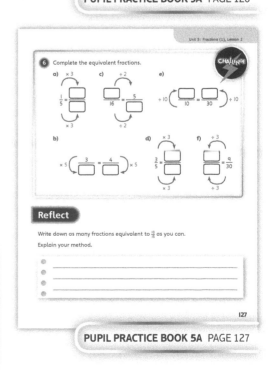

PUPIL PRACTICE BOOK 5A PAGE 127

After the lesson ⏸

- Can children find fractions equivalent to a unit fraction?
- Can children find fractions equivalent to a non-unit fraction?
- Can children find missing numerators/denominators in pairs of equivalent fractions?

Equivalent fractions – families of equivalent fractions

Learning focus

In this lesson, children further develop their understanding of equivalent fractions. They use their learning developed in the previous two lessons to find families of equivalent fractions for a given fraction.

Before you teach

- Can children recognise if two fractions are equivalent?
- Can children find fractions equivalent to a unit fraction?
- Can children find fractions equivalent to a non-unit fraction?

NATIONAL CURRICULUM LINKS

Year 5 Number – fractions (including decimals and percentages)

Identify, name and write equivalent fractions of a given fraction, represented visually, including tenths and hundredths.

ASSESSING MASTERY

Children can recognise and find families of equivalent fractions for a given fraction. They can recognise and explain the links between equivalent fractions using the correct mathematical vocabulary and can confidently represent equivalent fractions in concrete, pictorial and abstract ways.

COMMON MISCONCEPTIONS

Children may forget that you must multiply or divide both the top and bottom of a fraction by the same number in order to find an equivalent fraction, for example, stating that $\frac{30}{40} = \frac{10}{20}$. Ask:

- Can you represent both $\frac{30}{40}$ and $\frac{10}{20}$ on a fraction strip? Are the representations the same? Can the fractions be equivalent?

STRENGTHENING UNDERSTANDING

Where children struggle with finding families of equivalent fractions, ensure they break the question down to just identify one fraction at a time using their learning from the previous two lessons. Models and representations can continue to be used to provide support where needed.

GOING DEEPER

Children could investigate equivalent fractions further to develop their understanding and recognise that there are an infinite number of fractions that are equivalent to any given fraction.

KEY LANGUAGE

In lesson: equivalent fraction, numerator, denominator, fraction

Other language to be used by the teacher: multiply, divide, multiplication, division

STRUCTURES AND REPRESENTATIONS

Arrays, bar models, number lines

RESOURCES

Optional: blank paper (to fold), sheets of paper with fractions shaded, scissors

 In the eTextbook of this lesson, you will find interactive links to a selection of teaching tools.

Quick recap 🔎

Play equivalent fractions bingo. List some fractions on the board, for example $\frac{2}{8}, \frac{5}{25}, \frac{7}{14}, \frac{3}{30}, \frac{8}{10}, \frac{6}{9}$, and ask children to choose four of them. On the board, present children with a simplified fraction, such as $\frac{1}{4}$. If they have a fraction that is equivalent to this, they can cross it out. The first child to cross out all of their fractions wins.

Discover

Pair work

ASK

- Question **1** a) and **1** b): *What is the same and what is different about each of the flags? What other ways could you record the fractions shown? What fraction of each flag is shaded?*
- Question **1** b): *If you drew a vertical line through the middle of the flag that is the odd one out, what fraction would be coloured in now?*

IN FOCUS Question **1** a) is important as it offers the opportunity to begin discussing what is the same and what is different about the fractions shown in the picture. Children could be given photocopies of the flags and be encouraged to cut up the parts and overlap them to highlight the equivalencies, building on their previous learning. Question **1** b) encourages children to use their learning from the previous two lessons to explore other equivalent fractions.

PRACTICAL TIPS Children could be given pieces of paper to replicate the flags, that are shaded into different fractions. Challenge them to colour in and create their own flag where exactly one-quarter is shaded. Ask: *On how many of the different pieces of paper is this possible?* Encourage children to consider what links the fractions that are shown by the pieces of paper, and how they are different to those that do not work.

ANSWERS

Question **1** a): Fractions $\frac{1}{2}$, $\frac{2}{4}$ and $\frac{4}{8}$ are equivalent.

Question **1** b): The fraction on Luis's flag is the odd one out. The odd fraction out is $\frac{2}{3}$. Equivalent fractions to $\frac{2}{3}$ given in **Share** are $\frac{14}{21}$ and $\frac{24}{36}$ but pupils may give examples such as $\frac{4}{6}, \frac{6}{9}, \frac{8}{12}$, etc.

Share

Whole class teacher led

ASK

- Question **1** a): *Is there a pattern in the equivalent fractions? What are the next two equivalent fractions? How are the pictures and written fractions linked? Which part of the picture represents the numerator and which part represents the denominator?*
- Question **1** b): *Why is $\frac{2}{3}$ not equivalent to the other fractions? How did you find equivalent fractions for $\frac{2}{3}$?*

IN FOCUS Children should be given the opportunity to create the fractions they are discussing in questions **1** a) and **1** b) using pictures or concrete resources, such as folded paper, so they can link these to the abstract fractions. When explaining how each fraction is linked through multiplication and division, encourage children to use diagrams or other representations to show each relationship.

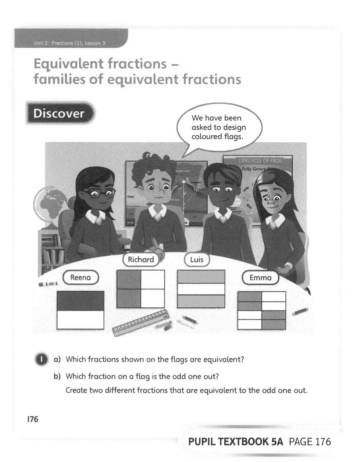

PUPIL TEXTBOOK 5A PAGE 176

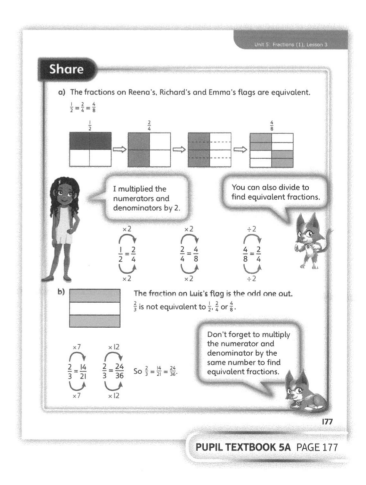

PUPIL TEXTBOOK 5A PAGE 177

Think together

Whole class teacher led (I do, We do, You do)

ASK

- Question ❶: *How many parts are there altogether? How many are shaded? What fraction of the diagram is shaded? How can you find equivalent fractions?*
- Question ❶: *Which fraction is the odd one out? How do you know?*
- Question ❷: *Can you use a diagram to prove that one of these statements is incorrect?*

IN FOCUS In question ❶, children are shown pictorial representations of four fractions, one of which does not belong in the family of equivalent fractions. They should describe what fraction each image shows and explain how they worked out which three were equivalent.
In question ❷, children should be encouraged to explain the mistake that has been made as well as identifying the incorrect statement.

STRENGTHEN Provide children with strips of paper to fold, which they can use to represent fractions. This will support their understanding and help them to find equivalent fractions.

DEEPEN In question ❸, challenge children to find other pairs of equivalent fractions that might not necessarily appear equivalent at first glance, such as $\frac{8}{10}$ and $\frac{12}{15}$.

ASSESSMENT CHECKPOINT Use question ❶ to assess whether children can confidently find families of equivalent fractions and use this to identify the odd one out. Use question ❷ to assess whether children can spot and explain common errors in finding equivalent fractions.

ANSWERS

Question ❶: The fractions are: a) $\frac{4}{10}$, b) $\frac{6}{15}$, c) $\frac{4}{15}$, d) $\frac{2}{5}$
Equivalent fractions are a), b) and d).
The odd one out is c) $\frac{4}{15}$.

Question ❷: The incorrect statement is b).

Question ❸ a): Yes, the statement is correct.

Question ❸ b): $\frac{10}{12} = \frac{20}{24} = \frac{15}{18}$
$\frac{12}{27} = \frac{8}{18}$
$\frac{10}{14} = \frac{20}{28} = \frac{15}{21}$

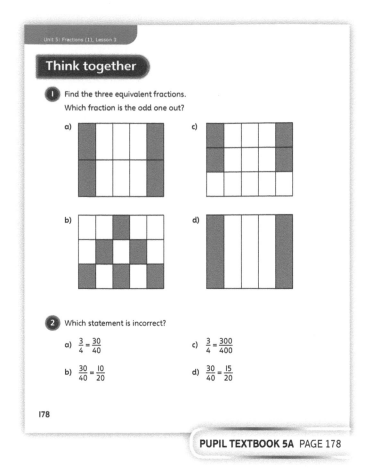

Think together

❶ Find the three equivalent fractions. Which fraction is the odd one out?

a)   c)

b)   d)

❷ Which statement is incorrect?

a) $\frac{3}{4} = \frac{30}{40}$ c) $\frac{3}{4} = \frac{300}{400}$

b) $\frac{30}{40} = \frac{10}{20}$ d) $\frac{30}{40} = \frac{15}{20}$

178

PUPIL TEXTBOOK 5A PAGE 178

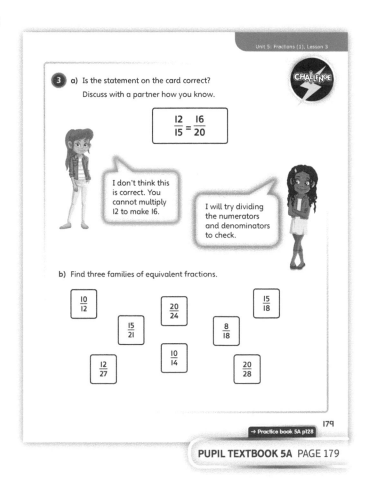

❸ a) Is the statement on the card correct? Discuss with a partner how you know.

CHALLENGE

$$\frac{12}{15} = \frac{16}{20}$$

I don't think this is correct. You cannot multiply 12 to make 16.

I will try dividing the numerators and denominators to check.

b) Find three families of equivalent fractions.

$\frac{10}{12}$ $\frac{20}{24}$ $\frac{15}{18}$

$\frac{15}{21}$ $\frac{8}{18}$

$\frac{12}{27}$ $\frac{10}{14}$ $\frac{20}{28}$

179

→ Practice book 5A p128

PUPIL TEXTBOOK 5A PAGE 179

Practice

WAYS OF WORKING Independent thinking

IN FOCUS In question ① b), refer back to how the different ways of dividing each diagram demonstrates equivalent fractions. Encourage children to use this understanding to find ways of expanding the thirds shown. In question ① c), ask children to consider the number of squares there are in total and to use their knowledge of equivalent fractions to work out how many they need to shade in. In questions ② and ③, children should consider and explain the relationship between equivalent fractions to identify the missing numbers.

STRENGTHEN Provide children with diagrams or other representations that they can use to represent the fractions and support their understanding.

DEEPEN In question ⑤, encourage children to explain each step of their thinking process. This will improve their reasoning skills and ensure they are not using a trial and error method to answer the question. Ask: *Can you create a similar question for a partner?*

ASSESSMENT CHECKPOINT Use questions ① to ③ to assess children's understanding of finding families of equivalent fractions for a given fraction. Use question ④ to check whether children can identify and explain common mistakes when finding equivalent fractions.

ANSWERS Answers for the **Practice** part of the lesson can be found in the *Power Maths* online subscription.

Reflect

WAYS OF WORKING Independent thinking

IN FOCUS The focus of this question is to assess children's understanding of finding equivalent fractions.

ASSESSMENT CHECKPOINT Children should be able to explain their methods and provide examples of questions for which they would use each method.

ANSWERS Answers for the **Reflect** part of the lesson can be found in the *Power Maths* online subscription.

After the lesson ⏸

- Can children find pairs of equivalent fractions?
- Can children find families of equivalent fractions?
- Can children find missing numbers in families of equivalent fractions?

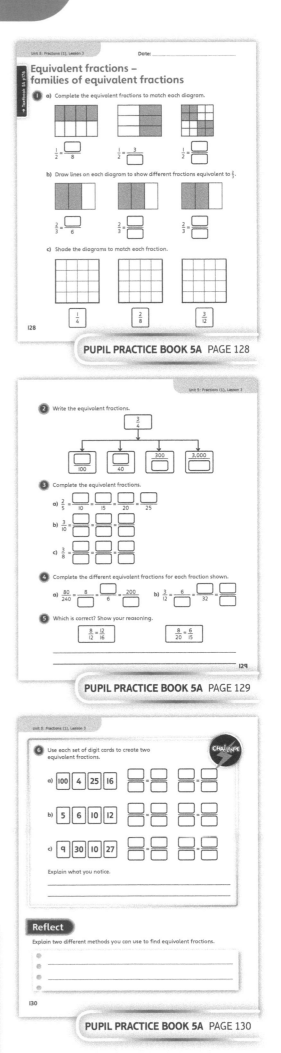

PUPIL PRACTICE BOOK 5A PAGE 128

PUPIL PRACTICE BOOK 5A PAGE 129

PUPIL PRACTICE BOOK 5A PAGE 130

Improper fractions to mixed numbers

Learning focus

In this lesson, children will use their understanding of fractions to recognise mixed number fractions. They will independently convert improper fractions to mixed numbers.

Before you teach

- How confident are children at adding fractions together?
- If children find adding fractions challenging, how will this impact the lesson and what will you do to help support their understanding?

NATIONAL CURRICULUM LINKS

Year 5 Number – fractions (including decimals and percentages)

Recognise mixed numbers and improper fractions and convert from one form to the other and write mathematical statements > 1 as a mixed number [for example, $\frac{2}{5} + \frac{4}{5} = \frac{6}{5} = 1\frac{1}{5}$].

ASSESSING MASTERY

Children can use their understanding of fractions, and pictorial and abstract representations, to recognise mixed number fractions and fluently convert from improper fractions to mixed numbers, explaining the links between both using the correct mathematical vocabulary.

COMMON MISCONCEPTIONS

Children may be inclined to record improper fractions as proper fractions, for example, recording $\frac{11}{3}$ as $\frac{3}{11}$. Show children pictorial or concrete representations of the two fractions and ask:

- *What is the same and what is different about these two fractions?*
- *Which of these fractions could be represented as $3\frac{2}{3}$?*

STRENGTHENING UNDERSTANDING

For children likely to find this lesson challenging, offer the opportunity to add unit fractions with the same denominator before the lesson begins. This could be done through folding paper into quarters. Ask children to colour one-quarter in. Ask: *How many more quarters would you need to add to this to make one whole? How many quarters would you need to add together to make two? Can you prove it?*

GOING DEEPER

Ask: *Is it always, sometimes or never true that a simplified fraction equals a different mixed number than the one it equalled before it was simplified? What about if it has been expanded? Can you prove your ideas?*

KEY LANGUAGE

In lesson: improper fraction, mixed number, fraction, numerator, denominator, convert, simplify

Other language to be used by the teacher: equivalent to

STRUCTURES AND REPRESENTATIONS

Bar model, number line

RESOURCES

Optional: paper (for folding), printed pictures of pots of paint with different unit fractions listed on them

 In the eTextbook of this lesson, you will find interactive links to a selection of teaching tools.

Quick recap

Recap fractions that are equivalent to one whole. Show children different representations, such as a bar model split into 5 equal parts with 5 parts shaded. Ask: *What fraction is shaded? What is that the same as?* Repeat for other fractions and ask children what they notice. They should know that when the numerator is equal to the denominator, then the fraction is equivalent to one whole.

Discover

Pair work

ASK

- Question ❶ a): *How many halves make one whole? What other way could you write this number? What other fractions could be used to make the same number?*

IN FOCUS Question ❶ a) allows children to relate their understanding of fractions to a context they should recognise. Begin by discussing how many half litres there are. Once children have agreed the number of half litres, move onto question ❶ b). Discuss how many half litres will be needed to make a litre. Once children have agreed that two-halves will make the whole, they will then be able to use the picture to say how many wholes there are and what fraction is still left.

PRACTICAL TIPS Children could be given printed pictures of pots of paint with $\frac{1}{2}$ litre labels on them, to replicate the scenario. Ensure children use the same unit fraction to create their mixed number fraction.

ANSWERS

Question ❶ a): Sofia has $\frac{5}{2}$ litres of paint.

Question ❶ b): Sofia has $2\frac{1}{2}$ litres of paint.

Share

WAYS OF WORKING Whole class teacher led

ASK

- Question ❶ b): *How do the number line and the bar model help demonstrate how the fraction converts to a mixed number? Which way of recording is easier to read? Why?*
- Questions ❶ a) and b): *What do you notice about the fraction and the whole number it represents? What is the same and what is different?*

IN FOCUS At this point in the lesson, it is important to ensure children can recognise both improper and mixed number fractions and can accurately describe the differences between them. This understanding could be drawn out in the discussion about how the two ways of representing the same number are similar and different.

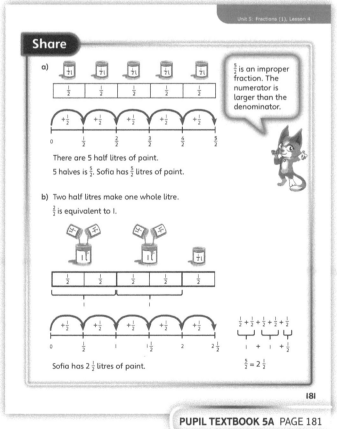

Think together

Whole class teacher led (I do, We do, You do)

ASK

- Question ❶: *How does the bar model help to represent the question? How can this fraction be written?*
- Question ❷: *Is there a way of finding how many wholes there are in an improper fraction?*
- Question ❸: *What do you notice about how the mixed numbers change? Is there a pattern?*

IN FOCUS Question ❶ supports children's understanding through the use of the bar model to demonstrate what number of thirds will create one whole. Children should be encouraged to explain the links between the pictorial and written representations of the numbers to help secure their conceptual understanding.

For question ❷, children should start to link their understanding of multiples to help them find how many whole numbers are within the improper fraction.

STRENGTHEN If children find converting the improper fractions in questions ❷ and ❸ challenging, ask: *Can you draw a bar model that represents each improper fraction? How many of each fraction makes one whole? How many wholes can be made with the improper fraction? What fraction is left over?*

DEEPEN Question ❸ should be used to deepen children's fluency with fractions by encouraging them to identify where they can use their ability to simplify fractions to make conversions easier, for example, simplifying $\frac{20}{6}$ to $\frac{10}{3}$ before converting to $3\frac{1}{3}$. Ask: *Could you simplify any of these fractions? Would that help you convert them to a mixed number more easily? Do you get the same mixed number after simplifying the fraction compared to the one you would have got before simplifying it?*

ASSESSMENT CHECKPOINT Question ❶ assesses children's ability to link their understanding of the different representations of fractions to help convert between improper fractions and mixed numbers.

Question ❸ assesses children's recognition of how changing the numerator and denominator in an improper fraction can have an effect on the mixed number. Look for children who recognise that if the numerator increases, then the mixed number also increases, but that if the denominator increases, the mixed number will decrease. Children should be able to explain why this is, if necessary using pictorial representations such as the bar model.

ANSWERS

Question ❶: $\frac{10}{3} = 1 + 1 + 1 + \frac{1}{3} = 3\frac{1}{3}$

Question ❷ a): $1\frac{1}{4}$ b): $3\frac{1}{4}$ c): $3\frac{3}{4}$ d): $10\frac{1}{4}$

Question ❸ a): $\frac{17}{6} = 2\frac{5}{6}$, $\frac{18}{6} = 3$, $\frac{19}{6} = 3\frac{1}{6}$, $\frac{20}{6} = 3\frac{2}{6} = 3\frac{1}{3}$, $\frac{21}{6} = 3\frac{3}{6} = 3\frac{1}{2}$, $\frac{22}{6} = 3\frac{4}{6} = 3\frac{2}{3}$, $\frac{23}{6} = 3\frac{5}{6}$

Question ❸ b): $\frac{24}{4} = 6$, $\frac{24}{5} = 4\frac{4}{5}$, $\frac{24}{6} = 4$, $\frac{24}{7} = 3\frac{3}{7}$, $\frac{24}{8} = 3$, $\frac{24}{9} = 2\frac{6}{9} = 2\frac{2}{3}$, $\frac{24}{10} = 2\frac{4}{10} = 2\frac{2}{5}$

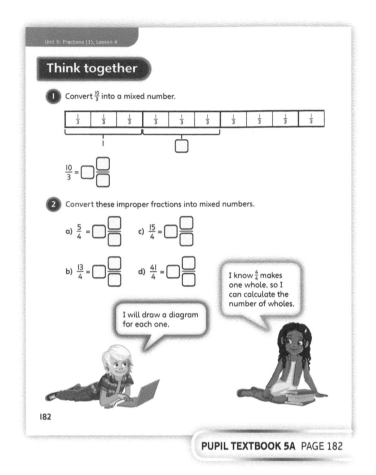

Think together

❶ Convert $\frac{10}{3}$ into a mixed number.

$\frac{10}{3} = \square\frac{\square}{\square}$

❷ Convert these improper fractions into mixed numbers.

a) $\frac{5}{4} = \square\frac{\square}{\square}$ c) $\frac{15}{4} = \square\frac{\square}{\square}$

b) $\frac{13}{4} = \square\frac{\square}{\square}$ d) $\frac{41}{4} = \square\frac{\square}{\square}$

I will draw a diagram for each one.

I know $\frac{4}{4}$ makes one whole, so I can calculate the number of wholes.

182

PUPIL TEXTBOOK 5A PAGE 182

❸ Complete each set.
What stays the same and what changes?
Explain the patterns of answers.

a) $\frac{17}{6} = \square\frac{\square}{\square}$ b) $\frac{24}{4} = \square\frac{\square}{\square}$

$\frac{18}{6} = \square\frac{\square}{\square}$ $\frac{24}{5} = \square\frac{\square}{\square}$

$\frac{19}{6} = \square\frac{\square}{\square}$ $\frac{24}{6} = \square\frac{\square}{\square}$

$\frac{20}{6} = \square\frac{\square}{\square}$ $\frac{24}{7} = \square\frac{\square}{\square}$

$\frac{21}{6} = \square\frac{\square}{\square}$ $\frac{24}{8} = \square\frac{\square}{\square}$

$\frac{22}{6} = \square\frac{\square}{\square}$ $\frac{24}{9} = \square\frac{\square}{\square}$

$\frac{23}{6} = \square\frac{\square}{\square}$ $\frac{24}{10} = \square\frac{\square}{\square}$

I wonder if some answers can be written in different ways.

I think I can simplify some of the fractions.

183

→ Practice book 5A p131

PUPIL TEXTBOOK 5A PAGE 183

Practice

WAYS OF WORKING Independent thinking

IN FOCUS Question ❶ develops children's independent understanding of the link between the pictorial representation provided by the bar model and converting improper fractions to mixed numbers. Children should be encouraged to complete the diagrams where necessary and use this to support their reasoning.

Question ❷ presents the fractions using different shapes in different orientations to develop children's fluency.

Question ❸ is important as it develops children's understanding of the effect that changing the numerator or denominator can have on a mixed number. Link this to children's earlier discussion and encourage them to predict what will happen to the mixed numbers in each column.

STRENGTHEN Encourage children to find more than one way to record the fractions. Ask: *How could you use your learning from the previous lesson to help you here? Could you expand or simplify the fraction? How would that help?*

DEEPEN For question ❺, deepen children's reasoning about the concepts covered in the question by asking: *Can you show your reasoning using a pictorial representation? Explain how your picture supports your ideas.*

ASSESSMENT CHECKPOINT Questions ❶ and ❷ assess children's ability to recognise pictorial representations of improper fractions and to use them to convert the fractions to a mixed number. Look for children explaining how the pictures support their conversions.

Question ❸ assesses children's recognition of how changing the numerator and denominator in an improper fraction affects the mixed number. Look for children who recognise that if the numerator increases, then the mixed number increases, but if the denominator increases, the mixed number actually decreases.

ANSWERS Answers for the **Practice** part of the lesson can be found in the *Power Maths* online subscription.

Reflect

WAYS OF WORKING Independent thinking

IN FOCUS When solving this question, children should be encouraged to share their reasoning about how they have converted the fractions.

ASSESSMENT CHECKPOINT Look for children accurately converting each improper fraction to a mixed number. Children may do this by individually converting each fraction; they may recognise the multiples and use this to help them; they may convert either the lowest fraction and add one more third each time to their mixed number; or they may start at the largest and subtract one-third from their mixed number.

ANSWERS Answers for the **Reflect** part of the lesson can be found in the *Power Maths* online subscription.

After the lesson ⏸

- Did children recognise how their understanding of multiples could be useful in this mathematical context?
- Were children fluently simplifying fractions when converting to mixed numbers? How will you continue to develop this method?

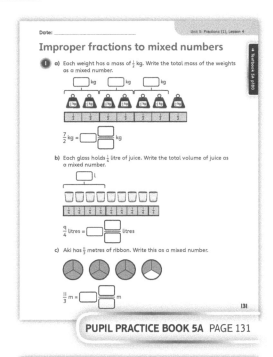

PUPIL PRACTICE BOOK 5A PAGE 131

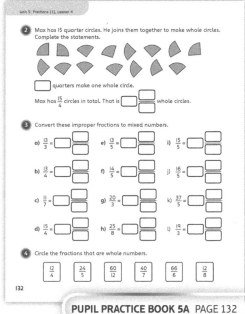

PUPIL PRACTICE BOOK 5A PAGE 132

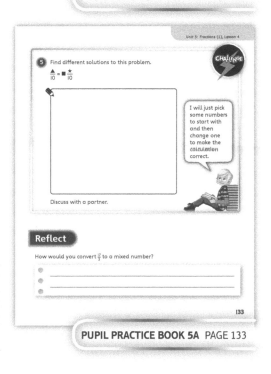

PUPIL PRACTICE BOOK 5A PAGE 133

Mixed numbers to improper fractions

Learning focus

In this lesson, children will use their understanding of fractions to recognise mixed numbers. They will independently convert mixed numbers to improper fractions.

Before you teach

- Are there any misconceptions regarding converting from improper fractions to mixed numbers that will need addressing in this lesson?
- What real-life contexts could you use to engage children in the learning during this lesson?

NATIONAL CURRICULUM LINKS

Year 5 Number – fractions (including decimals and percentages)

Recognise mixed numbers and improper fractions and convert from one form to the other and write mathematical statements > 1 as a mixed number [for example, $\frac{2}{5} + \frac{4}{5} = \frac{6}{5} = 1\frac{1}{5}$].

ASSESSING MASTERY

Children can use their understanding of fractions, and pictorial and abstract representations, to recognise mixed numbers and fluently convert from mixed numbers to improper fractions, explaining the links between both using the correct mathematical vocabulary.

COMMON MISCONCEPTIONS

When converting a mixed number to an improper fraction, children may simply add the whole number to the numerator of the mixed number, rather than adding what the whole number is worth, for example, they may convert $4\frac{5}{6}$ into $\frac{9}{6}$, rather than $\frac{29}{6}$. Ask:

- *What fraction does each whole represent? How do you know?*

STRENGTHENING UNDERSTANDING

For children who find the concept of converting a mixed number to an improper fraction challenging, it would be beneficial to offer them shapes cut up into equal fractions, for example, a hexagon cut up into equal sixths. Ask: *How many sixths do you need to make one whole hexagon? Finish this statement: One whole equals ☐ sixths. How many sixths would you need for three hexagons?*

Show children a mixed number represented as a whole and parts of shapes. Ask children to investigate how many of each fraction there are in the mixed number.

GOING DEEPER

Children could be encouraged to write their own word problems that require conversion from a mixed number to an improper fraction. This could be deepened further by asking children to write a problem that requires a fraction to be simplified or expanded first.

KEY LANGUAGE

In lesson: improper fraction, mixed number, numerator, denominator, equivalent fraction

Other language to be used by the teacher: convert

STRUCTURES AND REPRESENTATIONS

Fraction strips, bar model

RESOURCES

Optional: circles of paper, pictures of shapes cut into equal fractions, strips of paper

 In the eTextbook of this lesson, you will find interactive links to a selection of teaching tools.

Quick recap 🔄

Recap converting improper fractions to mixed numbers from the previous lesson. Write an improper fraction on the board, such as $\frac{15}{4}$. Ask: *How many quarters are in one whole? How many quarters are in two wholes? How many quarters are in three wholes? How many more quarters do we have?* Ask children to draw a diagram to show that $\frac{15}{4}$ is equivalent to $3\frac{3}{4}$. Repeat for other improper fractions.

Discover

WAYS OF WORKING Pair work

ASK

- Question ① a): *How many whole fruit tarts are there? What fraction of a fruit tart is on the other plate? What is the total amount of fruit tarts?*
- Question ① b): *How many quarters of a fruit tart are there on the full plates? How many quarters of a fruit tart is that in total? What about the other quarter?*

IN FOCUS Questions ① a) and ① b) encourage children to consider the number of unit fractions within a given mixed number. While solving this, children should be asked to represent the fruit tarts using concrete or pictorial representations to support their understanding of the conversion.

PRACTICAL TIPS Use circles of paper to represent the fruit tarts. Children could fold or cut them into quarters to represent the question and support their understanding. If they line the slices up separately, they can clearly see the improper fraction, whilst if they arrange them like the tarts in the **Discover** artwork, they can see the mixed number.

ANSWERS

Question ① a): There are $4\frac{1}{4}$ fruit tarts.

Question ① b): $4\frac{1}{4} = \frac{17}{4}$

PUPIL TEXTBOOK 5A PAGE 184

Share

WAYS OF WORKING Whole class teacher led

ASK

- Question ① a): *Where can you see the 4 whole tarts in the picture? Where is the other quarter?*
- Question ① b): *How many quarters are there in each whole tart? How many quarters are there in 4 whole tarts? There is one more quarter. How many quarters are there in total?*
- Question ① b): *How is Astrid's abstract method supported by the bar model?*

IN FOCUS Children should explain how the bar models link to the image of the tarts. They should use their explanation to then explore Dexter's and Astrid's comments.

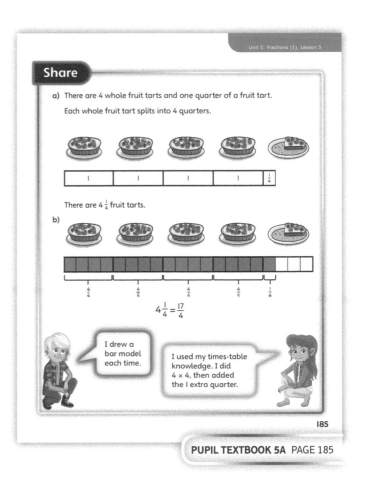

PUPIL TEXTBOOK 5A PAGE 185

Think together

Whole class teacher led (I do, We do, You do)

ASK

- Question **1** a): *How many fifths are there in one whole? How many fifths are there in two wholes? How many fifths are there in three wholes? How many fifths are there altogether?*
- Question **2**: *Are there only two solutions for B?*
- Question **3**: *If star is worth 1, then what is the mixed number that we are converting? What is the solution if the star is worth 2?*

IN FOCUS Question **1** a) uses a pictorial representation to support children's understanding of converting mixed numbers to improper fractions. Children should be given the opportunity to draw the model and split the wholes into fifths to support their understanding. Question **3** gives further practice of converting mixed numbers to improper fractions, whilst encouraging children to make connections between the values of the star and the triangle.

STRENGTHEN If children need extra support, provide them with strips of paper that they can use to represent the fractions and then divide up to support their conversions.

DEEPEN In question **3**, encourage children to explain what they notice and use this to find the value of the triangle if the star is 200. How can they check their answer? Encourage them to discuss and explore Flo and Ash's comments.

ASSESSMENT CHECKPOINT Use question **1** to assess whether children can use pictorial representations to support them in converting mixed numbers to improper fractions. Use question **2** to assess whether children can confidently convert mixed numbers to improper fractions.

ANSWERS

Question **1** a): $3\frac{4}{5} = \frac{19}{5}$

Question **1** b): $2\frac{2}{3} = \frac{8}{3}$

Question **2**: A = $1\frac{1}{4}$ or $\frac{5}{4}$ C = 3 or $\frac{12}{4}$

           B = $2\frac{1}{2}$ or $\frac{10}{4}$ D = 4 or $\frac{16}{4}$

Question **3**:

☆	▲
1	6
2	11
3	16
4	21
5	26
10	51

Each time you increase the value of the star by 1, the triangle increases by 5, because 1 star represents 5 fifths.

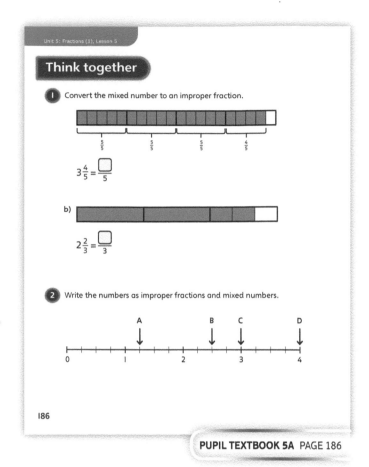

Think together

1 Convert the mixed number to an improper fraction.

$3\frac{4}{5} = \frac{\Box}{5}$

b)

$2\frac{2}{3} = \frac{\Box}{3}$

2 Write the numbers as improper fractions and mixed numbers.

186

PUPIL TEXTBOOK 5A PAGE 186

3 Bella is finding solutions to:

$\frac{☆}{5} = \frac{▲}{5}$

CHALLENGE

She chooses different numbers for ☆ and looks at the effect on ▲.

Complete the table and explain what happens to ▲ when you increase ☆ by 1.

☆	▲
1	
2	
3	
4	
5	
10	

I think the numbers will go up by 1 each time.

I'm not so sure. They go up by a different amount I think.

→ Practice book 5A p134

187

PUPIL TEXTBOOK 5A PAGE 187

Practice

IN FOCUS Questions ① and ② develop children's understanding of how mixed numbers can be represented in multiple ways. They also develop children's independent ability to convert mixed numbers to improper fractions.

Question ③ is important as it develops children's awareness that mixed numbers can appear on a number line or in a count.

STRENGTHEN For children who find question ⑤ challenging, ask: *How could you make it easier for yourself to use the fractions in the question? Could you convert the halves to quarters or the quarters to halves? How would that help?*

If children find question ⑥ challenging, ask: *Could you use a bar model to represent this? How would that representation help you solve this question?*

DEEPEN When solving question ④ a) to c), encourage children to deepen their understanding of the conversions they are doing by discussing the observable pattern. Ask: *What happens when you increase the numerator by 1 each time? What happens when those numbers decrease instead?*

ASSESSMENT CHECKPOINT Question ⑤ assesses children's ability to solve word problems involving mixed numbers. Look for children using their understanding of fractions from previous lessons to support their solutions, which might include converting the halves to quarters.

Question ⑥ assesses children's ability to reason and problem solve using all the skills they have learnt so far. Look for children recognising where they will need to expand or simplify a fraction before converting it.

ANSWERS Answers for the **Practice** part of the lesson can be found in the *Power Maths* online subscription.

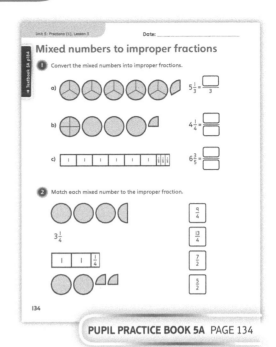

PUPIL PRACTICE BOOK 5A PAGE 134

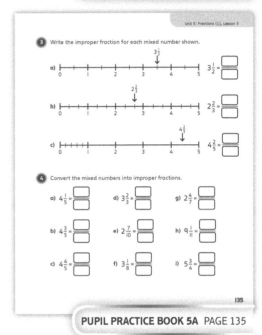

PUPIL PRACTICE BOOK 5A PAGE 135

Reflect

IN FOCUS This question offers a good opportunity to make a final assessment of children's understanding of how the two types of fractions are linked and can be represented. Children could be encouraged to write an explanation of how their picture proves their thinking.

ASSESSMENT CHECKPOINT Look for children representing their thinking using the pictorial representations from this unit, such as a bar model.

ANSWERS Answers for the **Reflect** part of the lesson can be found in the *Power Maths* online subscription.

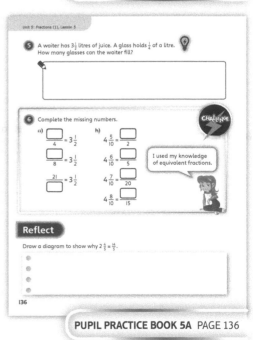

PUPIL PRACTICE BOOK 5A PAGE 136

After the lesson ⏸

- Were children as confident converting from a mixed number to an improper fraction as they were converting the other way around?
- If they were more confident at one type of conversion than the other, what were the barriers to their learning? How will you plan for this next time?

Compare fractions less than 1

Learning focus

In this lesson, children will learn to compare fractions based on their size. They will use pictorial representations to justify their comparisons and share their reasoning.

Before you teach

- How confident are children with the concept that a fraction is a part of a whole and that the denominator represents the total number of parts?
- Can children represent fractions using diagrams and other representations?
- Can children find equivalent fractions?

NATIONAL CURRICULUM LINKS

Year 5 Number – fractions (including decimals and percentages)

Compare and order fractions whose denominators are all multiples of the same number.

ASSESSING MASTERY

Children can understand and represent fractions fluently. They can recognise and explain how a greater denominator equates to a smaller part of a whole. They can use pictorial representations, such as bar models, to represent different fractions and compare them. They can use their knowledge of equivalent fractions from earlier in the unit to support them in comparing fractions.

COMMON MISCONCEPTIONS

Children may simply compare the numbers within a fraction rather than considering the fractions as being part of a whole, for example, they may think that $\frac{3}{10}$ is greater than $\frac{1}{2}$ because 3 is greater than 1 and 10 is greater than 2. Ask:

- *Can you draw a fraction strip to show $\frac{3}{10}$? How many parts is the whole divided into? How many of these parts are shaded to show $\frac{3}{10}$?*
- *Can you draw a fraction strip to show $\frac{1}{2}$? How many parts is the whole divided into? How many of these parts are shaded to show $\frac{1}{2}$?*
- *Which of these represents the greater part of the whole?*

STRENGTHENING UNDERSTANDING

If children find accurately drawing bar models a challenge in this lesson, it may help to give them a printed fraction wall to use. Ask: *How can the fraction wall help you compare fractions? Can you explain how to compare $\frac{3}{4}$ and $\frac{5}{8}$ using the fraction wall?*

GOING DEEPER

Children could be challenged to create their own real-life word problems which require the comparison of two or more fractions. Ask: *What could you change about the fractions in your problem to make the question more or less challenging?*

KEY LANGUAGE

In lesson: fraction, numerator, denominator, compare, less than (<), greater than (>), equal to (=)

Other language to be used by the teacher: comparison, sort

STRUCTURES AND REPRESENTATIONS

Bar models, number lines, fraction wall

RESOURCES

Optional: paper (for folding), printed fraction wall, printed number line split into tenths, fraction cards, blank fraction strips

 In the eTextbook of this lesson, you will find interactive links to a selection of teaching tools.

Quick recap ⟲

Show children a fraction on the board and ask them to find two fractions that are equivalent to it. Then increase the level of challenge by telling them one of the numerators has to be ... or one of the denominators has to be ....

Discover

WAYS OF WORKING Pair work

ASK

- Question ❶ a): *How many penalties did Danny take? How many did he score? What fraction of his penalties did Danny score?*
- Question ❶ a): *How many penalties did Jamilla take? How many did she score? What fraction of her penalties did she score?*
- Question ❶ b): *Who scored more penalties? Does this mean their score is better?*
- Question ❶ b): *Who missed more penalties? Does this mean their score is worse?*

IN FOCUS Use this activity to explore why you cannot simply compare the numerators or denominators in isolation when comparing fractions. In question ❶ b), prompt children to recognise that Danny scored more penalties and missed more penalties, so children cannot compare using this information alone.

PRACTICAL TIPS Provide children with a printed number line split up into tenths that they can use to represent and compare the fractions. Ask: *Jamilla did not take 10 penalties. How can we show her score on the number line?*

ANSWERS

Question ❶ a): 4 out of 5 is $\frac{4}{5}$. 7 out of 10 is $\frac{7}{10}$.

Question ❶ b): $\frac{4}{5}$ is the better score.

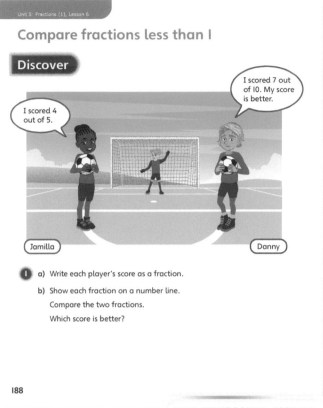

Compare fractions less than 1

Discover

❶ a) Write each player's score as a fraction.

b) Show each fraction on a number line.
Compare the two fractions.
Which score is better?

188

PUPIL TEXTBOOK 5A PAGE 188

Share

WAYS OF WORKING Whole class teacher led

ASK

- Question ❶ a): *How is 4 out of 5 the same as $\frac{4}{5}$? How is 7 out of 10 the same as $\frac{7}{10}$? What do the numbers in the fractions represent?*
- Question ❶ b): *Why can't you compare the fractions if the numerators and denominators are different? How do equivalent fractions help? What equivalent fractions can you see on the number line?*

IN FOCUS Use the comments from Ash, Sparks and Astrid as a discussion point with children. They need to understand the importance of finding equivalent fractions when comparing fractions with different numerators and denominators.

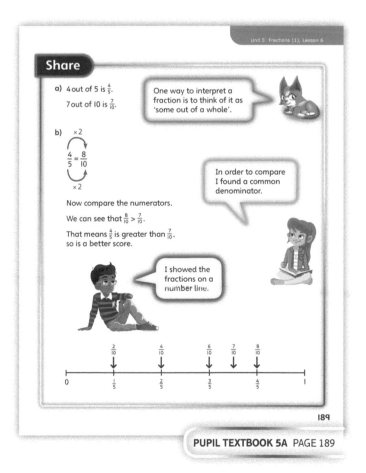

PUPIL TEXTBOOK 5A PAGE 189

Think together

Whole class teacher led (I do, We do, You do)

ASK

• Question ❶: *What can you use to represent the fractions? How can you use this to find equivalent fractions? Once you have found equivalent fractions, why is it easier to compare?*

• Question ❷: *What number would make this number sentence correct: $\frac{7}{18} = \frac{8}{9}$?*

• Question ❷: *If $\frac{?}{18}$ is bigger than $\frac{8}{9}$, what can you say about the missing number?*

• Question ❷: *How could you represent the fractions to support your thinking?*

IN FOCUS In question ❶, encourage children to use Astrid's method. Children may wish to use a number line to support this method. In question ❷, ask: *What do you know? What can you find out? How can you use equivalent fractions to help you?*

STRENGTHEN Provide children with strips of paper or blank number lines that they can use to represent fractions and support their understanding.

DEEPEN In question ❸, encourage children to explain why each method works and whether they only work for those fractions. Give them a pair of fractions and ask: *Which of the methods would you use to compare these?*

ASSESSMENT CHECKPOINT Use question ❶ to assess whether children can compare fractions using inequality signs. Use question ❷ to assess whether children can apply this understanding to work out missing numerators and denominators in given inequalities.

ANSWERS

Question ❶ a): $\frac{4}{5} > \frac{11}{15}$

Question ❶ b): $\frac{12}{20} = \frac{3}{5}$

Question ❶ c): $\frac{19}{50} < \frac{2}{5}$

Question ❷ a): $\frac{8}{9} < \frac{17}{18}$

Question ❷ b): $\frac{2}{3} > \frac{5}{9}$

Question ❷ c): $\frac{3}{4} > \frac{11}{16}$

Question ❷ d): $\frac{1}{2} < \frac{13}{24}$

Question ❸: Answers will vary dependent on understanding.

Think together

❶ Choose the correct symbol <, > or =.

a) $\frac{4}{5} \bigcirc \frac{11}{15}$

b) $\frac{12}{20} \bigcirc \frac{3}{5}$

c) $\frac{19}{50} \bigcirc \frac{2}{5}$

I will use equivalent fractions.

❷ Use each card once to complete the statements correctly.

| 1 | 3 | 16 | 17 |

a) $\frac{8}{9} < \frac{\square}{18}$

b) $\frac{2}{\square} > \frac{5}{9}$

c) $\frac{3}{4} > \frac{11}{\square}$

d) $\frac{\square}{2} < \frac{13}{24}$

190

PUPIL TEXTBOOK 5A PAGE 190

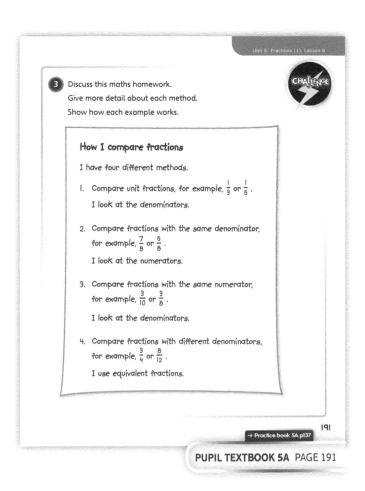

❸ Discuss this maths homework.
Give more detail about each method.
Show how each example works.

CHALLENGE

How I compare fractions

I have four different methods.

1. Compare unit fractions, for example, $\frac{1}{3}$ or $\frac{1}{5}$.
   I look at the denominators.

2. Compare fractions with the same denominator, for example, $\frac{7}{8}$ or $\frac{5}{8}$.
   I look at the numerators.

3. Compare fractions with the same numerator, for example, $\frac{3}{10}$ or $\frac{3}{8}$.
   I look at the denominators.

4. Compare fractions with different denominators, for example, $\frac{3}{4}$ or $\frac{8}{12}$.
   I use equivalent fractions.

191

→ Practice book 5A p137

PUPIL TEXTBOOK 5A PAGE 191

Practice

WAYS OF WORKING Independent thinking

IN FOCUS Question ① provides children with the opportunity to explore the different methods for comparing fractions and then to use visual representations to support their findings. Ask: *Could you have used a different method? Would your answers have been the same?* Question ② provides the opportunity for children to consolidate their understanding of comparing fractions by choosing the correct inequality signs. Children should be encouraged to choose the most appropriate method for each comparison and explain why they have chosen it.

STRENGTHEN Provide children with blank fraction strips, which they can divide into the correct number of parts, or blank number lines to support their understanding, if needed.

DEEPEN In question ⑤, encourage children to justify their choice of method for each part. Can they use different methods and then identify which is most efficient and why? Can they create a question like this for a partner?

THINK DIFFERENTLY Question ③ highlights the importance of the size of the whole when comparing fractions. Children should already know that $\frac{5}{5}$ is equivalent to $\frac{10}{10}$ and should use this fact when explaining why Bella's bar models are incorrect.

ASSESSMENT CHECKPOINT Use questions ① and ② to assess whether children can confidently and accurately compare fractions using inequality signs. Use question ③ to assess whether children understand the importance of the size of the whole when comparing fractions.

ANSWERS Answers for the **Practice** part of the lesson can be found in the *Power Maths* online subscription.

Reflect

WAYS OF WORKING Pair work

IN FOCUS Children are asked to summarise their learning in their own words whilst working with a partner.

ASSESSMENT CHECKPOINT Children should be able to explain different methods for comparing fractions and provide examples for when each method could be used.

ANSWERS Answers for the **Reflect** part of the lesson can be found in the *Power Maths* online subscription.

After the lesson ⏸

- Can children compare fractions with the same numerator?
- Can children compare fractions with the same denominator?
- Can children compare fractions using equivalent fractions?

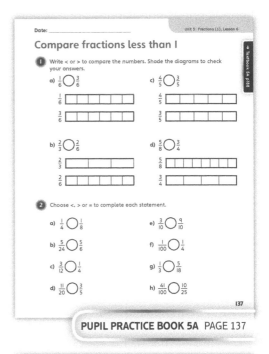

PUPIL PRACTICE BOOK 5A PAGE 137

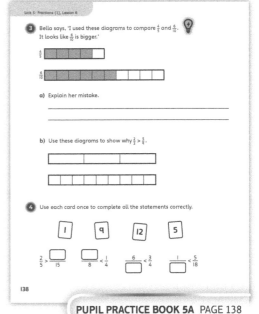

PUPIL PRACTICE BOOK 5A PAGE 138

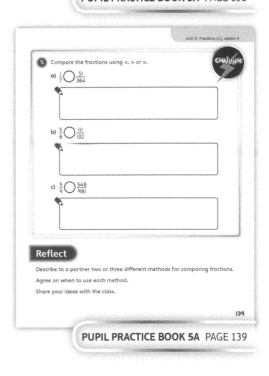

PUPIL PRACTICE BOOK 5A PAGE 139

Order fractions less than I

Learning focus

In this lesson, children will use their knowledge of comparing fractions to order them based on their size. They will use pictorial representations to justify their comparisons and share their reasoning.

Before you teach

- Can children find equivalent fractions?
- Can children compare fractions where the numerators or denominators are equal?
- Can children use their knowledge of equivalent fractions to compare fractions?

NATIONAL CURRICULUM LINKS

Year 5 Number – fractions (including decimals and percentages)
Compare and order fractions whose denominators are all multiples of the same number.

ASSESSING MASTERY

Children can understand and represent fractions fluently. They can recognise and explain how a greater denominator equates to a smaller part of a whole. They can use pictorial representations such as bar models to represent different fractions and order them.

COMMON MISCONCEPTIONS

Children may simply compare or order isolated parts of the fractions, rather than considering each fraction as part of a whole. For example, they may think that $\frac{3}{10} > \frac{2}{6} > \frac{1}{2}$ because 3 > 2 > 1 and 10 > 6 > 2. Ask:
- *Can you find equivalent fractions where the number on the bottom is the same for each of the fractions? What do the equivalent fractions tell you about how much of the whole they represent? Which fraction takes up the greater part of the whole? Which fraction takes up the least part of the whole?*

STRENGTHENING UNDERSTANDING

If children find accurately drawing bar models a challenge in this lesson, it may help to give them a printed fraction wall to use. Ask: *How can a fraction wall help you compare and order fractions?*

GOING DEEPER

Encourage children to look for alternative ways of ordering fractions rather than relying solely on the use of equivalent fractions. For example, if ordering $\frac{2}{5}$, $\frac{11}{15}$ and $\frac{2}{9}$, children could use the fact that the numerators are the same in $\frac{2}{5}$ and $\frac{2}{9}$ to find that $\frac{2}{9}$ is smaller than $\frac{2}{5}$, then use the fact that $\frac{2}{5}$ is less than one-half and $\frac{11}{15}$ is greater than one-half to order them.

KEY LANGUAGE

In lesson: fraction, numerator, denominator, compare, order, greater than (>), less than (<), equal to (=)

Other language to be used by the teacher: comparison, sort

STRUCTURES AND REPRESENTATIONS

Bar model, number lines

RESOURCES

Optional: paper (for folding), printed fraction wall, blank number lines, fraction cards

 In the eTextbook of this lesson, you will find interactive links to a selection of teaching tools.

Quick recap 🔎

Give children pairs of fractions and ask them to compare them using <, > or =. Use different examples which use different methods, for example, the same numerator, the same denominator, or the need to find equivalent fractions. Encourage children to explain their reasoning throughout and represent the fractions using diagrams where appropriate.

Discover

WAYS OF WORKING Pair work

ASK

- Question ① a) and ① b): *Which of the fractions are easiest to compare? Which of the fractions are hardest to compare?*
- Question ① b): *How many eighths are equivalent to one-half? How many sixteenths are equivalent to one-half? How does this help you order the fractions?*

IN FOCUS This activity introduces children to the idea of using the position of a fraction relative to one half to compare and order them. Encourage children to discuss why this is useful when comparing fractions.

PRACTICAL TIPS Provide children with strips of paper or a blank number line to represent the fractions and support them in identifying which fractions are less than one-half and which are greater.

ANSWERS

Question ① a): $\frac{5}{8}$, $\frac{9}{16}$ and $\frac{15}{16}$ are greater than half.

Question ① b): $\frac{1}{8} < \frac{5}{16} < \frac{9}{16} < \frac{5}{8} < \frac{15}{16}$

Order fractions less than 1

Discover

① a) Which of the fractions are greater than $\frac{1}{2}$?

b) Order the fractions from least to greatest.

192

Share

WAYS OF WORKING Whole class teacher led

ASK

- Question ① a): *Why is the middle of the number line labelled with $\frac{4}{8}$ and with $\frac{8}{16}$? What are both of these fractions equivalent to?*
- Question ① a): *How does knowing this help you work out which fractions are greater than one-half?*
- Question ① b): *Why did we need to find equivalent fractions to write the fractions in order?*

IN FOCUS This activity provides an opportunity for children to explore different methods for ordering fractions. In question ① b), they should recognise that knowing which fractions are less than $\frac{1}{2}$ and which are greater makes it easier to compare and order them.

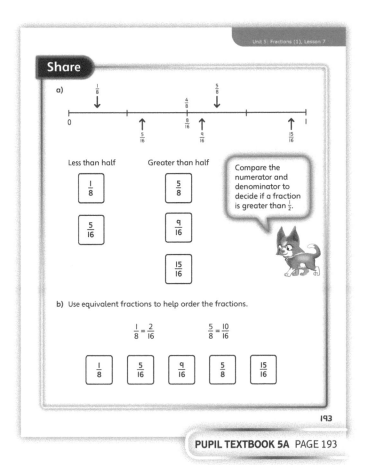

193

Think together

Think together

WAYS OF WORKING Whole class teacher led (I do, We do, You do)

ASK

• Question **1** a): *How many twelfths are equivalent to one-half? How many hundredths are equivalent to one-half? How many two-hundredths are equivalent to one-half?*
• Question **1** b): *How many twentieths are equivalent to one-tenth? How many hundredths are equivalent to one-tenth? How many sixtieths are equivalent to one-tenth?*
• Question **2**: *Why do you think Dexter is going to sort the fractions into those which are greater than and less than one-half?*
• Question **3** a) and **3** b): *How many possible answers are there? How do you know?*

IN FOCUS Question **1** provides an opportunity for children to secure their understanding of fractions being less than or greater than a given amount, which will support them when ordering fractions. In question **2**, children should be encouraged to recognise the importance of comparing fractions to one-half and see how this helps them when ordering fractions.

STRENGTHEN Provide children with strips of paper or blank number lines to represent the fractions and support their understanding.

DEEPEN In question **3**, children use their knowledge of fractions as a number to identify fractions in a given region of a number line. They should use their knowledge of equivalent fractions to support them.

ASSESSMENT CHECKPOINT Use question **1** to assess whether children can confidently compare fractions to another given fraction. Use question **2** to assess whether children can use this knowledge to order a list of fractions.

ANSWERS

Question **1** a): $\frac{5}{12}$ and $\frac{99}{200}$ are less than one-half.

Question **1** b): $\frac{1}{20}$ and $\frac{5}{60}$ are less than one-tenth.

Question **2**: $\frac{1}{6} < \frac{2}{6} < \frac{5}{12} < \frac{3}{6} < \frac{2}{3} < \frac{5}{6}$

Question **3** a): Answers will vary, e.g.
$$\frac{1}{4}, \frac{2}{8}, \frac{1}{3}, \frac{2}{6}, \frac{2}{5}, \frac{4}{10}$$

Question **3** b): Answers will vary, e.g.
$$\frac{3}{5} < \frac{7}{10} < \frac{4}{5}$$

PUPIL TEXTBOOK 5A PAGE 194

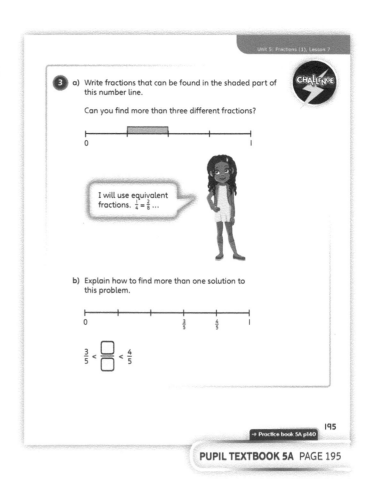

PUPIL TEXTBOOK 5A PAGE 195

Practice

WAYS OF WORKING Independent thinking

IN FOCUS Questions **1** and **2** provide opportunities for children to consolidate their learning on comparing fractions. Children are asked to sort fractions into those greater and less than one-half in question **1**, and one-tenth in question **2**. In question **3**, children can then build on this understanding to order sets of fractions.

STRENGTHEN If children find this particularly challenging, provide them with blank number lines or a fraction wall to support their understanding.

DEEPEN In question **5**, children use their knowledge of fractions as a number to identify fractions in a given region of a number line. They should use their knowledge of equivalent fractions to support them. Ask: *How many possible answers are there? How do you know?*

THINK DIFFERENTLY In question **4**, encourage children to explain how they know which question has more possible answers. How do they know they have found all the answers? Can they think of a question which would have more or fewer possible answers?

ASSESSMENT CHECKPOINT Use questions **1** and **2** to assess whether children can confidently compare fractions to another given fraction. Use question **3** to assess whether children can use this knowledge to order a list of fractions.

ANSWERS Answers for the **Practice** part of the lesson can be found in the *Power Maths* online subscription.

Reflect

WAYS OF WORKING Independent thinking

IN FOCUS Children can only use each digit card once to make the three fractions. This should prompt them to compare the fractions using a bar model and demonstrate their ability to compare and order fractions. They should be able to explain how to use the numerator and denominator to accurately compare and order fractions.

ASSESSMENT CHECKPOINT Look for children who use the methods and representations covered in the lesson fluently and with confidence. Their reasoning should show fluent understanding of what numerators and denominators represent and, if using a bar model, why it is important to ensure the bars are the same size.

ANSWERS Answers for the **Reflect** part of the lesson can be found in the *Power Maths* online subscription.

After the lesson ⏸

- Can children confidently compare and order fractions with different denominators?
- Can children fluently explain why their bar models needed to be the same size when using them to compare fractions?

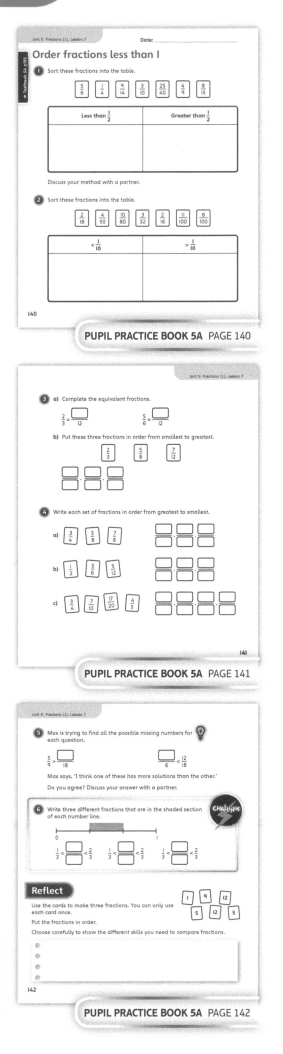

PUPIL PRACTICE BOOK 5A PAGE 140

PUPIL PRACTICE BOOK 5A PAGE 141

PUPIL PRACTICE BOOK 5A PAGE 142

231

Compare and order fractions greater than I

Learning focus

In this lesson, children will learn to compare mixed numbers and order them based on their size. They will use pictorial representations to justify their comparisons and share their reasoning.

NATIONAL CURRICULUM LINKS

Year 5 Number – fractions (including decimals and percentages)
Compare and order fractions whose denominators are all multiples of the same number.

ASSESSING MASTERY

Children can understand and represent mixed fractions fluently. They can recognise and explain how a larger denominator equates to a smaller part of a whole. They can use pictorial representations, such as bar models, to represent different fractions, and to compare and order them.

COMMON MISCONCEPTIONS

Children may order improper fractions based on the numbers in the fraction, rather than the number the fraction represents. For example, they may say that $\frac{8}{2}$ is smaller than $\frac{16}{15}$ because 8 and 2 are smaller than 16 and 15. Ask:
- *Can you draw a picture of each of these fractions?*
- *Using the pictures, can you explain why $\frac{8}{2}$ is actually larger than $\frac{16}{15}$?*

STRENGTHENING UNDERSTANDING

It is important to make sure that children are confident converting between mixed numbers and improper fractions. Give children opportunities to convert between the two types of fractions through activities such as using card shapes cut up into fractions, such as rectangles cut up into quarters. Ask: *Can you find $\frac{17}{4}$? Can you show me how many complete rectangles that fraction would make? Can you write it as a mixed number?*

Encourage children to draw the two fraction types and discuss the similarities and differences between them.

GOING DEEPER

Children could be asked to write their own real-life word problem that requires the person solving it to compare and order improper fractions or mixed numbers. Ask: *Can you write a word problem that has two or three steps to solve it?*

KEY LANGUAGE

In lesson: mixed number, improper fraction, whole, fraction, numerator, denominator, compare, order, greater than (>), less than (<), equal to (=), closer, further

STRUCTURES AND REPRESENTATIONS

Bar model, number line

RESOURCES

Mandatory: dice

Optional: paper (for folding), number cards, printed fraction wall, card shapes cut into fractions, blank number line

 In the eTextbook of this lesson, you will find interactive links to a selection of teaching tools.

Quick recap ⟳

Recap converting between mixed numbers and improper fractions. Draw three boxes on the board to represent a mixed number with all digits missing. Roll a dice three times and fill in the missing digits of the mixed number. Ask: *What mixed number have we made? What is this mixed number as an improper fraction?* Repeat with two boxes converting from improper fractions to mixed numbers.

Discover

Pair work

ASK

- Question ① a): *What is different about these numbers to those you compared in the previous lesson? What part of the number will you compare first? Why? Can you draw a diagram to support your comparison?*
- Question ① b): *How is this question similar and different to question ① a)? Can you use a pictorial representation to prove your solution?*

IN FOCUS Question ① a) is important as it challenges children to compare two mixed numbers for the first time. Question ① b) uses mixed numbers where the whole numbers are the same as each other. Discuss with children whether the numbers in question ① a) or the numbers in question ① b) are easier to compare.

Question ① b) challenges children to find the larger number, rather than the smallest.

PRACTICAL TIPS Children could be given fictional 'tape measures' that have centimetres (not to scale) between which there are intervals of different sizes. For example, a tape measure that measures in halves, one that measures in thirds, and so on. Give children two mixed numbers that have different denominators and ask them to find out which is greatest and which is the least, and to prove it. The tape measures could be easily made by printing five or six fraction walls, cutting each fraction strip off and combining it with its matching strip.

	0 cm		1 cm		2 cm		3 cm
Halves							
Quarters							

ANSWERS

Question ① a): The shop is closer than the café.

Question ① b): The castle is further than the beach.

Share

Whole class teacher led

ASK

- Question ① a): *Why is 3 and a bit always greater than 2 and a bit? Why does it make sense to compare the integer part of the mixed numbers first?*
- Question ① b): *If the integers are the same, how can you compare the mixed numbers?*

IN FOCUS Children are given the opportunity to consider mixed numbers and their position in the number system. This will support them in comparing and ordering mixed numbers. In question ① a), they should recognise that the integer part of a mixed number gives an indication of the size of the mixed numbers and so this is a sensible place to start when comparing and ordering. In question ① b), where the integer parts are the same, they should realise that they can compare the fractional parts in the same way as they have done in previous lessons to compare the mixed numbers.

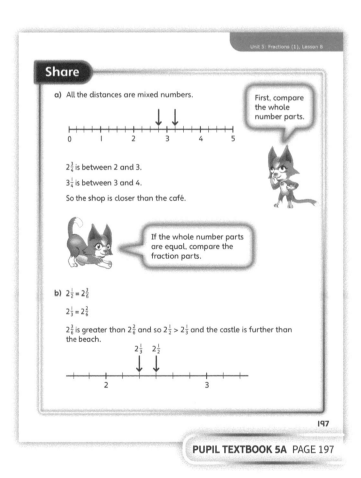

Compare and order fractions greater than 1

Discover

café $3\frac{1}{4}$ miles
shop $2\frac{3}{4}$ miles
beach $2\frac{1}{3}$ miles
castle $2\frac{1}{2}$ miles
garage $3\frac{1}{2}$ miles
pier $1\frac{3}{4}$ miles

① a) Which is closer, the café or the shop?

b) Which is further, the beach or the castle?

196

PUPIL TEXTBOOK 5A PAGE 196

Share

a) All the distances are mixed numbers.

First, compare the whole number parts.

$2\frac{3}{4}$ is between 2 and 3.

$3\frac{1}{4}$ is between 3 and 4.

So the shop is closer than the café.

If the whole number parts are equal, compare the fraction parts.

b) $2\frac{1}{2} = 2\frac{3}{6}$

$2\frac{1}{3} = 2\frac{2}{6}$

$2\frac{3}{6}$ is greater than $2\frac{2}{6}$ and so $2\frac{1}{2} > 2\frac{1}{3}$ and the castle is further than the beach.

197

PUPIL TEXTBOOK 5A PAGE 197

Think together

Whole class teacher led (I do, We do, You do)

ASK

- Question **1**: *All of the fractions have the same denominator. Does this mean you can compare them straight away? What do you need to do first? If you convert the mixed numbers to improper fractions or convert the improper fractions to mixed numbers, do you get the same answer?*
- Question **2**: *Which are the smaller two fractions? Which are the greater two fractions? How can you decide which is smallest and which is greatest?*

IN FOCUS Question **1** provides children with an opportunity to explore comparing and ordering mixed numbers and improper fractions. They should be encouraged to explore converting the improper fraction to a mixed number as well as converting the two mixed numbers to improper fractions. This will help them to recognise that both methods give the same answer. They should realise that comparing the improper fractions in this example is made easier because the denominators are the same. In question **2**, children should compare the integer parts first to recognise that they only actually need to compare two pairs of fractions, rather than comparing all the fractions at once.

STRENGTHEN Provide children with strips of paper or blank number lines to represent fractions and support their understanding.

DEEPEN In question **3**, children should be encouraged to discuss the pros and cons of both methods of comparing improper fractions. They should discuss and come up with examples of fractions that are suited to each of the two methods.

ASSESSMENT CHECKPOINT Use questions **1** and **2** to assess whether children can confidently and accurately compare and order mixed numbers and improper fractions.

ANSWERS

Question **1**: From smallest to greatest: $1\frac{1}{2}, 3\frac{1}{2}, \frac{9}{2}$

Question **2** a): $2\frac{3}{4}, 2\frac{7}{8}, 3\frac{3}{8}, 3\frac{3}{4}$

Question **2** b): $1\frac{1}{5}, 1\frac{1}{4}, 2\frac{3}{10}$

Question **3**: $5\frac{3}{4} > \frac{21}{4}$

$\frac{16}{3} < 5\frac{1}{2}$

$1\frac{1}{2} > \frac{11}{8}$

$\frac{25}{6} < 4\frac{2}{3}$

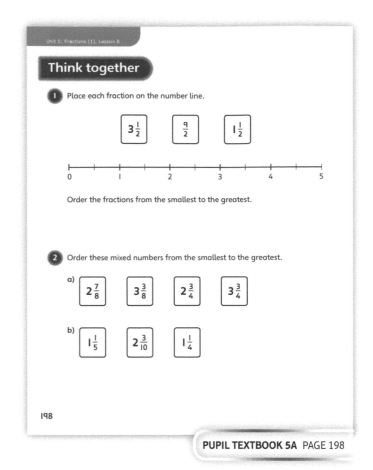

PUPIL TEXTBOOK 5A PAGE 198

PUPIL TEXTBOOK 5A PAGE 199

Practice

WAYS OF WORKING Independent thinking

IN FOCUS Question ② develops understanding of comparing mixed numbers and improper fractions with different denominators. The scaffolding of pictorial representations is removed in question ④. Encourage children to use pictures to support their transition into the use of abstract representation.

Question ③ presents the concept of comparing and ordering fractions in a word problem. This will challenge children to use their understanding of the vocabulary in the question to order the fractions successfully.

STRENGTHEN If children find comparing the fractions in question ⑤ challenging, ask: *If you are comparing two mixed numbers, what part should you look at first? How can you make a comparison where something is missing from one of the fractions? How could you use a picture to help you compare a mixed number and an improper fraction?*

DEEPEN Questions ⑥ a) and ⑥ b) deepen children's problem-solving and reasoning skills, developing their understanding of fractions that lie between two given mixed numbers or improper fractions. Encourage children to show their understanding of this concept by asking: *Can you create your own challenge, similar to these, for a partner to solve? How can you make it more challenging?*

ASSESSMENT CHECKPOINT Questions ①, ② and ③ assess children's ability to compare mixed numbers and improper fractions using different representations to support their thinking. Look for children's explanations referencing the pictorial and abstract representations and using the links between them to prove their comparisons.

Questions ⑤ and ⑥ assess children's reasoning and problem-solving ability. Look for children's fluency in their reasoning and the use of the methods learnt in this unit. They should be able to explain what strategies are useful and why.

ANSWERS Answers for the **Practice** part of the lesson can be found in the *Power Maths* online subscription.

Reflect

WAYS OF WORKING Independent thinking

IN FOCUS This final question is a good opportunity to assess children's confidence with the strategies they have been using in this lesson. Children should be able to fluently explain how to convert one of the fractions into the form of the other, and how they would then go about comparing them.

ASSESSMENT CHECKPOINT Look for children's understanding of the methods they have been learning and their use of the mathematical vocabulary linked to the concepts.

ANSWERS Answers for the **Reflect** part of the lesson can be found in the *Power Maths* online subscription.

After the lesson ⏸

- Were children as confident comparing and ordering mixed numbers and improper fractions as they were with regular fractions?
- How could you link comparing and ordering fractions to other areas of the curriculum?

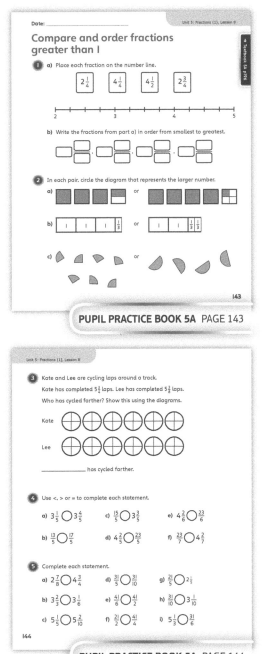

PUPIL PRACTICE BOOK 5A PAGE 143

PUPIL PRACTICE BOOK 5A PAGE 144

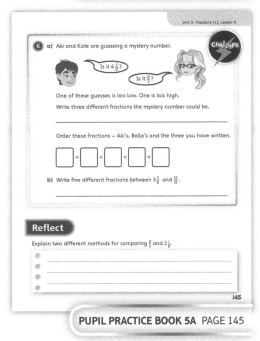

PUPIL PRACTICE BOOK 5A PAGE 145

End of unit check

Don't forget the unit assessment grid in your *Power Maths* online subscription.

WAYS OF WORKING Group work adult led

IN FOCUS

- Question **1** assesses children's ability to recognise and find equivalent fractions. Question **2** assesses children's ability to convert an improper fraction to a mixed number, and their understanding of simplifying.
- Question **3** assesses children's ability to compare and order fractions. Children may use a picture or their understanding of equivalent fractions to find a common denominator to help them compare and order the fractions.
- Question **4** assesses children's ability to convert a mixed number to an improper fraction.
- Question **5** assesses children's ability to find a collection of equivalent fractions when given various numerators or denominators.
- Question **6** is a SATs-style question which assesses children's ability to compare and order mixed number and improper fractions.

ANSWERS AND COMMENTARY

Children who have mastered the concepts in this unit will be able to fluently convert between mixed numbers and improper fractions and find equivalent fractions. They will reliably complete number sequences involving the different fraction types and will be able to compare and order different fractions, including those with different denominators.

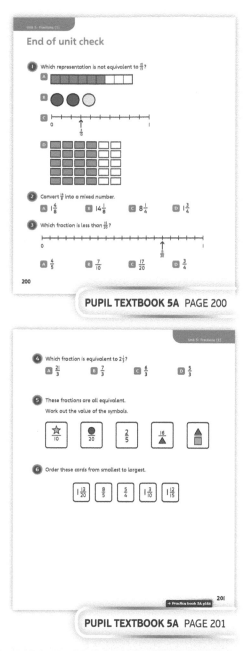

PUPIL TEXTBOOK 5A PAGE 200

PUPIL TEXTBOOK 5A PAGE 201

Q	A	WRONG ANSWERS AND MISCONCEPTIONS	STRENGTHENING UNDERSTANDING
1	C	Choosing any other option may suggest children lack fluency with different ways of representing equivalent fractions.	To convert between the two types of fraction, ask: • *Can you show a drawing of the mixed number $2\frac{1}{4}$?* • *Can you share your drawings of the two wholes into quarters?* • *How many quarters are there in total? Show this as an improper fraction.*
2	D	Choosing A may suggest children are using any digit in the 10s in the numerator as their whole number, ignoring the denominator.	
3	B	Choosing any other option may indicate that children find comparing/ordering fractions challenging.	
4	B	Choosing C might indicate that children have forgotten to add the 1 after multiplying 2 by 3.	
5	☆ = 4 ○ = 8 △ = 40 □ = 100	In the last part, children will need to carry their answer from the previous part through to find the numerical values of the square.	
6	$\frac{5}{4}, 1\frac{3}{10}, \frac{8}{5}, 1\frac{12}{20},$ $1\frac{12}{15},$ or $\frac{5}{4}, 1\frac{3}{10},$ $1\frac{12}{20}, \frac{8}{5}, 1\frac{12}{15}$	Children should identify that two of the fractions are equivalent and so can go either way round.	

My journal

WAYS OF WORKING Independent thinking

ANSWERS AND COMMENTARY Children should recognise the link between the numerator and denominator they choose, namely that the relationship between the two denominators should be the inverse relationship to that between the two numerators. For example, if the numerator they choose is double 6, then the denominator they choose needs to be half of 10. Ask:

• *What do you know about the fractions? How will this help you complete them?*
• *Could you find a solution that works if you put 5 as the numerator in the first fraction?*

Power check

WAYS OF WORKING Independent thinking

ASK

• *How much more confident do you feel about finding equivalent fractions? Why?*
• *Can you explain the similarities and differences between improper fractions and mixed numbers?*
• *What has been the most useful skill you have learnt in this unit? Why?*

Power play

WAYS OF WORKING Pair work

IN FOCUS Use this **Power play** to assess whether children can reliably and fluently find equivalent fractions for any given fraction, including mixed numbers and improper fractions. To deepen children's thinking in this concept, ask:

• *Are there certain numbers, or types of number, that are better or easier to use to create equivalent fractions?*
• *Are some numbers better used as a denominator than a numerator or vice versa?*

ANSWERS AND COMMENTARY Children should be able to use their understanding of improper fractions and mixed numbers to fluently find equivalent fractions. If asked, children should be able to prove their equivalent fractions using a picture of shaded 2D shapes or a fraction strip. If children find this activity challenging, it will be beneficial to offer them support with finding equivalent fractions and converting between mixed numbers and improper fractions. If dice are not available then children could use spinners labelled 1 to 6 (which they can make with a paper clip and some paper).

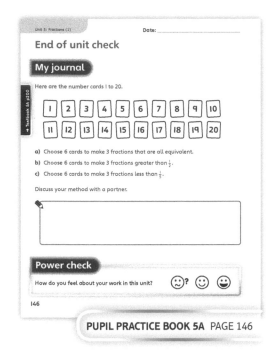

PUPIL PRACTICE BOOK 5A PAGE 146

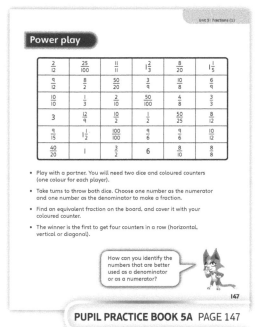

PUPIL PRACTICE BOOK 5A PAGE 147

After the unit ⏸

• Are there any misconceptions regarding fractions from this unit that will impact on the following unit, Fractions (2)?
• What concrete representations of fractions did you include in your teaching? How would you develop their use in the future?

Strengthen and **Deepen** activities for this unit can be found in the *Power Maths* online subscription.

Unit 6
Fractions ②

Mastery Expert tip! 'To develop fluency and confidence in adding and subtracting fractions and mixed numbers, encourage children to see the use of fractions in different areas of the curriculum. For example, in geography, children might add fractional distances between two or more places on a map, or find the difference between them.'

Don't forget to watch the Unit 6 video!

WHY THIS UNIT IS IMPORTANT

This unit introduces children to adding and subtracting related fractions by finding a common denominator. The particular focus is on examples where one number is a multiple of another. This is the first time children will have met such a concept and visual representations of fractions should help children grasp this fundamental topic. Children will extend their knowledge to adding and subtracting simple mixed numbers as well as proper and improper fractions. It is important that children develop their confidence and flexibility with fractions so that they are confident exploring the most efficient methods in problem solving with fractions. These skills will be key for Year 6 and beyond as children add together or subtract any two or more fractions.

WHERE THIS UNIT FITS

→ Unit 5: Fractions (1)
→ **Unit 6: Fractions (2)**
→ Unit 7: Multiplication and division (2)

In this unit, children extend their understanding of fractions by focusing on addition and subtraction of related fractions and mixed numbers. Children will continue to develop their confidence in reasoning and problem solving and will also explore different methods for addition and subtraction of fractions.

Before they start this unit, it is expected that children:
• can find factors and multiples of numbers using multiplication facts
• can find equivalent fractions and convert between improper fractions and mixed numbers.

ASSESSING MASTERY

Children who have mastered this unit are able to confidently add and subtract fractions and mixed numbers using formal written methods. They can fluently convert between mixed numbers and improper fractions, solve word problems and display their reasoning when explaining the use of different methods.

COMMON MISCONCEPTIONS	STRENGTHENING UNDERSTANDING	GOING DEEPER
Children may simply add or subtract the numerators and denominators when adding or subtracting mixed numbers, or fractions with different denominators.	Start with fractions that have the same denominator to strengthen understanding. Explain that the denominators must be the same before adding or subtracting the numerators, using fraction strips for support.	Encourage children to solve problems using their knowledge of finding equivalent fractions. For example, $3\frac{4}{5} + \boxed{} = \boxed{} + 5\frac{1}{15}$
Children may, when subtracting mixed numbers, focus on the fractional parts and subtract the smaller fraction from the larger fraction, regardless of the order in which the fractions occur.	Show children the numbers on a fraction strip so they can see why it is not always possible to subtract the smaller fraction from the larger fraction. It may be useful to encourage children to use a number line to 'count on' to carry out a subtraction using an alternative method.	Encourage children to explain the most efficient method depending on the numbers in the question.

Unit 6: Fractions ❷

UNIT STARTER PAGES

Introduce this unit using teacher-led discussion. Allow children time to discuss questions in pairs or small groups and share ideas as a whole class. Encourage children to use representations to visualise the fractions.

STRUCTURES AND REPRESENTATIONS

Fraction wheels: Circles divided into equal parts can represent fractions, show equivalence and support the addition and subtraction of proper fractions, improper fractions and mixed numbers.

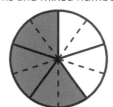

$\frac{3}{5}$ is equivalent to $\frac{6}{10}$

$\frac{3}{5} = \frac{6}{10}$

Fraction strips: These models represent proper fractions, improper fractions and mixed numbers. They can demonstrate operations involving fractions and support conversion between improper fractions and mixed numbers. Fraction strips can also be used with a number line.

$\frac{1}{3} + \frac{7}{12} = \frac{4}{12} + \frac{7}{12}$

Number lines: These models support children in converting between improper fractions and mixed numbers and can represent addition and subtraction.

$1\frac{3}{4} - 1\frac{1}{4} = \frac{1}{2}$

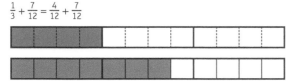

KEY LANGUAGE

There is some key language that children will need to know as part of the learning in this unit:

→ fraction, whole, part, equal parts, equivalent

→ add, sum, total, subtract, difference

→ divide, multiply, multiple

→ numerator, denominator, common denominator

→ simplify, convert

→ proper fraction, improper fraction, mixed number, equivalent fraction

→ method, multi-step, efficient

In this unit we will …

⚡ Add and subtract fractions with the same denominator

⚡ Add and subtract fractions, including mixed numbers, where one denominator is a multiple of the other

⚡ Solve word problems involving fractions

How can you add these two fractions?

$\frac{1}{4} + \frac{3}{8}$

We will need some maths words. Do you know what they all mean?

add · subtract · proper fraction

improper fraction · convert

equivalent fraction · mixed number

denominator · numerator

whole · common denominator

We need to be able to convert between mixed numbers and improper fractions. Use your skills to convert $2\frac{1}{3}$ into an improper fraction.

Add and subtract fractions

Learning focus

In this lesson, children will recap their knowledge of adding and subtracting fractions by adding and subtracting fractions with the same denominator. They will use their knowledge of fractions equivalent to one whole to subtract from a whole amount.

Before you teach

- Can children represent fractions using fraction strips and other representations?
- Do children understand that when the numerator and denominator of a fraction are equal, the fraction is equivalent to one whole?

NATIONAL CURRICULUM LINKS

Year 5 Number – fractions (including decimals and percentages)

Add and subtract fractions with the same denominator and denominators that are multiples of the same number.

ASSESSING MASTERY

Children can confidently add and subtract fractions with the same denominator. They understand why you add/subtract the numerators and the denominators do not change. They can use their knowledge of equivalent fractions to subtract from whole amounts.

COMMON MISCONCEPTIONS

Children may add or subtract both the numerators and denominators, leading to an incorrect answer, for example $\frac{1}{2} + \frac{1}{2} = \frac{2}{4}$. Ask:

- *Do you need to add both the numerators and add both the denominators? Can you draw a fraction strip to help you?*

STRENGTHENING UNDERSTANDING

Provide children with strips of paper to represent fractions to support their understanding of adding and subtracting fractions with the same denominator.

GOING DEEPER

Encourage children to not only notice when questions such as $\frac{1}{2} + \frac{1}{2} = \frac{2}{4}$ have been answered incorrectly, but also to consider why this answer cannot be true as $\frac{2}{4}$ is equivalent to $\frac{1}{2}$.

KEY LANGUAGE

In lesson: fraction, common denominator, equivalent, whole, denominator, numerator

Other language to be used by the teacher: subtraction, parts

STRUCTURES AND REPRESENTATIONS

Fraction walls, fraction strips

RESOURCES

Optional: paper (for folding), strips of paper, concrete objects such as pens and pencils

 In the eTextbook of this lesson, you will find interactive links to a selection of teaching tools.

Quick recap 🔍

Show a fraction on the board and ask children to draw a diagram to represent the fraction. Encourage them to explain how their diagram represents the fraction, including where the numerator and denominator can be seen in their diagram. Repeat for other fractions.

Discover

Unit 6: Fractions (2), Lesson 1

Add and subtract fractions

Discover

WAYS OF WORKING Pair work

ASK

- Question ① a): *What can you see in each panel? What can you see three of in each panel? What can you see two of in each panel? What can you see five of in each panel?*

IN FOCUS This activity provides children with an opportunity to use unitising to support their understanding of adding and subtracting fractions. In the bottom right panel, children may at first refer to these as 'slices', but should be encouraged to consider what fraction of the whole pizza each slice represents, leading them to find that 3 sixths + 2 sixths = 5 sixths, which will then support them in question ① b).

PRACTICAL TIPS Provide children with different objects that they can use to replicate the scenario. If they have 3 pencils and another 4 pencils, how many do they have in total? If they have 3 one-sevenths and another 4 one-sevenths, how much do they have altogether?

ANSWERS

Question ① a): 3 elephants + 2 elephants = 5 elephants
3 ducks + 2 ducks = 5 ducks
3 tens + 2 tens = 5 tens
3 sixths + 2 sixths = 5 sixths

Question ① b): $\frac{3}{8} + \frac{2}{8} = \frac{5}{8}$

$\frac{3}{11} + \frac{2}{11} = \frac{5}{11}$

$\frac{3}{12} + \frac{2}{12} = \frac{5}{12}$

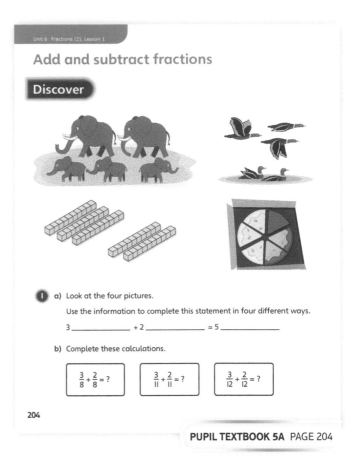

① a) Look at the four pictures.
Use the information to complete this statement in four different ways.

3 _____ + 2 _____ = 5 _____

b) Complete these calculations.

$\frac{3}{8} + \frac{2}{8} = ?$ $\frac{3}{11} + \frac{2}{11} = ?$ $\frac{3}{12} + \frac{2}{12} = ?$

204

PUPIL TEXTBOOK 5A PAGE 204

Share

WAYS OF WORKING Whole class teacher led

ASK

- Question ① a): *Why does it matter that each panel contained the same thing? Why did none of the panels contain three elephants and two ducks? Why wouldn't that work?*
- Question ① b): *Why doesn't the denominator change in the answer?*

IN FOCUS Use this activity to explore with children the idea of unitising in more detail. In question ① a), encourage them to recognise that, if they are counting elephants, it is just the numbers that they add, they do not add the elephants together. Similarly, if they are counting sixths, it is just the numerators they add together, not the denominators.

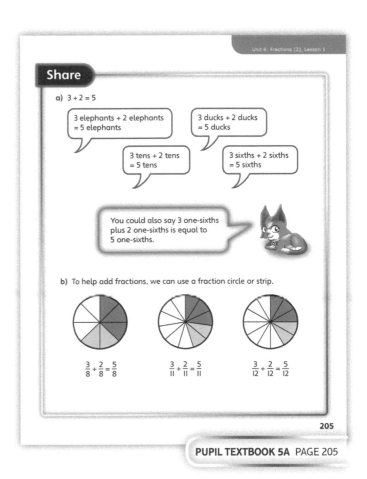

Unit 6: Fractions (2), Lesson 1

Share

a) 3 + 2 = 5

3 elephants + 2 elephants = 5 elephants

3 ducks + 2 ducks = 5 ducks

3 tens + 2 tens = 5 tens

3 sixths + 2 sixths = 5 sixths

You could also say 3 one-sixths plus 2 one-sixths is equal to 5 one-sixths.

b) To help add fractions, we can use a fraction circle or strip.

$\frac{3}{8} + \frac{2}{8} = \frac{5}{8}$ $\frac{3}{11} + \frac{2}{11} = \frac{5}{11}$ $\frac{3}{12} + \frac{2}{12} = \frac{5}{12}$

205

PUPIL TEXTBOOK 5A PAGE 205

Think together

WAYS OF WORKING Whole class teacher led (I do, We do, You do)

ASK

- Question **1**: *What does each part of the circle represent? Does this change if you put the parts together? Why does Lee think this diagram shows tenths?*
- Question **2**: *What is the same about the questions in each part? What is different?*
- Question **2**: *If 4 ones add 6 ones is equal to 10 ones, then what is four-elevenths add six-elevenths?*

IN FOCUS These questions provide an opportunity for children to explore addition and subtraction of fractions with the same denominator by unitising. Question **1** quickly addresses the common misconception that the denominators are also added together, as children should recognise that each part of the circle still represents one-fifth.

STRENGTHEN In question **1**, provide children with copies of the diagram that they can cut out. They should place the one-fifth in the other circle and see that there are three-fifths of the whole shaded.

DEEPEN In question **3**, children use their understanding of fractions equivalent to one whole to subtract from whole amounts. Ask them what they notice and provide them with increasingly challenging examples to see if their method works.

ASSESSMENT CHECKPOINT Use questions **1** and **2** to assess whether children can add and subtract fractions with the same denominator. Use question **3** to assess whether children can use fractions equivalent to one whole to subtract from whole amounts.

ANSWERS

Question **1**: $\frac{2}{5} + \frac{1}{5} = \frac{3}{5}$, not $\frac{3}{10}$

Lee added the denominators when he should only add the numerators.

Question **2** a): $\frac{4}{11} + \frac{6}{11} = \frac{10}{11}$

Question **2** b): $\frac{8}{13} - \frac{7}{13} = \frac{1}{13}$

Question **2** c): $\frac{9}{17} + \frac{6}{17} = \frac{15}{17}$

Question **2** d): $\frac{17}{19} - \frac{7}{19} = \frac{10}{19}$

Question **3** a): $\frac{3}{7} + \frac{4}{7} = 1$

Question **3** b): $1 - \frac{7}{8} = \frac{1}{8}$

Question **3** c): $\frac{7}{9} + \frac{2}{9} = 1$

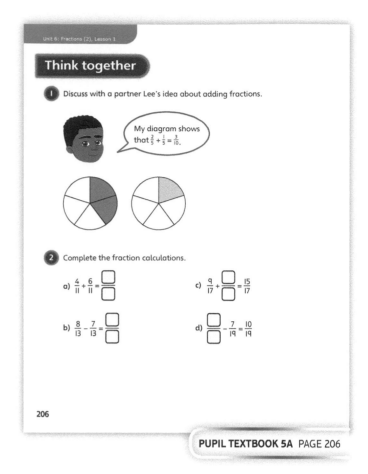

PUPIL TEXTBOOK 5A PAGE 206

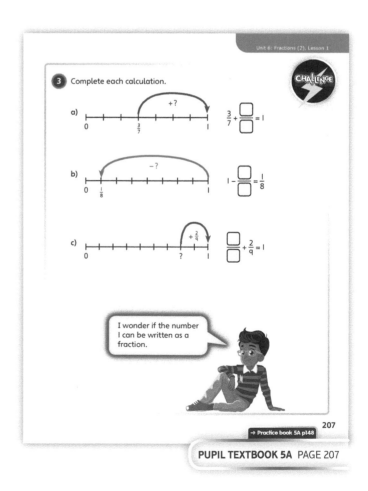

PUPIL TEXTBOOK 5A PAGE 207

Practice

WAYS OF WORKING Independent thinking

IN FOCUS These questions provide the opportunity for children to consolidate their understanding of adding and subtracting fractions with varying levels of scaffolding. Questions ❶ and ❷ provide them with pictorial support, before moving on to more abstract representations. Encourage them to consider what they are counting in each case and link back to the idea of unitising to support their thinking.

STRENGTHEN Provide children with strips of paper to represent the fractions and support their understanding.

DEEPEN Children should be encouraged to spot and explain anything they notice in question ❸. When answering question ❼, children may initially think they cannot do this because the denominators are not the same. However, they should recognise that the first two fractions total one whole and then use this to show that the calculation is correct.

ASSESSMENT CHECKPOINT Use questions ❶ and ❷ to assess whether children can add and subtract fractions with the same denominator where pictorial representations are given to support them. Use questions ❸ and ❹ to assess whether children can add and subtract fractions with the same denominator in the abstract form.

ANSWERS Answers for the **Practice** part of the lesson can be found in the *Power Maths* online subscription.

Reflect

WAYS OF WORKING Independent thinking

IN FOCUS This activity will demonstrate children's understanding of the concept taught in the lesson.

ASSESSMENT CHECKPOINT Children should recognise that the fractions they are adding or subtracting should have a denominator of 20, and the numerators should have a sum or a difference of 13.

ANSWERS Answers for the **Reflect** part of the lesson can be found in the *Power Maths* online subscription.

After the lesson ⏸

- Can children add and subtract fractions with the same denominator using pictorial representations to support them?
- Can children add and subtract fractions with the same denominator abstractly?
- Can children subtract fractions from the whole?

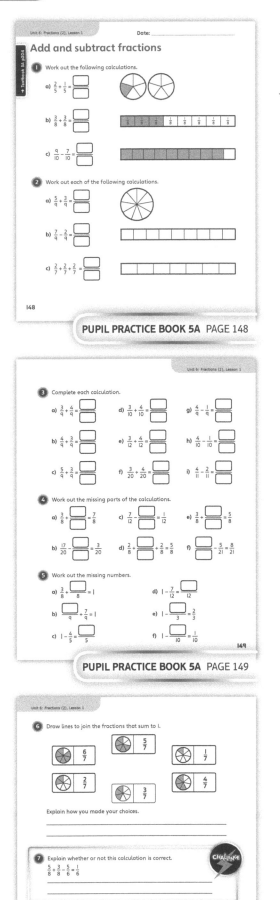

PUPIL PRACTICE BOOK 5A PAGE 148

PUPIL PRACTICE BOOK 5A PAGE 149

PUPIL PRACTICE BOOK 5A PAGE 150

Add fractions within 1

Learning focus

In this lesson, children will use their knowledge from the previous lesson, along with their knowledge of equivalent fractions, to add fractions where one denominator is a multiple of the other, where the answer does not exceed one whole.

Before you teach ⏸

- Can children add fractions with the same denominator?
- Can children find equivalent fractions?
- Can children identify common denominators?

NATIONAL CURRICULUM LINKS

Year 5 Number – fractions (including decimals and percentages)
Add and subtract fractions with the same denominator and denominators that are multiples of the same number.

ASSESSING MASTERY

Children can confidently approach questions involving fractions where one denominator is a multiple of the other. They use their knowledge of common denominators and equivalent fractions to rewrite the question so that both fractions have the same denominator, and then use this to find the answer.

COMMON MISCONCEPTIONS

Children may multiply the denominators to find a common denominator rather than spotting that one denominator is a multiple of the other. Say:
- *Look at the denominators. What are the multiples of both numbers? What is the smallest multiple of both numbers?*

Children may add the denominators as well as the numerators. Ask:
- *Do you need to add both the numerators <u>and</u> add both the denominators? Can you draw on a fraction strip to help you?*

STRENGTHENING UNDERSTANDING

Provide pictorial representations such as those on the **Share** section in the textbook, to aid children's understanding of why they need to first find a common denominator before they can add the fractions.

Provide children with strips of paper that they can fold and shade to represent the fractions in the question. They could represent the two fractions separately and then cut them out and lay them over a blank fraction strip to work out the total, if needed.

GOING DEEPER

Challenge children to write addition questions in context that give a desired answer.

KEY LANGUAGE

In lesson: fraction, **common denominator**, equivalent, denominator, total, multiple

Other language to be used by the teacher: numerator, addition, divide, split, parts, multiply

STRUCTURES AND REPRESENTATIONS

Fraction walls, fraction strips, bar models

RESOURCES

Optional: paper (for folding), blank number lines

 In the eTextbook of this lesson, you will find interactive links to a selection of teaching tools.

Quick recap 🔎

Give children a blank number line split into eight intervals, with the start and end points labelled 0 and 1 respectively. Ask children to label the intervals on the number line in their simplest form, then count along the number line as a class: *'zero, one-eighth, one-quarter, three-eighths, one-half…'* This can be used to support children's understanding of how one-eighth more than one-quarter is three-eighths.

Discover

Unit 6: Fractions (2), Lesson 2

WAYS OF WORKING Pair work

ASK

- Question ① a): *Is $\frac{3}{8}$ more or less than half of the course? How do you know?*
- Question ① a): *How can you draw a diagram to show three-eighths?*
- Question ① b): *How can you show $\frac{1}{4}$ more on the same diagram? How many eighths is $\frac{1}{4}$ equivalent to?*
- Question ① b): *How much have you shaded in total?*

IN FOCUS In question ① b), children are given the opportunity to recognise why they cannot simply add the numerators and the denominators. They can use pictorial representations of the fractions to aid their understanding of this.

PRACTICAL TIPS Provide children with strips of paper that they can fold and shade to represent the fractions in the question. They could represent the two fractions separately and then cut them out and lay them over a blank fraction strip to work out the total, if needed.

ANSWERS

Question ① a):

Accept diagrams that show $\frac{3}{8}$ successfully shaded.

Question ① b): $\frac{3}{8} + \frac{1}{4} = \frac{5}{8}$

The diagram should have $\frac{5}{8}$ shaded.

Share

WAYS OF WORKING Whole class teacher led

ASK

- Question ① a): *What is the same about each diagram? What is different?*
- Question ① b): *Why do you need to find a common denominator?*
- Question ① b): *Why is the answer to both additions the same?*

IN FOCUS In question ① b), children explore the addition of fractions where one denominator is a multiple of the other, using different diagrams as pictorial representations. Children should recognise that, in order to shade in $\frac{1}{4}$ of a diagram that is split into 8 equal parts, they need to use their knowledge of equivalent fractions.

Add fractions within I

Discover

> The first $\frac{3}{8}$ of the course is climbing.

> The next $\frac{1}{4}$ of the course is crawling.

Bella

① a) Draw a diagram with $\frac{3}{8}$ shaded.

b) Now shade $\frac{1}{4}$ more on the same diagram.
What total fraction have you shaded?

208

PUPIL TEXTBOOK 5A PAGE 208

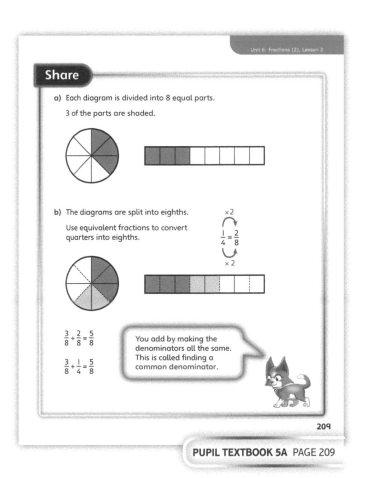

Share

a) Each diagram is divided into 8 equal parts.

3 of the parts are shaded.

b) The diagrams are split into eighths.

Use equivalent fractions to convert quarters into eighths.

$$\frac{1}{4} = \frac{2}{8} \quad (\times 2)$$

$$\frac{3}{8} + \frac{2}{8} = \frac{5}{8}$$

$$\frac{3}{8} + \frac{1}{4} = \frac{5}{8}$$

> You add by making the denominators all the same. This is called finding a common denominator.

209

PUPIL TEXTBOOK 5A PAGE 209

Think together

Whole class teacher led (I do, We do, You do)

ASK

- Question **1**: *Which of the fractions in each number sentence is easiest to shade? Which fraction in each number sentence do you need to find an equivalent fraction for? Why?*
- Question **2**: *How can you help Dexter decide which fraction to convert? Can you split two equal parts into six equal parts? Can you split six equal parts into two equal parts? How does that help you decide?*

IN FOCUS These questions provide opportunity for children to consolidate their learning on adding fractions where one denominator is the multiple of the other. In question **2**, they should use diagrams and other representations to support them in deciding which fraction to convert in each number sentence.

STRENGTHEN Provide children with strips of paper to fold and represent the fractions to support their understanding. Ask: *Which fraction are you going to shade first? Why? How can you show the other fraction on the same strip?*

DEEPEN In question **3**, children recognise and explain the common misconception that both the numerators and the denominators are added together. They should be able to use fraction strips or pictorial representations to recognise the mistake, explain why it is wrong and explain why Aki's answer cannot be correct.

ASSESSMENT CHECKPOINT Use question **1** to assess whether children can add fractions where one denominator is a multiple of the other, where pictorial representations are given for support. Use question **2** to assess whether children can work more abstractly when completing these calculations, and decide which of the two fractions they need to convert.

ANSWERS

Question **1** a): $\frac{7}{10} + \frac{1}{5} = \frac{9}{10}$

Question **1** b): $\frac{3}{5} + \frac{1}{10} = \frac{7}{10}$

Question **2** a): $\frac{1}{2} + \frac{1}{6} = \frac{4}{6}$

Question **2** b): $\frac{1}{6} + \frac{1}{3} = \frac{3}{6}$

Question **2** c): $\frac{1}{2} + \frac{1}{10} = \frac{6}{10}$

Question **2** d): $\frac{1}{2} + \frac{1}{100} = \frac{51}{100}$

Question **3**: This circle shows $\frac{1}{4} + \frac{1}{2}$.

Aki should have converted $\frac{1}{2}$ into quarters.
$\frac{1}{4} + \frac{2}{4} = \frac{3}{4}$

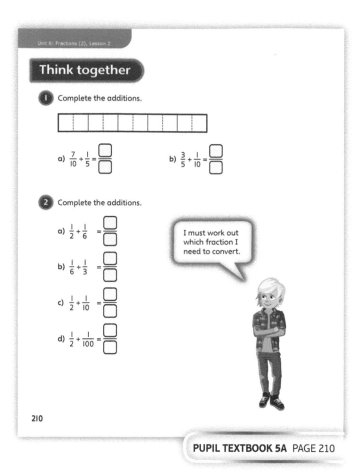

PUPIL TEXTBOOK 5A PAGE 210

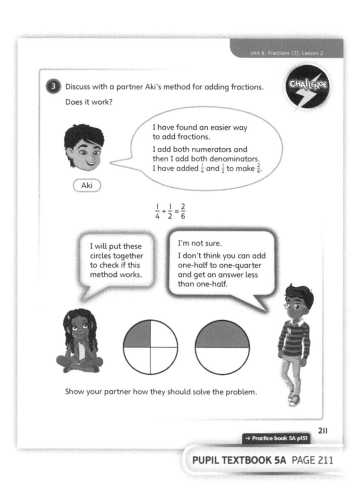

PUPIL TEXTBOOK 5A PAGE 211

Practice

WAYS OF WORKING Independent thinking

IN FOCUS In question ❶, children are given pictorial representations to support their understanding of adding fractions where one denominator is a multiple of the other. In question ❷, children are required to add fractions where pictorial representations are less useful, because of the number of parts needed, and they will demonstrate their ability to apply the abstract method. In question ❸, they then apply this skill in context as they answer a word problem. Ask: *What do you know? What do you want to find out? What can you use to find this?*

STRENGTHEN Provide children with strips of paper to support their understanding in earlier questions. Encourage them to explain each step of their thinking, so that they can apply this logic when working more abstractly.

DEEPEN In question ❹ b), children are required to add more than two fractions and should be encouraged to explore whether it matters what order they do this in. In question ❺, they need to use their knowledge of inverse operations alongside their knowledge of fractions to find missing numbers.

ASSESSMENT CHECKPOINT Use question ❶ to assess whether children can add fractions where one denominator is a multiple of the other when pictorial representations are given for support. Use question ❸ to assess whether children can apply this in context. Use question ❷ to assess whether children can abstractly add fractions.

ANSWERS Answers for the **Practice** part of the lesson can be found in the *Power Maths* online subscription.

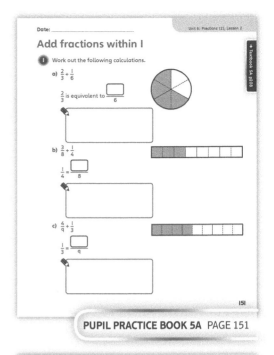

PUPIL PRACTICE BOOK 5A PAGE 151

PUPIL PRACTICE BOOK 5A PAGE 152

Reflect

WAYS OF WORKING Independent thinking

IN FOCUS Children are exposed to a common misconception in adding fractions. Explaining the mistake will demonstrate their understanding of adding fractions where one denominator is a multiple of the other.

ASSESSMENT CHECKPOINT Children should be able to not only recognise that the answer is wrong, but also explain the mistake that has been made and find the correct answer.

ANSWERS Answers for the **Reflect** part of the lesson can be found in the *Power Maths* online subscription.

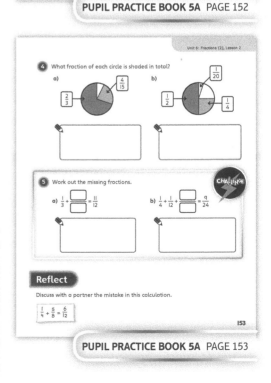

PUPIL PRACTICE BOOK 5A PAGE 153

After the lesson ⏸

- Can children use pictorial representations to add fractions where one denominator is a multiple of the other?
- Can children identify the common denominator needed to add a pair of fractions?
- Can children abstractly add fractions where one denominator is a multiple of the other?

Add fractions with a total greater than 1

Learning focus

In this lesson, children will add fractions with a sum greater than 1. Children will convert mixed numbers and improper fractions and achieve answers in their simplest form.

NATIONAL CURRICULUM LINKS

Year 5 Number – fractions (including decimals and percentages)

Add and subtract fractions with the same denominator and denominators that are multiples of the same number.

Recognise mixed numbers and improper fractions and convert from one form to the other and write mathematical statements > 1 as a mixed number [for example, $\frac{2}{5} + \frac{4}{5} = \frac{6}{5} = 1\frac{1}{5}$].

ASSESSING MASTERY

Children can confidently find equivalent fractions with the same denominator and use this to add fractions. Children can convert improper fractions into mixed numbers and vice-versa, knowing which representation is required at each stage in the solution process.

COMMON MISCONCEPTIONS

Children may add the numerators and place the result over the largest denominator, for example, $\frac{4}{5} + \frac{9}{10} = \frac{13}{10}$. They have not converted the fractions so they both have the same denominator before performing the addition. Ask:
- *What is a common denominator? Do all fractions need to have the same denominator before adding?*

When converting an improper fraction to a mixed number, children may not understand that the value of the numerator needs to be divided by the denominator, for example, $\frac{19}{15} = 19 \div 15 = 1$ remainder $4 = 1\frac{4}{15}$. Children may confuse fractions with the rules for place value and divide the numerator by 10, splitting 19 into 1 remainder 9. Ask:
- *How many parts are in one whole? How many parts are left over?*

STRENGTHENING UNDERSTANDING

The **Textbook** models how to draw fraction circles to represent the fractions involved. Some children will benefit from using concrete representations such as plates or circles which they can cut and manipulate. By moving the parts to form whole circles (each representing one whole), children can see the remaining parts.

GOING DEEPER

Ask children to explore and explain different methods. Encourage children to decide which method is most efficient depending on the context. Children may represent methods in more than one way, for example, fraction strips and fraction circles; this may be extended to number lines.

KEY LANGUAGE

In lesson: fraction, denominator, equivalent, total, improper fraction, mixed number, convert

Other language to be used by the teacher: whole, numerator, simplify, equal parts, sum

STRUCTURES AND REPRESENTATIONS

Fraction walls, fraction strips, fraction shapes

RESOURCES

Optional: fraction strips, fraction circles, scissors, analogue clocks, blank clock faces, blank number lines

 In the eTextbook of this lesson, you will find interactive links to a selection of teaching tools.

Quick recap

Write the fractions $\frac{1}{5}, \frac{3}{10}, \frac{7}{15}, \frac{11}{20}, \frac{1}{2}$ and $\frac{2}{3}$ on the board. Ask: *Which two fractions have a sum of $\frac{10}{15}$? Which two fractions have a difference of $\frac{1}{20}$? Which two fractions have a sum of $\frac{3}{4}$?*

Discover

WAYS OF WORKING Pair work

ASK

- Question ① a): *How long did Reena spend on her English homework?*
- Question ① a): *How can you represent $\frac{3}{4}$? Is there enough of the diagram left to show $\frac{1}{2}$? What can you do?*
- Question ① b): *How long did Reena spend on her maths homework? Do you think the answer is more or less than one hour?*

IN FOCUS Question ① b) introduces children to adding fractions with a total above 1 through a real-life context. They should use their understanding of time to recognise that Reena has spent longer than 1 hour on her homework, so her total time is above 1.

PRACTICAL TIPS Provide children with a blank clock face on paper that they can shade in to represent the times and work out the total time spent on the homework.

ANSWERS

Question ① a):

Question ① b): $\frac{3}{4} + \frac{1}{2} = 1\frac{1}{4}$, or we can write this as a mixed number $\frac{5}{4}$.

PUPIL TEXTBOOK 5A PAGE 212

Share

WAYS OF WORKING Whole class teacher led

ASK

- Question ① a): *How does this show $\frac{3}{4}$ of an hour?*
- Question ① a): *Is Sparks correct? How do you know?*
- Question ① b): *Is there enough room on the clock to shade another half an hour? How do you know? Why do you need a second clock?*

IN FOCUS In question ① b), children explore the addition of the two fractions using fractions represented on a clock face to aid their understanding. They should understand and be able to explain why the answer goes above 1, and use their knowledge of converting improper fractions and mixed numbers to give their answer as a mixed number.

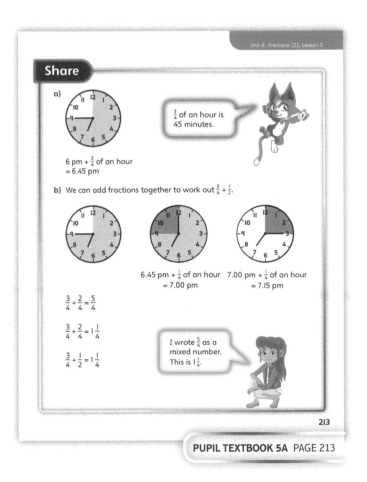

PUPIL TEXTBOOK 5A PAGE 213

Think together

WAYS OF WORKING Whole class teacher led (I do, We do, You do)

ASK

• Question **1**: *Do you need another circle? How do you know? What should you split the second circle into? How many parts are shaded in total? What is this as a mixed number?*
• Question **2**: *Which answers do you think are greater than 1? Why do you think this? How can you check your answers? Were you correct?*

IN FOCUS Children are given the opportunity to practise adding fractions with a total greater than 1. Pictorial representations are used in question **1**, but children need to recognise that they need to draw an additional whole in order to represent the additions.

STRENGTHEN Provide children with strips of paper or blank number lines that they can use to represent the calculations and support them in finding their answers.

DEEPEN In question **3** a), children use their knowledge of adding fractions with a total greater than 1 and their knowledge of inverse operations to find missing numbers in calculations. In question **3** b), children should be encouraged to find multiple different answers and explain what they notice.

ASSESSMENT CHECKPOINT Use questions **1** and **2** to assess whether children can confidently and accurately add fractions with a total greater than 1.

ANSWERS

Question **1** a): $\frac{5}{6} + \frac{1}{2} = \frac{8}{6}$ or $1\frac{2}{6}$ or $1\frac{1}{3}$

Question **1** b): $\frac{3}{4} + \frac{7}{8} = \frac{13}{8}$ or $1\frac{5}{8}$

Question **2** a): No, $\frac{2}{3} + \frac{1}{9} = \frac{7}{9}$

Question **2** b): Yes, $\frac{2}{3} + \frac{4}{9} = \frac{10}{9}$

Question **3** a): $\frac{3}{4} + \frac{7}{20} = 1\frac{2}{20}$ or $1\frac{1}{10}$

$\quad\quad\quad\quad \frac{4}{5} + \frac{7}{15} = 1\frac{4}{15}$

$\quad\quad\quad\quad \frac{11}{18} + \frac{4}{9} = 1\frac{1}{18}$

$\quad\quad\quad\quad \frac{4}{7} + \frac{11}{21} = 1\frac{2}{21}$

Question **3** b): Various answers, for example $\frac{7}{8} + \frac{5}{16} = 1\frac{3}{16}$

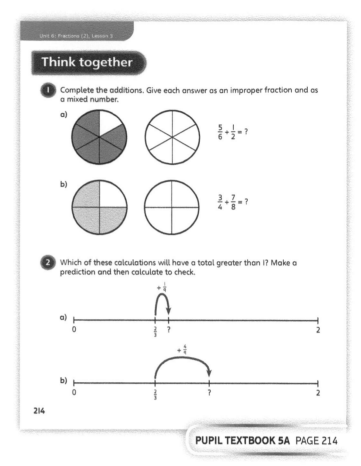

PUPIL TEXTBOOK 5A PAGE 214

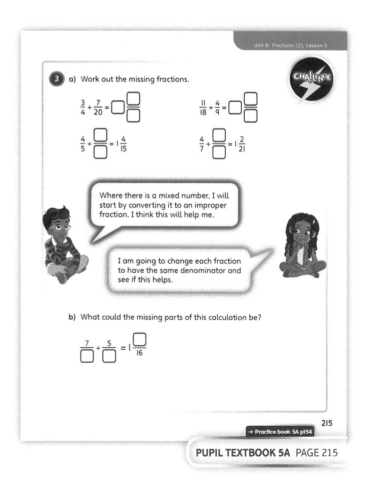

PUPIL TEXTBOOK 5A PAGE 215

Practice

WAYS OF WORKING Independent thinking

IN FOCUS Question ❶ aims to consolidate children's understanding of the need to find a common denominator to add fractions where the answer exceeds 1. Question ❸ encourages children to find the common denominator independently. Number lines are used as pictorial representations of the additions. This question presents a problem with numerators that are the same. Many children think the fractions do not need to be converted in this case, so the question will challenge this misconception.

STRENGTHEN Encourage children to find a common denominator of the fractions and, when diagrams are not given, to make their own representations.

DEEPEN Question ❻ can be explored further by giving children more complex missing number problems (for example, $\square + \frac{2}{3} = 1\frac{1}{6} + \square$) and asking them to find more than one solution.

ASSESSMENT CHECKPOINT Can children convert between improper fractions and mixed numbers? Are they achieving a final answer in its simplest form? Can they produce accurate diagrams, if required, which help them to find the answer as a mixed number? Can they identify the calculation from a picture and independently find fractions with the same denominator, while confidently converting between improper fractions and mixed numbers?

ANSWERS Answers for the **Practice** part of the lesson can be found in the *Power Maths* online subscription.

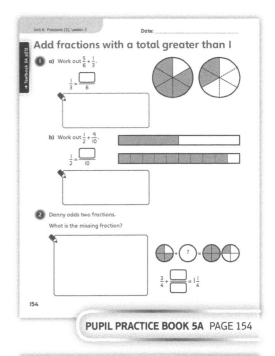

PUPIL PRACTICE BOOK 5A PAGE 154

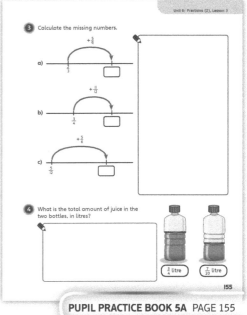

PUPIL PRACTICE BOOK 5A PAGE 155

Reflect

WAYS OF WORKING Pair work

IN FOCUS This question tests children's understanding of the steps needed to add two fractions together. Encourage them to explain the mistake and the correct process as well as calculating the correct answer.

ASSESSMENT CHECKPOINT This **Reflect** activity will assess children's ability to add two fractions where the answer exceeds 1. Look for children who can spot the mistake and explain the correct steps needed, showing understanding of finding a common denominator and converting an improper fraction to a mixed number.

ANSWERS Answers for the **Reflect** part of the lesson can be found in the *Power Maths* online subscription.

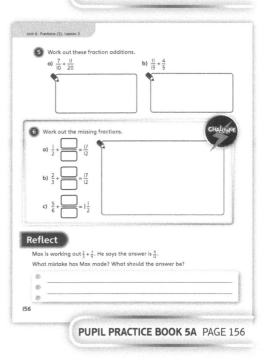

PUPIL PRACTICE BOOK 5A PAGE 156

After the lesson ⏸

- Can children add fractions by finding a common denominator and explain which fraction needs to be converted?
- Can children explain how to convert between improper fractions and mixed numbers?
- Can children explain how to find missing fractions in calculations?

251

Add to a mixed number

Learning focus

In this lesson, children continue their learning on adding fractions as they add to a mixed number. They start by adding only wholes or only parts, before adding other mixed numbers where one denominator is a multiple of the other.

Before you teach

- Can children add fractions where one denominator is a multiple of the other and the total is less than 1?
- Can children add fractions where one denominator is a multiple of the other and the total is above 1?

NATIONAL CURRICULUM LINKS

Year 5 Number – fractions (including decimals and percentages)

Add and subtract fractions with the same denominator and denominators that are multiples of the same number.

Recognise mixed numbers and improper fractions and convert from one form to the other and write mathematical statements > 1 as a mixed number [for example, $\frac{2}{5} + \frac{4}{5} = \frac{6}{5} = 1\frac{1}{5}$].

ASSESSING MASTERY

Children can add mixed numbers and improper fractions by first adding the whole numbers and then adding the parts. Children will be efficient in identifying a common denominator and finding equivalent fractions when dealing with questions involving fractions where one denominator is a multiple of the other.

COMMON MISCONCEPTIONS

Children may add the wholes, the numerators and the denominators, for example, in the calculation $4\frac{2}{5} + 2\frac{3}{10}$, children may answer $6\frac{5}{15}$. Remind them that fractions must have a common denominator before they can be added. Ask:

- *Do the fractions have the same denominator? What do we need to do to make the denominators the same?*
- *Do you need to add the numerators <u>and</u> add the denominators?*
- *Can you draw a fraction strip to help you?*

STRENGTHENING UNDERSTANDING

Children who are finding it difficult to add mixed numbers should use part-whole models to partition the mixed numbers into wholes and parts, so that they can consider these in isolation before combining them to find the total.

GOING DEEPER

Help children to gain further understanding of adding mixed numbers by presenting questions with missing numbers and placing these problems in a real-life context. Encourage children to draw their own diagrams and model their workings on a number line. This will deepen their ability to communicate their knowledge.

KEY LANGUAGE

In lesson: common denominator, whole, part, mixed number, fraction part, equivalent

Other language to be used by the teacher: simplify, proper fraction

STRUCTURES AND REPRESENTATIONS

Fraction strips, fraction diagrams, bar models, part-whole models

RESOURCES

Optional: blank number lines, fraction strips, laminated part-whole models

 In the eTextbook of this lesson, you will find interactive links to a selection of teaching tools.

Quick recap

Recap representing mixed numbers and ask children to explain them in words. For example, show $3\frac{2}{5}$ on the board and ask children to draw a diagram to represent this. Using their diagram, encourage them to explain that $3\frac{2}{5}$ is made up of 3 wholes and two fifths. Repeat for other fractions.

Discover

ASK

- Question **1** b): *Who has more water? How do you know?*
- Question **1** b): *Will all of Jamie's water fit in Andy's bucket? How do you know?*
- Question **1** b): *Will the total number of full buckets change? Will the amount in the partly-filled bucket change?*

IN FOCUS In question **1** b), children consider adding to a mixed number in a real-life context. They should use their existing understanding of fractions to recognise that Jamie's water will fit in Andy's bucket, because $\frac{1}{4}$ is less than $\frac{1}{2}$, so the number of wholes will not change.

PRACTICAL TIPS Provide children with diagrams of buckets with intervals showing quarters on the bucket (make sure the bucket diagrams have straight sides so all the quarters are visually equal). Ask them to shade in Andy's water, then shade in Jamie's water. Ask: *How much water is there in total?*

ANSWERS

Question **1** a): $3\frac{1}{2} = 3 + \frac{1}{2}$

Question **1** b): $3\frac{1}{2} + \frac{1}{4} = 3\frac{3}{4}$

Share

ASK

- Question **1** a): *Why is $3\frac{1}{2}$ the same as $3 + \frac{1}{2}$? How does the diagram show this?*
- Question **1** b): *Why can you add the fraction parts separately? Will this change the answer?*

IN FOCUS In question **1** b), children recognise that they can use their existing knowledge of adding fractions to support them when adding to a mixed number.

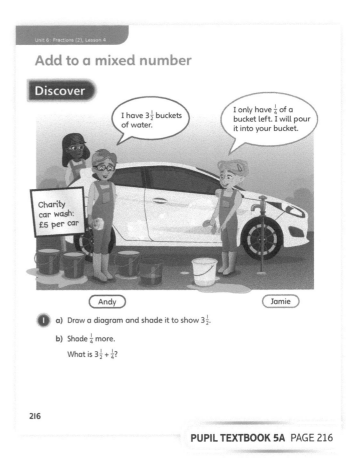

PUPIL TEXTBOOK 5A PAGE 216

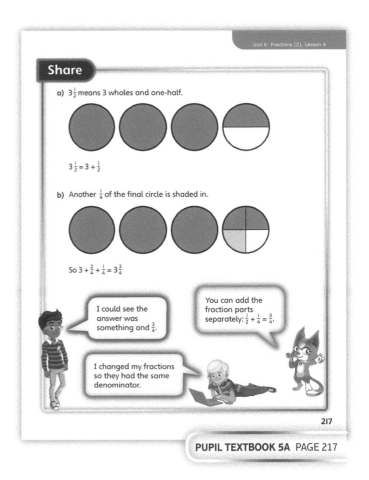

PUPIL TEXTBOOK 5A PAGE 217

Think together

Whole class teacher led (I do, We do, You do)

ASK

- Question **1**: *Will the whole change? Will the fraction change? How do you know?*
- Question **2**: *Which part is going to change? How do you know?*
- Question **3**: *Can you think of similar calculations for a partner to answer?*

IN FOCUS Questions **1** and **2** introduce children to adding to a mixed number by first considering adding only wholes or only fractions. They should recognise which part of the mixed number is going to change by considering whether they are adding wholes or fractions.

STRENGTHEN Provide children with a part-whole model that they can use to partition the mixed numbers into parts and wholes. This will help them to break the questions down further and support them in working out the answers.

DEEPEN In question **3**, encourage children to consider whether the order in which they add the fraction changes the answer. Ask: *Is one way easier than the other?*

ASSESSMENT CHECKPOINT Use questions **1** and **2** to assess whether children can add wholes or add fractions to a mixed number.

ANSWERS

Question **1** a): $2\frac{1}{4} + \frac{1}{2} = 2\frac{3}{4}$

Question **1** b): $2\frac{1}{4} + 3 = 5\frac{1}{4}$

Question **2** a): $5\frac{1}{3} + \frac{1}{3} = 5\frac{2}{3}$; $5\frac{1}{3} + 3 = 8\frac{1}{3}$

Question **2** b): $3\frac{1}{7} + \frac{4}{7} = 3\frac{5}{7}$; $3\frac{1}{7} + 4 = 7\frac{1}{7}$

Question **2** c): $8\frac{3}{10} + \frac{2}{5} = 8\frac{7}{10}$; $8\frac{3}{10} + 2 = 10\frac{3}{10}$

Question **3** a): $3\frac{1}{2} + \frac{1}{2} = 4$

Question **3** b): $3\frac{1}{4} + \frac{1}{4} = 3\frac{1}{2}$

Question **3** c): $3\frac{1}{4} + 2 + \frac{3}{4} = 6$

Question **3** d): $3\frac{1}{4} + \frac{1}{2} + \frac{1}{4} = 4$

Question **3** e): $3\frac{1}{4} + \frac{3}{4} + \frac{1}{2} = 4\frac{1}{2}$

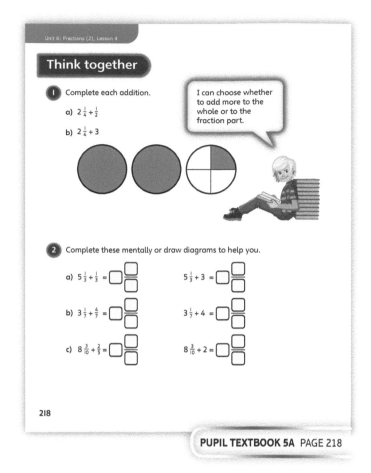

PUPIL TEXTBOOK 5A PAGE 218

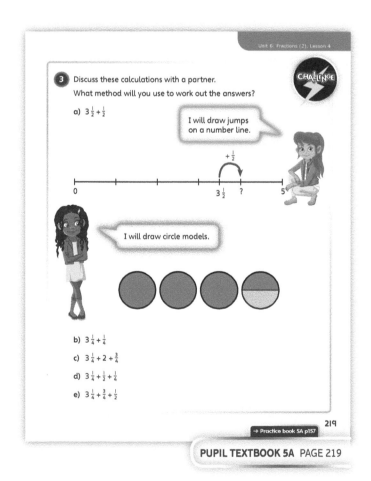

PUPIL TEXTBOOK 5A PAGE 219

Practice

IN FOCUS These questions provide an opportunity for children to consolidate adding to a mixed number by first adding only wholes or parts, before adding another mixed number. Children should be encouraged to consider which part of the mixed number is going to change: the whole, the fraction or both. Questions ❶ and ❷ provide pictorial representations before the questions become more abstract.

STRENGTHEN Provide children with a part-whole model that they can use to partition the mixed numbers into parts and wholes. This will help them to break down the questions further and support them in working out the answers.

DEEPEN In question ❻, ask: Did you need to work out the answers to match them? Why or why not? In question ❺, discuss the methods children used. Ask: *Did you work from left to right or do it another way? If so, why?*

ASSESSMENT CHECKPOINT Use questions ❶ and ❷ to assess whether children can add to a mixed number where pictorial representations are given for support. Use questions ❸ and ❹ to assess whether children can work abstractly to add to a mixed number.

ANSWERS Answers for the **Practice** part of the lesson can be found in the *Power Maths* online subscription.

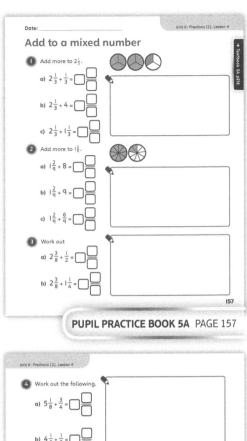

PUPIL PRACTICE BOOK 5A PAGE 157

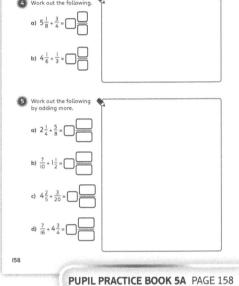

PUPIL PRACTICE BOOK 5A PAGE 158

Reflect

WAYS OF WORKING Pair work

IN FOCUS Children should first consider whether the answer will cross the whole or not and be able to explain why. They should use the most appropriate method for each addition.

ASSESSMENT CHECKPOINT Children should be able to add $\frac{1}{2}$ to each value using an appropriate method and explain their choice of method.

ANSWERS Answers for the **Reflect** part of the lesson can be found in the *Power Maths* online subscription.

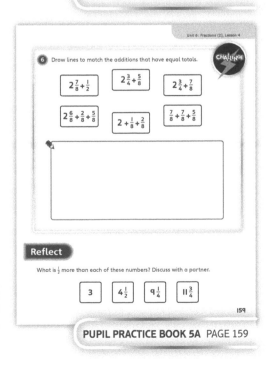

PUPIL PRACTICE BOOK 5A PAGE 159

After the lesson ⏸

- Can children add a whole number to a mixed number?
- Can children add a proper fraction to a mixed number?
- Can children add two mixed numbers?

Add two mixed numbers

Learning focus

In this lesson, children build on their understanding from the previous lesson to add pairs of mixed numbers. They partition the mixed numbers into parts and wholes to support their calculations and give their answers in the simplest form, where necessary.

Before you teach

- Can children partition mixed numbers into parts and wholes?
- Can children add a whole number to a mixed number?
- Can children add a fraction to a mixed number?

NATIONAL CURRICULUM LINKS

Year 5 Number – fractions (including decimals and percentages)

Add and subtract fractions with the same denominator and denominators that are multiples of the same number.

Recognise mixed numbers and improper fractions and convert from one form to the other and write mathematical statements > 1 as a mixed number [for example, $\frac{2}{5} + \frac{4}{5} = \frac{6}{5} = 1\frac{1}{5}$].

ASSESSING MASTERY

Children can partition a mixed number into parts and wholes and use this to add pairs of mixed numbers. They can find a common denominator where appropriate to complete the addition and give their answers in their simplest form, where necessary.

COMMON MISCONCEPTIONS

Children may add the wholes, the numerators and the denominators, for example, in the calculation $4\frac{2}{5} + 2\frac{3}{10}$, children may answer $6\frac{5}{15}$. Remind them that fractions must have a common denominator before they can be added. Ask:

- *Do the fractions have the same denominator? What do we need to do to make the denominators the same?*
- *Do you need to add the numerators <u>and</u> add the denominators?*
- *Can you draw a fraction strip to help you?*

STRENGTHENING UNDERSTANDING

Children who find it difficult to add mixed numbers should use part-whole models to partition the mixed numbers into wholes and parts, so that they can consider these in isolation before combining them to find the total.

GOING DEEPER

Help children to gain a deeper understanding of adding mixed numbers by presenting questions with missing numbers and placing these problems in a real-life context. Encourage children to draw their own diagrams and model their workings on a number line. This will deepen their ability to communicate their knowledge.

KEY LANGUAGE

In lesson: common denominator, whole, part, mixed number, fraction part, equivalent

Other language to be used by the teacher: simplify, proper fraction

STRUCTURES AND REPRESENTATIONS

Fraction strips, fraction diagrams, bar models, part-whole models, number lines

RESOURCES

Optional: strips of paper, blank number lines, fraction strips, laminated part-whole models

 In the eTextbook of this lesson, you will find interactive links to a selection of teaching tools.

Quick recap 🔍

Show children a calculation that involves adding to a mixed number and ask them whether the whole will change, the fraction will change, or both. Ask them to explain their reasoning. For example, in $2\frac{3}{5} + 5$ only the whole will change, in $2\frac{3}{5} + \frac{1}{5}$ only the fraction will change, and in $2\frac{3}{5} + \frac{4}{5}$ both will change because $\frac{3}{5} + \frac{4}{5} > 1$.

Discover

Add two mixed numbers

Discover

WAYS OF WORKING Pair work

ASK

• Question ① a): *Where can you see Toshi's journey?*
• Question ① a): *What mixed numbers do you need to represent on fraction strips? How many wholes have you drawn in total? What else have you drawn?*
• Question ① b): *How can you work out the total distance?*

IN FOCUS In question ① b), children are required to add two mixed numbers. They focus on first adding the wholes, then the parts, and then combining these to find the total.

PRACTICAL TIPS Provide children with strips of paper or blank number lines that can be used to represent the fractions to support them in calculating the answer.

ANSWERS

Question ① a):

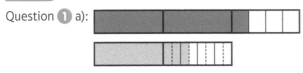

① a) Draw fraction strips for each part of Toshi's drive to work.

b) Find the total distance of Toshi's journey.

Question ① b): $2\frac{1}{4} + 1\frac{3}{8} = 3\frac{5}{8}$

220

Share

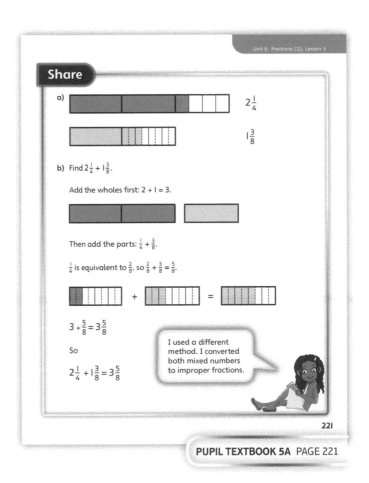

WAYS OF WORKING Whole class teacher led

ASK

• Question ① a): *How does the fraction strip show $2\frac{1}{4}$? How does the fraction strip show $1\frac{3}{8}$? Why is it important that the wholes are the same size?*
• Question ① b): *How many wholes are there altogether? How can you add $\frac{1}{4}$ and $\frac{3}{8}$? If Toshi has driven 3 whole kilometres and another $\frac{5}{8}$ of a kilometre, how far has Toshi driven in total?*

IN FOCUS In question ① a), children explore adding two mixed numbers using fraction strips to support their understanding. Using their knowledge of addition being commutative, they should recognise that you can add the wholes and parts separately, and combine them to find the total distance in question ① b).

Share

a)

$2\frac{1}{4}$

$1\frac{3}{8}$

b) Find $2\frac{1}{4} + 1\frac{3}{8}$.

Add the wholes first: $2 + 1 = 3$.

Then add the parts: $\frac{1}{4} + \frac{3}{8}$.

$\frac{1}{4}$ is equivalent to $\frac{2}{8}$, so $\frac{2}{8} + \frac{3}{8} = \frac{5}{8}$.

$3 + \frac{5}{8} = 3\frac{5}{8}$

So

$2\frac{1}{4} + 1\frac{3}{8} = 3\frac{5}{8}$

> I used a different method. I converted both mixed numbers to improper fractions.

221

Think together

Whole class teacher led (I do, We do, You do)

ASK

- Question **1**: *How many wholes are there altogether? What fractions do you need to add together? What common denominator should you use? What is the total?*
- Question **2**: *How many wholes are there? What fractions do you need to add together? How can you convert an improper fraction to a mixed number? What is the total?*

IN FOCUS Here children get to practise adding mixed numbers together, including an example where the fractional parts cross the whole in question **2**. Children should use their knowledge of adding fractions, along with their previous learning of adding to a mixed number to break question **3** down into more manageable parts.

STRENGTHEN If children are finding this concept difficult, encourage them to draw a part-whole model and partition the mixed numbers into parts and wholes to help them break down the questions further.

DEEPEN In question **3**, ask children to explore the different methods and discuss which one they prefer and why. Ask: *Is one method always better than the other, or does it depend on the question?*

ASSESSMENT CHECKPOINT Use questions **1** and **2** to assess whether children can add mixed numbers using pictorial representations for support.

ANSWERS

Question **1**: $1\frac{2}{5} + 1\frac{3}{10} = 2\frac{7}{10}$

Question **2**: $2\frac{3}{4} + \frac{5}{8} = 3\frac{3}{8}$

Question **3** a): $2\frac{1}{4} + 1\frac{3}{20} = 3\frac{8}{20}$ or $3\frac{2}{5}$

Question **3** b): $4\frac{6}{15}$ or $4\frac{2}{5}$

PUPIL TEXTBOOK 5A PAGE 222

PUPIL TEXTBOOK 5A PAGE 223

Practice

WAYS OF WORKING Independent thinking

IN FOCUS Children are given opportunity to consolidate their understanding of adding mixed numbers with varying levels of scaffolding. Questions ❶ and ❷ use pictorial representations to support children and also guide them, step by step, in how to approach answering the questions. They should break each question down so that they first add the wholes, then the parts, before combining them to find the answer. Question ❺ presents the addition of mixed numbers in the form of a word problem, and questions ❻ and ❼ move fully to the abstract form.

STRENGTHEN If children are finding this concept difficult, encourage them to draw part-whole models and partition the mixed numbers into parts and wholes to help them break down the questions further.

DEEPEN In question ❼, children combine their knowledge of inverse operations with their new learning to find a missing number in a calculation. Encourage them to compare their methods with a partner. Can they think of any other methods that could be used to solve this?

ASSESSMENT CHECKPOINT Use questions ❶ and ❷ to assess whether children can add mixed numbers where pictorial representations are given to support them. Use questions ❸, ❹ and ❻ to assess whether children can work more abstractly to add mixed numbers.

ANSWERS Answers for the **Practice** part of the lesson can be found in the *Power Maths* online subscription.

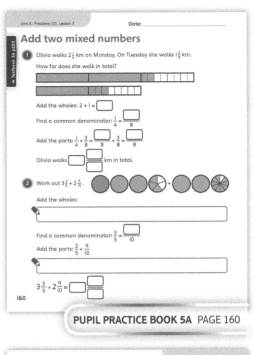

PUPIL PRACTICE BOOK 5A PAGE 160

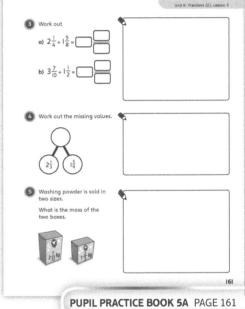

PUPIL PRACTICE BOOK 5A PAGE 161

Reflect

WAYS OF WORKING Independent thinking

IN FOCUS Children should explore completing the calculation both by partitioning into wholes and parts and by converting to improper fractions. They should be able to discuss the pros and cons of each method, and explain which one they prefer and why.

ASSESSMENT CHECKPOINT Can children explain how to subtract mixed numbers using an appropriate method and how calculations differ depending on the size of the fractional parts?

ANSWERS Answers for the **Reflect** part of the lesson can be found in the *Power Maths* online subscription.

After the lesson ⏸

- Can children add two mixed numbers where the fractions do not cross the whole?
- Can children add two mixed numbers where the fractions cross the whole?
- Can children add mixed numbers by converting to improper fractions?

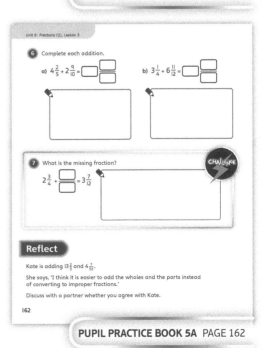

PUPIL PRACTICE BOOK 5A PAGE 162

Subtract fractions within 1

Learning focus

In this lesson, children build on their knowledge of adding fractions as they begin to subtract fractions within 1, where one denominator is a multiple of the other. They use their knowledge of equivalent fractions to find a common denominator and complete the subtractions.

Before you teach

- Can children subtract fractions with the same denominator?
- Can children identify what common denominator to use?
- Can children find equivalent fractions?

NATIONAL CURRICULUM LINKS

Year 5 Number – fractions (including decimals andpercentages)

Add and subtract fractions with the same denominator and denominators that are multiples of the same number.

Recognise mixed numbers and improper fractions and convert from one form to the other and write mathematical statements > 1 as a mixed number [for example, $\frac{2}{5} + \frac{4}{5} = \frac{6}{5} = 1\frac{1}{5}$].

ASSESSING MASTERY

Children can confidently approach questions involving subtracting fractions where one denominator is a multiple of the other. They can use their knowledge of common denominators and equivalent fractions to rewrite the question, so that both fractions have the same denominator, and then use this to find the answer.

COMMON MISCONCEPTIONS

Children may multiply the denominators to find a common denominator rather than spotting that one denominator is a multiple of the other. Ask:

- *Look at the denominators. What are the multiples of both denominators? What is the smallest multiple of both denominators?*

Children may subtract the denominators as well as the numerators. Ask:
- *Do you need to subtract the numerators <u>and</u> subtract the denominators?*
- *Can you draw a fraction strip to help you?*

STRENGTHENING UNDERSTANDING

Provide pictorial representations to aid children's understanding of why they need to first find a common denominator before they can subtract the fractions.

GOING DEEPER

Challenge children to write subtraction questions in context that give a desired answer.

KEY LANGUAGE

In lesson: fraction, common denominator, equivalent, denominator, difference, multiple

Other language to be used by the teacher: numerator, subtraction, divide, split, parts, multiply

STRUCTURES AND REPRESENTATIONS

Fraction walls, fraction strips, bar models

RESOURCES

Optional: paper (for folding)

 In the eTextbook of this lesson, you will find interactive links to a selection of teaching tools.

Quick recap

In order to recap prior learning on subtracting fractions with the same denominator, show children two representations of fractions with the same denominator, such as two fraction strips each folded or divided up into 6 equal parts, one with 1 part shaded and the other with 5 parts shaded. Ask: *How much more of the second fraction strip is shaded than the first?* Repeat for other diagrams.

Discover

WAYS OF WORKING Pair work

ASK

- Question **1** a): *What is half a tonne? How do you write this as a fraction?*
- Question **1** a): *Do you know any fractions that are equivalent to one-half?*
- Question **1** b): *How can you subtract $\frac{3}{10}$ from $\frac{1}{2}$? What do you need to do first?*

IN FOCUS In question **1** b), children are introduced to subtracting fractions in a real-life context, where one denominator is a multiple of the other. They should use their knowledge from adding fractions to recognise that they first need to find a common denominator before they can work out the answer.

PRACTICAL TIPS Provide children with strips of paper to represent the fractions and support them in working out the answer.

ANSWERS

Question **1** a):

Question **1** b): $\frac{1}{2} - \frac{3}{10} = \frac{2}{10}$

Subtract fractions within 1

Discover

1 a) Draw and shade a diagram to show $\frac{1}{2}$.
Convert your diagram to show $\frac{1}{2}$ as tenths.

b) Calculate $\frac{1}{2} - \frac{3}{10}$.

224

Share

WAYS OF WORKING Whole class teacher led

ASK

- Question **1** a): *How does the fraction strip show one-half? How has the diagram been changed to show tenths? How many tenths are equivalent to one-half?*
- Question **1** b): *Why are the two calculations the same? How do you know? Can you write $\frac{2}{10}$ in another way?*

IN FOCUS In question **1** b), children see subtraction represented in a fraction strip. They should notice that the $\frac{3}{10}$ has been crossed out, because it has been removed, and there is $\frac{2}{10}$ remaining.

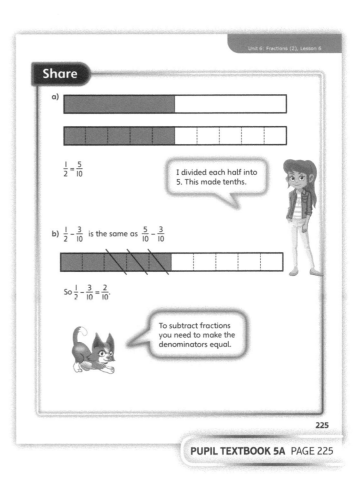

261

Think together

WAYS OF WORKING Whole class teacher led (I do, We do, You do)

ASK

• Question **1**: *What common denominator will you use? Why?*
• Question **2**: *What is the same about the subtractions? What is different? What do you notice?*

IN FOCUS Children are given the opportunity to practise subtracting fractions where one denominator is a multiple of the other. Question **1** provides pictorial support, whilst question **2** shows only the abstract representations. They should draw on their knowledge from earlier in the unit, when adding fractions, to support them in identifying a common denominator and use this to subtract the fractions.

STRENGTHEN Provide children with strips of paper to support them in representing the fractions and calculating the answers. Ask: *Why don't you need to draw both fractions? Why do you only need to represent the first one?*

DEEPEN In question **3**, children look at subtraction in the context of difference on a number line. They could be encouraged to find the smallest and greatest difference they can in question **3** a), and consider how many possible solutions there are in question **3** b).

ASSESSMENT CHECKPOINT Use question **1** to assess whether children can subtract fractions where one denominator is a multiple of the other when pictorial representations are given to support them. Use question **2** to assess whether children can work more abstractly to subtract fractions.

ANSWERS

Question **1** a): $\frac{1}{8}$

Question **1** b): $\frac{4}{9}$

Question **1** c): $\frac{1}{6}$

Question **2** a): $\frac{1}{3}$

Question **2** b): $\frac{3}{8}$

Question **2** c): $\frac{9}{20}$

Question **2** d): $\frac{49}{100}$

Question **2** e): $\frac{1}{6}$

Question **2** f): $\frac{1}{8}$

Question **2** g): $\frac{3}{20}$

Question **2** h): $\frac{7}{100}$

Question **3** a): Answers are dependent on the fractions chosen, for example the difference between $\frac{3}{25}$ and $\frac{38}{50}$ is $\frac{16}{25}$.

Question **3** b): For example, $\frac{1}{6} + \frac{1}{3} = \frac{1}{2}$

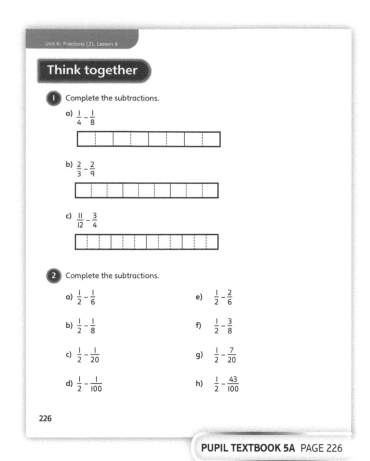

PUPIL TEXTBOOK 5A PAGE 226

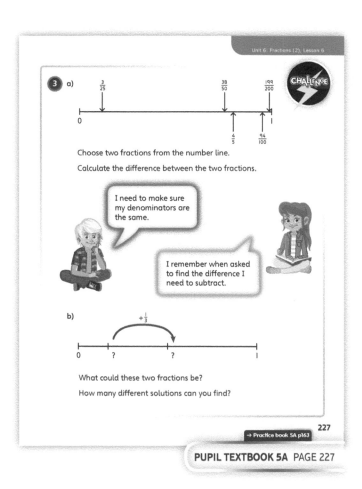

PUPIL TEXTBOOK 5A PAGE 227

Practice

WAYS OF WORKING Independent thinking

IN FOCUS Children are given the opportunity to consolidate their learning and subtract fractions where one denominator is a multiple of the other. Pictorial representations are provided to support children in answering question **1**. Question **2** focuses on abstract calculations. In questions **3** and **4**, children apply their learning in different contexts to solve problems.

STRENGTHEN Provide children with strips of paper to represent the fractions and support them in completing the subtractions.

DEEPEN In question **5**, ask: *Does it matter which order you complete the subtractions? Why or why not?* Ask them to experiment to recognise that it does matter because subtraction is not commutative.

ASSESSMENT CHECKPOINT Use question **1** to assess whether children can subtract fractions where one denominator is a multiple of the other when pictorial representations are given to support them. Use question **2** to assess whether children can work more abstractly to subtract fractions. Use questions **3** and **4** to assess whether children can apply these skills in contexts.

ANSWERS Answers for the **Practice** part of the lesson can be found in the *Power Maths* online subscription.

Reflect

WAYS OF WORKING Independent thinking

IN FOCUS Children demonstrate their learning by creating questions that have a given answer. Encourage them to be as creative as possible and add a context to their question.

ASSESSMENT CHECKPOINT Children should at least be able to provide three abstract subtractions that give an answer of $\frac{1}{4}$. They should be able to think creatively to use more difficult fractions, or apply a context to their question.

ANSWERS Answers for the **Reflect** part of the lesson can be found in the *Power Maths* online subscription.

After the lesson ⏸

- Can children identify a common denominator for a pair of fractions?
- Can children use equivalent fractions to rewrite a subtraction?
- Can children subtract fractions where one denominator is a multiple of the other?

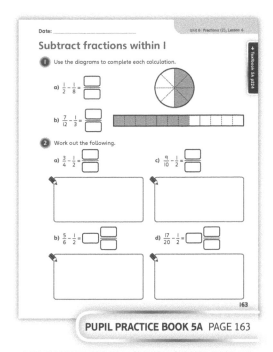

PUPIL PRACTICE BOOK 5A PAGE 163

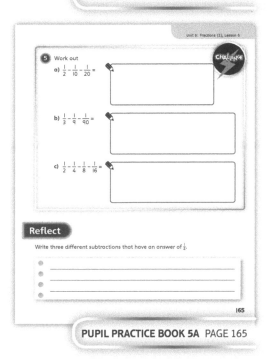

PUPIL PRACTICE BOOK 5A PAGE 164

PUPIL PRACTICE BOOK 5A PAGE 165

Subtract from a mixed number

Learning focus

In this lesson, children will subtract a fraction from a mixed number where one denominator is a multiple of the other. The calculations do not cross the whole.

Before you teach

- Can children subtract fractions that have the same denominator and fractions with different denominators?
- Can children convert between mixed numbers and improper fractions?
- Can children compare fractions by making the denominators equal?

NATIONAL CURRICULUM LINKS

Year 5 Number – fractions (including decimals and percentages)

Add and subtract fractions with the same denominator and denominators that are multiples of the same number.

Recognise mixed numbers and improper fractions and convert from one form to the other and write mathematical statements > 1 as a mixed number [for example, $\frac{2}{5} + \frac{4}{5} = \frac{6}{5} = 1\frac{1}{5}$].

ASSESSING MASTERY

Children can subtract a proper fraction from a mixed number where the fraction parts have related denominators. They can write equivalent fractions and use the 'crossing-out' method to subtract equal parts.

COMMON MISCONCEPTIONS

Children may focus on subtracting the fractional parts and forget about the whole number. For example, in the calculation $2\frac{3}{4} - \frac{1}{2}$, children may give an answer of $\frac{1}{4}$. Children need to understand that the whole number will still be in the answer. Show children the mixed numbers on fraction strips to help secure this understanding. Ask:

- *What number are you starting at? How much are you subtracting? What does the subtraction look like on fraction strips?*

STRENGTHENING UNDERSTANDING

If children find subtracting a fraction from a mixed number challenging, recap subtraction of proper fractions where one denominator is a multiple of the other. Strengthen their ability to visualise what the subtraction looks like by drawing different types of representations, such as fraction strips, number lines or fraction circles. When visualising subtraction with mixed numbers, ensure children can see why they should focus on subtracting the fractions and how the whole number stays the same.

GOING DEEPER

Give children some missing number problems that require subtraction, even though the question may be an addition, for example, $\boxed{} + 2\frac{2}{5} = 6\frac{11}{20}$. Have children explore problems with more than one calculation and ask them to draw fraction strips and compare them to help find the solution. For example, *Adam cycles $5\frac{7}{8}$ kilometres. Bethany cycles $\frac{1}{2}$ a kilometre less than Adam. Carl cycles $\frac{1}{4}$ a kilometre less than Bethany. How many kilometres does Carl cycle?*

KEY LANGUAGE

In lesson: common denominator, subtract, difference

Other language to be used by the teacher: equivalent, mixed number, improper fraction

STRUCTURES AND REPRESENTATIONS

Fraction strips, fraction shapes, number lines

RESOURCES

Optional: fraction strips, fraction shapes (circles)

 In the eTextbook of this lesson, you will find interactive links to a selection of teaching tools.

Quick recap

Show representations of mixed numbers to the class. Ask: *What is represented? How do you know? What is this mixed number made up of?* Repeat for other mixed numbers.

Discover

Unit 6: Fractions (2), Lesson 7

Subtract from a mixed number

Discover

WAYS OF WORKING Pair work

ASK

- Question **1** a): *How much water is in Amelia's watering can to start with? How much water does she use? How can you work out how much water is left?*
- Question **1** b): *How much water does Amelia have left? How much water does Danny have? How can you work out 'how much more'?*

IN FOCUS In question **1** a), children subtract a fraction from a mixed number, where the denominators of the fractions are the same. This introduces children to the 'crossing-out' method, subtracting equal parts to work out the answer.

Question **1** b) involves subtracting a fraction from a mixed number where one denominator is a multiple of the other. Children will need to find a common denominator and write an equivalent fraction, before subtracting equal parts.

PRACTICAL TIPS Use diagrams such as fraction circles or fraction strips to illustrate the original fraction and then 'cross out' the equal parts. In question **1** b), make sure children have found an equivalent fraction before subtracting the parts.

ANSWERS

Question **1** a): Amelia has $2\frac{7}{10}$ litres of water left.

Question **1** b): Amelia has $2\frac{3}{10}$ litres more water than Danny.

a) Amelia waters some flowers.

She uses $\frac{2}{10}$ of a litre of water from her watering can.

How much water does she have left?

b) How much more water than Danny does Amelia have now?

228

PUPIL TEXTBOOK 5A PAGE 228

Share

WAYS OF WORKING Whole class teacher led

ASK

- Question **1** a): *What do you notice about the denominators? What calculation do you need to do?*
- Question **1** b): *What do you notice about the denominators this time? Look at the diagram. Which part represents the difference?*

IN FOCUS In question **1** a), show children the diagrams of $2\frac{9}{10}$ and ask how they can find the amount Amelia has left. As the denominators are the same, children can easily subtract $\frac{2}{10}$ from $2\frac{9}{10}$ by crossing out two equal parts.

For question **1** b), discuss how to subtract $\frac{2}{5}$ from $2\frac{7}{10}$. Show children the fraction strips and point out the difference they are looking for. Discuss the need to find a common denominator and then an equivalent fraction before subtracting the fractions. (Point out that this is the same process as with addition.)

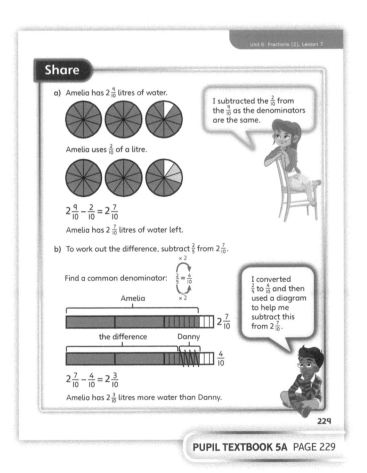

Share

a) Amelia has $2\frac{9}{10}$ litres of water.

Amelia uses $\frac{2}{10}$ of a litre.

$2\frac{9}{10} - \frac{2}{10} = 2\frac{7}{10}$

Amelia has $2\frac{7}{10}$ litres of water left.

I subtracted the $\frac{2}{10}$ from the $\frac{9}{10}$ as the denominators are the same.

b) To work out the difference, subtract $\frac{2}{5}$ from $2\frac{7}{10}$.

Find a common denominator: $\frac{2}{5} = \frac{4}{10}$

I converted $\frac{2}{5}$ to $\frac{4}{10}$ and then used a diagram to help me subtract this from $2\frac{7}{10}$.

$2\frac{7}{10} - \frac{4}{10} = 2\frac{3}{10}$

Amelia has $2\frac{3}{10}$ litres more water than Danny.

229

PUPIL TEXTBOOK 5A PAGE 229

Think together

WAYS OF WORKING Whole class teacher led (I do, We do, You do)

ASK

- Question ❶: *Can you subtract the fractions? What is a common denominator of the fractions?*
- Question ❷: *What calculation do you need to do? What is a common denominator of the fractions?*
- Question ❸: *Can you work out the common denominators?*

IN FOCUS Question ❶ requires children to subtract a proper fraction from a mixed number. There are diagrams of the starting fraction to support children's workings. The question progresses from fractions with the same denominator to fractions where one denominator is a multiple of the other.

Question ❷ looks at subtracting a fraction from a mixed number, where a diagram is provided but there is no support for the method. Encourage children to find a common denominator and show their method.

In question ❸, children subtract different fractions from a particular mixed number. Encourage children to find a common denominator and to discuss the patterns they notice.

STRENGTHEN Encourage children to use diagrams for support and to cross out the fractions they are subtracting.

DEEPEN After completing question ❸, encourage children to explore these concepts by presenting questions with a different starting number. Ask children to subtract proper fractions from this number and look for patterns. For example, start with $3\frac{23}{24}$ and subtract $\frac{1}{2}$, $\frac{2}{3}$, $\frac{3}{4}$, $\frac{5}{6}$, $\frac{7}{8}$ and $\frac{11}{12}$.

ASSESSMENT CHECKPOINT Can children subtract proper fractions from mixed numbers? Do they know how to determine a common denominator and find an equivalent fraction before subtracting? In question ❸, are children able to achieve the correct answers, identify a pattern and verbally describe the pattern in their own words?

ANSWERS

Question ❶ a): $3\frac{7}{8} - \frac{2}{8} = 3\frac{5}{8}$

Question ❶ b): $3\frac{7}{8} - \frac{3}{4} = 3\frac{7}{8} - \frac{6}{8} = 3\frac{1}{8}$

Question ❷: $4\frac{5}{6} - \frac{1}{3} = 4\frac{3}{6} = 4\frac{1}{2}$ pizzas left

Question ❸ a): $4\frac{11}{12} - \frac{11}{12} = 4$

$4\frac{11}{12} - \frac{5}{6} = 4\frac{1}{12}$

$4\frac{11}{12} - \frac{3}{4} = 4\frac{2}{12} = 4\frac{1}{6}$

$4\frac{11}{12} - \frac{2}{3} = 4\frac{3}{12} = 4\frac{1}{4}$

$4\frac{11}{12} - \frac{1}{2} = 4\frac{5}{12}$

Question ❸ b): Children spot patterns such as: 'As the numerator in the fraction that is being subtracted decreases, the numerator in the answer increases.'

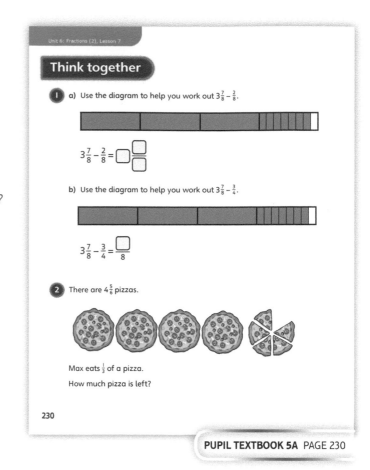

PUPIL TEXTBOOK 5A PAGE 230

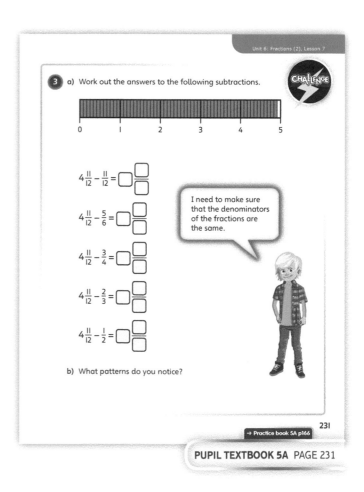

PUPIL TEXTBOOK 5A PAGE 231

Practice

WAYS OF WORKING Independent thinking

IN FOCUS In question ②, the fractions within the problems have one denominator that is a multiple of the other. Diagrams and support are given. Question ② c) looks at subtracting a whole number from a mixed number and subtracting a fraction, which results in a whole number.

Question ③ gives children a real-life context and asks them to subtract a fraction from a mixed number to work out how much is left. Diagrams are shown but no support is provided for the method. Encourage children to show their method step-by-step, with the equals signs underneath each other.

Question ④ looks at subtracting fractions from mixed numbers with no support and no context. Encourage children to look for common denominators and to show their method.

STRENGTHEN Encourage children to represent the fractions on fraction strips and to cross out the fractions to show a subtraction. Encourage children to use the structure presented in questions ① and ② to carry out their calculations.

DEEPEN Question ⑤ can be extended by giving children other missing number problems which involve subtracting a fraction from a mixed number. Ask questions where one denominator is a multiple of the other, rather than the denominators being the same.

ASSESSMENT CHECKPOINT Can children confidently subtract a proper fraction from a mixed number, knowing that the fraction and the fractional part of the whole number must have the same denominator? Can they identify that one denominator is a multiple of the other and, therefore, only one fraction needs to be converted to an equivalent fraction?

ANSWERS Answers for the **Practice** part of the lesson can be found in the *Power Maths* online subscription.

Reflect

WAYS OF WORKING Pair work

IN FOCUS This checks children's understanding of their method for subtracting a fraction from a mixed number. Encourage children to explain how they know that the whole number will not change, by finding a common denominator and comparing the fractional parts. Children should be able to explain that, since $\frac{3}{5}$ is greater than $\frac{3}{10}$, the answer is not going to be less than 2.

ASSESSMENT CHECKPOINT Can children explain why $\frac{3}{5}$ is larger than $\frac{3}{10}$, without the need for a diagram? Can children explain why it is possible to subtract the proper fraction from the fractional part of the mixed number without affecting the whole number?

ANSWERS Answers for the **Reflect** part of the lesson can be found in the *Power Maths* online subscription.

After the lesson ⏸

- Can children subtract a proper fraction from a mixed number?
- Can children recognise when one denominator is a multiple of the other and, therefore, change only one of the fractions?
- Can children show their solution process with clear steps and give the correct answer in its simplest form?

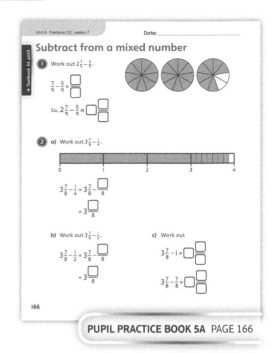

PUPIL PRACTICE BOOK 5A PAGE 166

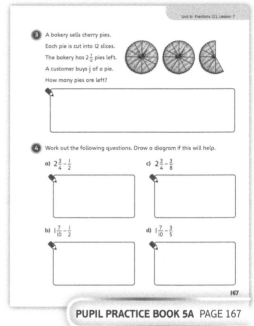

PUPIL PRACTICE BOOK 5A PAGE 167

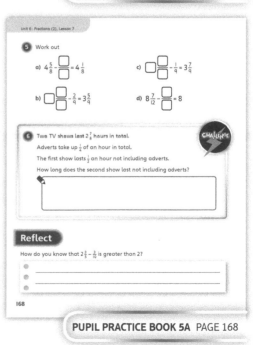

PUPIL PRACTICE BOOK 5A PAGE 168

Subtract from a mixed number – breaking the whole

Learning focus

In this lesson, children will subtract fractions from mixed numbers where the subtraction crosses the whole. They will split wholes into parts to subtract like fractions.

Before you teach

- Can children subtract fractions that have the same denominator?
- Can children subtract fractions with different denominators?
- Can children compare fractions by making the denominators equal?

NATIONAL CURRICULUM LINKS

Year 5 Number – fractions (including decimals and percentages)

Add and subtract fractions with the same denominator and denominators that are multiples of the same number.

Recognise mixed numbers and improper fractions and convert from one form to the other and write mathematical statements > 1 as a mixed number [for example, $\frac{2}{5} + \frac{4}{5} = \frac{6}{5} = 1\frac{1}{5}$].

ASSESSING MASTERY

Children can subtract a proper fraction from a mixed number where the subtraction crosses the whole. They can rewrite the mixed number by dividing one whole into parts and then subtract the required parts.

COMMON MISCONCEPTIONS

Children may, when subtracting a proper fraction from a mixed number, subtract the fractional parts without considering the order in which they are arranged. For example, in the calculation $3\frac{1}{4} - \frac{7}{8}$, children may calculate $\frac{7}{8} - \frac{1}{4} = \frac{5}{8}$ and then write the whole number at the front, obtaining an incorrect answer of $3\frac{5}{8}$. Children need to understand that, when subtracting, the order the numbers are presented in the question is important and the smaller fraction is not always subtracted from the larger fraction. Show children the mixed numbers and fractions in a diagram to help avoid this misconception and to secure understanding. Ask:

- *What does the start number look like? What fraction is being subtracted? When estimating the answer, what is the whole number?*

STRENGTHENING UNDERSTANDING

Some children may find it difficult to rewrite a mixed number as a whole number and a fractional part with a larger numerator, for example, $4\frac{1}{4} = 3\frac{5}{4}$. Demonstrate this using fraction circles or fraction strips and write an accompanying mathematical sentence, such as $4\frac{1}{4} = 3 + \frac{4}{4} + \frac{1}{4} = 3\frac{5}{4}$. Cross out parts on a diagram to help children understand why they need to make this change.

GOING DEEPER

Give children problems where they have to carry out more than one calculation, building on the previous lesson. This could include missing number problems such as $\boxed{} + \frac{3}{5} = 1\frac{2}{5} - \frac{7}{10}$.

KEY LANGUAGE

In lesson: common denominator, equivalent, whole, method

Other language to be used by the teacher: mixed number, convert, efficient, simplify, simplest form, numerator, denominator

STRUCTURES AND REPRESENTATIONS

Fraction strips, fraction shapes, bar models, number lines

RESOURCES

Optional: fraction shapes (circles), fraction strips, scissors, blank number lines, blank part-whole models

 In the eTextbook of this lesson, you will find interactive links to a selection of teaching tools.

Quick recap 🔁

As a class, practise counting on and back in mixed numbers on a number line. For example, show a number line from 2 to 4 where each whole is split into fifths. Ask children to count on, for example $2, 2\frac{1}{5}, 2\frac{2}{5}, \ldots$, and then back $4, 3\frac{4}{5}, 3\frac{3}{5}, \ldots$, with a specific focus on accuracy when counting back through a whole.

Discover

Unit 6: Fractions (2), Lesson 8

WAYS OF WORKING Pair work

ASK

- Question ➊ a): *What distance is the fun run? How many whole kilometres are there? What fraction of a kilometre is there?*
- Question ➊ a): *When Toshi has run 2 kilometres, does the whole number of kilometres or the fraction part change? How do you know?*
- Question ➊ b): *When Toshi has run a further $\frac{3}{4}$ kilometres, what changes now? How do you know?*

IN FOCUS In question ➊ b), children explore subtracting from a mixed number and breaking the whole in a real-life context. They should use their learning from earlier in the unit on adding to a mixed number, to support their understanding and develop their reasoning.

PRACTICAL TIPS Provide children with fraction strips or a blank number line to represent the question and support them in finding the answer.

ANSWERS

Question ➊ a): Toshi has $2\frac{1}{2}$ km left to run.

Question ➊ b): Toshi has $1\frac{3}{4}$ km left to run to complete the race.

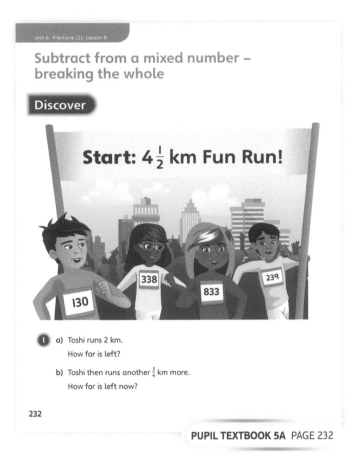

Subtract from a mixed number – breaking the whole

Discover

Start: $4\frac{1}{2}$ km Fun Run!

➊ a) Toshi runs 2 km.
How far is left?

b) Toshi then runs another $\frac{3}{4}$ km more.
How far is left now?

232

PUPIL TEXTBOOK 5A PAGE 232

Share

WAYS OF WORKING Whole class teacher led

ASK

- Question ➊ a): *How do you know the number line shows $4\frac{1}{2}$? Why have the jumps been drawn on? What does each jump represent? How much further does Toshi have left to run?*
- Question ➊ b): *What do the three smaller jumps represent? Why have we crossed over the whole? How much further has Toshi got left to run?*

IN FOCUS Children focus on subtracting from a mixed number using a number line to support them. In question ➊ b), they should understand that they need to cross the whole because $\frac{3}{4}$ is greater than $\frac{1}{2}$.

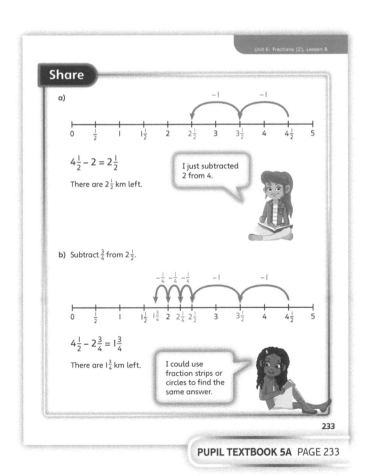

Share

a)

$4\frac{1}{2} - 2 = 2\frac{1}{2}$

There are $2\frac{1}{2}$ km left.

I just subtracted 2 from 4.

b) Subtract $\frac{3}{4}$ from $2\frac{1}{2}$.

$4\frac{1}{2} - 2\frac{3}{4} = 1\frac{3}{4}$

There are $1\frac{3}{4}$ km left.

I could use fraction strips or circles to find the same answer.

233

PUPIL TEXTBOOK 5A PAGE 233

Think together

Whole class teacher led (I do, We do, You do)

ASK

- Question **1**: *Why has Dexter also split one of the wholes into fifths? How does this help?*
- Question **1**: *Which method do you find easier, using a fraction strip or a number line? Why?*
- Question **2**: *What is the same about the calculations? What is different? Which do you find easier? Which method did you use?*

IN FOCUS Children are given the opportunity to practise subtracting from a mixed number by crossing the whole. Question **1** provides scaffolding and alternative methods for children to choose from. In question **2**, they can then use their preferred method as they answer abstract calculations. They should be encouraged to consider the similarities and differences between the set of questions, and what they notice in the answers.

STRENGTHEN Provide children with a part-whole model to partition the mixed numbers into parts and wholes to support them with breaking down the subtraction into more manageable steps. Encourage them to flexibly partition. For example, $\frac{3}{5} > \frac{1}{5}$, so when working out $3\frac{1}{5} - \frac{3}{5}$, they can partition $3\frac{1}{5}$ into 2 and $1\frac{1}{5}$ or 2 and $\frac{6}{5}$.

DEEPEN In question **3**, children combine their learning on equivalent fractions to answer more challenging subtraction questions. Ask: *Do you need to draw a fraction strip or number line every time? Can you think of any other methods? Which do you think is the most efficient?*

ASSESSMENT CHECKPOINT Use question **1** to assess whether children can subtract from a mixed number where pictorial representations are given to support them. Use question **2** to assess whether children can complete abstract subtractions. Question **3** can be used to assess whether children can subtract from a mixed number where they need to first find a common denominator.

ANSWERS

Question **1** a): $3\frac{1}{5}$ is $2\frac{6}{5}$. So $3\frac{1}{5} - \frac{3}{5} = 2\frac{6}{5} - \frac{3}{5} = 2\frac{3}{5}$.

Question **2** a): $4\frac{6}{7}$

Question **2** b): $3\frac{6}{7}$

Question **2** c): $4\frac{4}{7}$

Question **2** d): $3\frac{4}{7}$

Question **3** a): $2\frac{4}{5}$

Question **3** b): $2\frac{2}{3}$

$2\frac{7}{12}$

$2\frac{17}{20}$

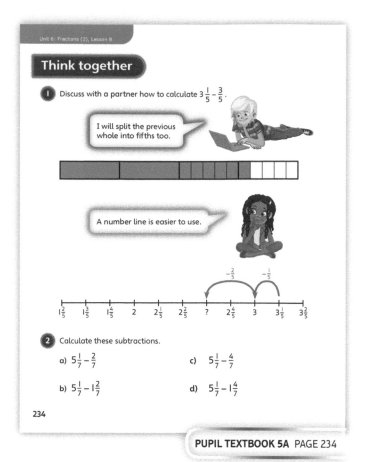

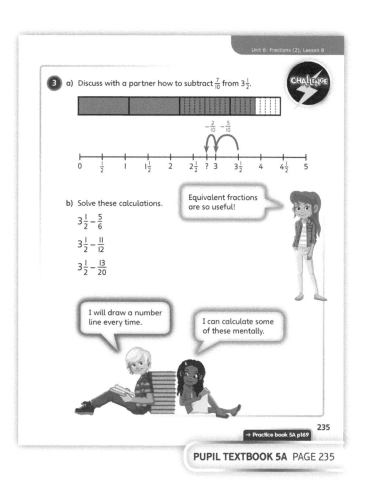

Practice

WAYS OF WORKING Independent thinking

IN FOCUS Questions **1** and **2** aim to consolidate children's understanding of subtracting a proper fraction from a mixed number, providing structure for a method and diagrams for support. In question **2**, children work out the missing numerator, given the answer. Provide time for discussion of the different methods children use to find the missing numbers.

Question **5** presents a problem in a real-life context. Children will identify the fractions and determine the calculations required to find the remaining fraction. Encourage children to draw a diagram to represent the sandwiches and to identify how much Danny eats before finding out how much is left.

STRENGTHEN When subtracting fractions, encourage children to represent the mixed number on a diagram or with fraction strips. Physically dividing one whole into equal parts and then crossing out the required number of parts will aid retention of the method used. Once children have completed question **3**, ask them which method they prefer and allow them to use this in further subtractions.

DEEPEN Provide children with missing number questions, similar in format to question **6**. Questions of this type involve a problem-solving element and allow children to explore different paths to reach an answer.

THINK DIFFERENTLY Question **5** develops children's understanding of fractions in the context of a real-life problem. It will be valuable to encourage children to explain their answers and develop their written reasoning.

ASSESSMENT CHECKPOINT Can children subtract a fraction from a mixed number by rewriting the mixed number and finding a common denominator? Are children's diagrams accurate and their written workings fluent? Can they solve problems by subtracting fractions?

ANSWERS Answers for the **Practice** part of the lesson can be found in the *Power Maths* online subscription.

Reflect

WAYS OF WORKING Pair work

IN FOCUS This question tests children's understanding of subtracting a fraction from a mixed number where the fraction they are subtracting is greater than the fractional part in the mixed number. Children may need to use a diagram to help them see why the answer is less than 2, or they may choose to find a common denominator and compare the fractional parts. Children should be able to explain that, since $\frac{9}{10}$ is greater than $\frac{2}{5}$, the answer is going to be less than 2.

ASSESSMENT CHECKPOINT Can children explain how to subtract a fraction from a mixed number and do they know why the size of the fraction they are subtracting will affect the whole number? Can children confidently explain why the answer will be less than 2?

ANSWERS Answers for the **Reflect** part of the lesson can be found in the *Power Maths* online subscription.

After the lesson ⏸

- Can children recognise when subtracting a fraction from a mixed number is going to affect the whole number?
- Can they rewrite a mixed number to help subtract a fraction?

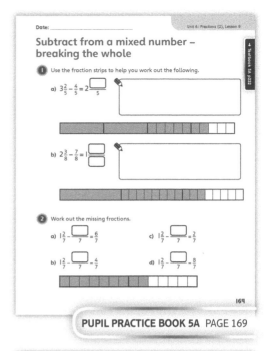

PUPIL PRACTICE BOOK 5A PAGE 169

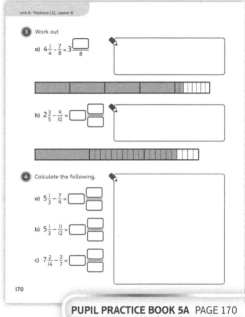

PUPIL PRACTICE BOOK 5A PAGE 170

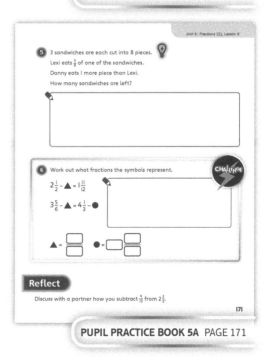

PUPIL PRACTICE BOOK 5A PAGE 171

Subtract two mixed numbers

Learning focus

In this lesson, children will subtract mixed numbers by subtracting the wholes and parts separately. The lesson also incorporates the concept of rewriting a mixed number in order to subtract the fractional part.

Before you teach

- Can children subtract fractions with different denominators?
- Can children explain why you can write, for example, $3\frac{1}{3}$ as $2 + \frac{4}{3}$?
- Are children familiar with adding mixed numbers by adding the wholes and parts separately?

NATIONAL CURRICULUM LINKS

Year 5 Number – fractions (including decimals and percentages)

Add and subtract fractions with the same denominator and denominators that are multiples of the same number.

Recognise mixed numbers and improper fractions and convert from one form to the other and write mathematical statements > 1 as a mixed number [for example, $\frac{2}{5} + \frac{4}{5} = \frac{6}{5} = 1\frac{1}{5}$].

ASSESSING MASTERY

Children can subtract mixed numbers by rewriting the mixed number and subtracting the wholes and parts separately.

COMMON MISCONCEPTIONS

When subtractions involve mixed numbers, children may subtract the wholes and subtract the fractions separately but incorrectly subtract the smaller fraction from the larger fraction, regardless of their order within the question. For example, in the calculation $5\frac{1}{3} - 2\frac{5}{6}$, children may work out $5 - 2 = 3$ and $\frac{5}{6} - \frac{1}{3} = \frac{3}{6}$, obtaining an incorrect answer of $3\frac{3}{6}$. Children need to understand that the order of the fractions does matter when subtracting. Provide a visual representation of the subtraction on a fraction strip to help avoid this misconception. Ask:

- *What does the start number look like on a fraction strip? What are you subtracting?*

STRENGTHENING UNDERSTANDING

Ensure children are confident rewriting a mixed number so they can subtract a fraction (as in the previous lesson) before looking at subtracting the wholes and fractions separately. In demonstrations, always draw a diagram to accompany the written process to help children visualise the subtraction. Encourage children to apply these skills when completing individual work.

GOING DEEPER

Present children with verbal problems involving subtraction with mixed numbers. For example, ask: *Find the difference between $1\frac{7}{8}$ and $3\frac{1}{4}$;* or: *How much more is $5\frac{2}{5}$ than $2\frac{7}{10}$?* Questions like this will help to incorporate key language and will also assess whether children know which number is the starting number when they are subtracting.

KEY LANGUAGE

In lesson: common denominator, whole, parts, method, subtract

Other language to be used by the teacher: equivalent, mixed number, convert, simplify, difference

STRUCTURES AND REPRESENTATIONS

Fraction strips

RESOURCES

Optional: fraction shapes (circles), fraction strips, blank number lines

 In the eTextbook of this lesson, you will find interactive links to a selection of teaching tools.

Quick recap 🔎

Show children fractions represented in uncommon forms, for example, $4\frac{7}{5}$. Ask them to rewrite the fraction as a mixed number and as an improper fraction, using their knowledge of wholes.

Discover

WAYS OF WORKING Pair work

ASK

- Question ① a): *What has Kate done first? Can you draw a diagram to represent the calculation?*
- Question ① b): *What do you notice about the fractional parts?*

IN FOCUS Question ① a) will help children to see that, when they are subtracting mixed numbers, they may be able to subtract the whole numbers and subtract the fractions separately.

Question ① b) introduces one way to deal with subtraction of mixed numbers where the fraction in the number to be subtracted is larger than the fraction in the other mixed number. Children will need to be confident rewriting a mixed number in a different form.

PRACTICAL TIPS Guide children into the question by asking everyone to draw a diagram showing $3\frac{3}{4}$. Some children will use fraction circles or shapes, others will represent the fraction on a fraction strip or number line. After children have performed the subtraction on their diagram, give them time to share their diagrams and discuss as a class the different representations.

ANSWERS

Question ① a): $3\frac{3}{4} - 1\frac{1}{2} = 3\frac{3}{4} - 1\frac{2}{4} = 2\frac{1}{4}$.

Kate is correct.

Question ① b): $3\frac{1}{2} - 1\frac{3}{4} = 1\frac{3}{4}$

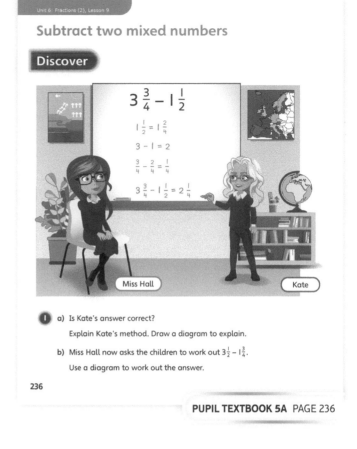

Subtract two mixed numbers

Discover

$$3\frac{3}{4} - 1\frac{1}{2}$$

$1\frac{1}{2} = 1\frac{2}{4}$

$3 - 1 = 2$

$\frac{3}{4} - \frac{2}{4} = \frac{1}{4}$

$3\frac{3}{4} - 1\frac{1}{2} = 2\frac{1}{4}$

Miss Hall   Kate

① a) Is Kate's answer correct?

Explain Kate's method. Draw a diagram to explain.

b) Miss Hall now asks the children to work out $3\frac{1}{2} - 1\frac{3}{4}$.

Use a diagram to work out the answer.

236

PUPIL TEXTBOOK 5A PAGE 236

Share

WAYS OF WORKING Whole class teacher led

ASK

- Question ① a): *What calculation do you need to do? Can you easily subtract $\frac{3}{4}$?*
- Question ① b): *What is the common denominator of the fractions? Can you use the same method as in part a)? How do you know? What method can you use to help?*

IN FOCUS Question ① a) asks children to look at the workings of a subtraction involving mixed numbers and determine whether the answer is correct. The method shows the wholes and fractional parts being subtracted separately. Fraction strips model the steps of this method, crossing out the whole and the required number of equal parts. As the fractional part of the first mixed number is larger than the fractional part of the second mixed number, the subtraction does not require wholes of the first mixed number to be crossed, as in the previous lesson.

Question ① b) requires $3\frac{1}{2}$ to be rewritten as $2\frac{6}{4}$ before subtracting. Use the diagram to explain this visually to children.

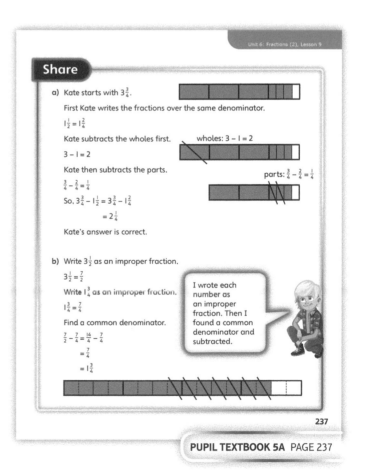

Share

a) Kate starts with $3\frac{3}{4}$.

First Kate writes the fractions over the same denominator.

$1\frac{1}{2} = 1\frac{2}{4}$

Kate subtracts the wholes first.   wholes: $3 - 1 = 2$

$3 - 1 = 2$

Kate then subtracts the parts.   parts: $\frac{3}{4} - \frac{2}{4} = \frac{1}{4}$

$\frac{3}{4} - \frac{2}{4} = \frac{1}{4}$

So, $3\frac{3}{4} - 1\frac{1}{2} = 3\frac{3}{4} - 1\frac{2}{4}$

$= 2\frac{1}{4}$

Kate's answer is correct.

b) Write $3\frac{1}{2}$ as an improper fraction.

$3\frac{1}{2} = \frac{7}{2}$

Write $1\frac{3}{4}$ as an improper fraction.

$1\frac{3}{4} = \frac{7}{4}$

Find a common denominator.

$\frac{7}{2} - \frac{7}{4} = \frac{14}{4} - \frac{7}{4}$

$= \frac{7}{4}$

$= 1\frac{3}{4}$

I wrote each number as an improper fraction. Then I found a common denominator and subtracted.

237

PUPIL TEXTBOOK 5A PAGE 237

Think together

WAYS OF WORKING Whole class teacher led (I do, We do, You do)

ASK

- Question **1**: *What fraction is equivalent to $\frac{4}{5}$ with a denominator of 10? Can you subtract the wholes and parts separately?*
- Question **2**: *What is the common denominator of the fractions? Can you easily subtract the wholes and parts? Why not? Can you rewrite $4\frac{2}{9}$? Now can you subtract the wholes and parts separately?*
- Question **3**: *What has Max done first? Do you need to divide a whole and then rewrite the mixed number before subtracting? How do you know?*

IN FOCUS Question **1** requires children to subtract mixed numbers. They are encouraged to find a common denominator and use the method of subtracting the wholes and parts separately.

Question **2** uses the same method but children need to rewrite the mixed number in order to subtract the fractional part. Diagrams and structured steps support children with this concept.

Question **3** introduces an alternative method and asks children to use this to carry out two further calculations. Children will find a common denominator and draw diagrams to support their workings.

STRENGTHEN In question **3**, encourage children to write down each step of their calculation, aligning the steps underneath each other as modelled in questions **1** and **2**. Children can use diagrams for support and cross out the fractions on their diagram to show their subtractions.

DEEPEN Question **3** demonstrates Max's method. Ask children to describe and demonstrate three different ways of answering a subtraction question involving mixed numbers that cross the whole. Encourage children to show and explain their steps progressively, using clear diagrams where necessary.

ASSESSMENT CHECKPOINT Can children subtract mixed numbers by subtracting the wholes and parts separately? Can they subtract mixed numbers by rewriting the first mixed number, then subtracting the wholes and parts separately? Can they subtract the fractions using written strategies and draw diagrams, if required, for support? Do they select the method that is most efficient for the question being asked? Can they use alternative methods to carry out a subtraction involving mixed numbers?

ANSWERS

Question **1**: $3\frac{4}{5} = 3\frac{8}{10}$; subtract the wholes: $3 - 1 = 2$;
subtract the parts: $\frac{8}{10} - \frac{7}{10} = \frac{1}{10}$
$3\frac{4}{5} - 1\frac{7}{10} = 2\frac{1}{10}$

Question **2**: $2\frac{1}{3} = 2\frac{3}{9}$; $4\frac{2}{9} - 2\frac{3}{9} = 3\frac{11}{9} - 2\frac{3}{9} = 1\frac{8}{9}$

Question **3** a): $3\frac{7}{10} - 1 = 2\frac{7}{10}$;
$2\frac{7}{10} - \frac{1}{2} = 2\frac{7}{10} - \frac{5}{10} = 2\frac{2}{10} = 2\frac{1}{5}$

Question **3** b): $10\frac{3}{8} - 4 = 6\frac{3}{8}$;
$6\frac{3}{8} - \frac{11}{16} = 6\frac{6}{16} - \frac{11}{16} = 5\frac{22}{16} - \frac{11}{16} = 5\frac{11}{16}$

1 Work out $3\frac{4}{5} - 1\frac{7}{10}$.

$3\frac{4}{5} - 1\frac{7}{10} = \boxed{}\dfrac{\boxed{}}{10}$

I will try to work out each one using different methods.

2 Work out $4\frac{2}{9} - 2\frac{1}{3}$.

$4\frac{2}{9} - 2\frac{1}{3} = \boxed{}\dfrac{\boxed{}}{\boxed{}}$

238

PUPIL TEXTBOOK 5A PAGE 238

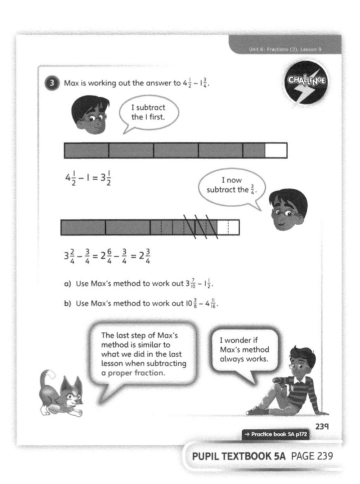

3 Max is working out the answer to $4\frac{1}{2} - 1\frac{3}{4}$.

CHALLENGE

I subtract the 1 first.

$4\frac{1}{2} - 1 = 3\frac{1}{2}$

I now subtract the $\frac{3}{4}$.

$3\frac{2}{4} - \frac{3}{4} = 2\frac{6}{4} - \frac{3}{4} = 2\frac{3}{4}$

a) Use Max's method to work out $3\frac{7}{10} - 1\frac{1}{2}$.

b) Use Max's method to work out $10\frac{3}{8} - 4\frac{11}{16}$.

The last step of Max's method is similar to what we did in the last lesson when subtracting a proper fraction.

I wonder if Max's method always works.

239

→ Practice book 5A p172

PUPIL TEXTBOOK 5A PAGE 239

Practice

WAYS OF WORKING Independent thinking

IN FOCUS In question ③, children will need to rewrite the mixed number before they can subtract the fractional parts. The question has a methodical structure and a diagram for support.

Question ⑥ gives children an opportunity to reason. It encourages them to compare the wholes and parts to estimate an answer without doing the full calculations.

Question ⑦ gives a real-life context and asks children to solve a problem that requires them to both add and subtract with mixed numbers. Children may draw a diagram to help them decide which calculations they need to do. Children should see that the towns could be ordered A B C or B A C.

STRENGTHEN When rewriting mixed numbers and subtracting the wholes and parts, encourage children to represent the calculations on a diagram to support their calculations.

DEEPEN Question ⑥ can be explored further by giving children other calculations and asking how they know whether the answer will be below a certain number.

ASSESSMENT CHECKPOINT Can children subtract mixed numbers by subtracting the wholes and parts separately? Can children identify when it is necessary to rewrite the mixed number? (Questions ③ and ⑤ will help to assess their ability to do this.) Can children solve problems involving addition and subtraction of mixed numbers?

ANSWERS Answers for the **Practice** part of the lesson can be found in the *Power Maths* online subscription.

Reflect

WAYS OF WORKING Pair work

IN FOCUS This question tests children's understanding of subtraction where the fraction in the mixed number they are subtracting is greater than the fraction in the mixed number they are subtracting from. It addresses a common mistake: subtracting the smaller fraction from the larger fraction regardless of their order within the question. Children should be able to explain that, since $\frac{3}{4}$ is greater than $\frac{1}{12}$, the first mixed number needs to be rewritten.

ASSESSMENT CHECKPOINT Can children explain how to subtract mixed numbers and do they know why the order of the fractions within the question is important?

ANSWERS Answers for the **Reflect** part of the lesson can be found in the *Power Maths* online subscription.

After the lesson ⏸

- Can children subtract mixed numbers by subtracting the wholes and subtracting the parts separately?
- Can they recognise when they are going to need to rewrite the calculation when subtracting with mixed numbers?
- Can they rewrite a mixed number to help subtract a fraction?

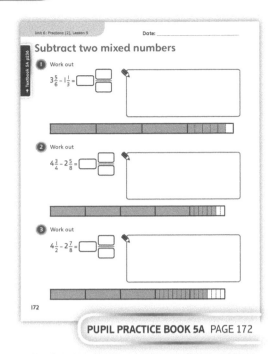

PUPIL PRACTICE BOOK 5A PAGE 172

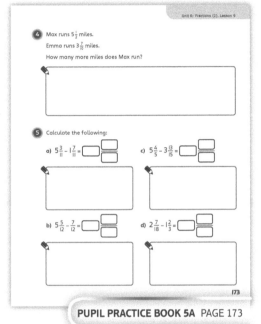

PUPIL PRACTICE BOOK 5A PAGE 173

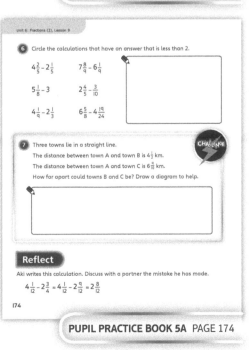

PUPIL PRACTICE BOOK 5A PAGE 174

Solve fraction problems

Learning focus

In this lesson, children will add and subtract mixed numbers in the context of word problems. They will practise solving problems by adding and subtracting wholes and parts.

Before you teach

- Can children find and use common denominators?
- Are they confident using improper fractions and mixed numbers?
- Can they solve a range of addition and subtraction calculations with fractions and mixed numbers, using an appropriate method?

NATIONAL CURRICULUM LINKS

Year 5 Number – fractions (including decimals and percentages)

Add and subtract fractions with the same denominator and denominators that are multiples of the same number.

ASSESSING MASTERY

Children can interpret the given information to create an appropriate calculation to solve the problem. Children can apply their knowledge of fractions to give the correct answer, stating it in the correct context.

COMMON MISCONCEPTIONS

Children may not be able to identify what calculation or operation is needed. Make problems as practical as possible, with resources, images and models; this may help children to recognise what is happening and, therefore, what operation is required. If children are not confident with number order, provide opportunities to talk through the problem. Ask:

- *What is the problem asking you to find? Are there key words in the question that tell you what you need to do, such as 'total' or 'difference'?*

STRENGTHENING UNDERSTANDING

Putting mixed numbers into a part-whole model may help children to see, quickly and easily, how many wholes and fractions are in each amount. Children can then determine what can be added or subtracted in its current form and what needs changing. Pictorial representations such as bar models and fraction strips will help to strengthen children's understanding of adding and subtracting mixed numbers.

GOING DEEPER

Ask children to show their working out in a variety of ways – fraction strips, bar models, number lines, and so on. Some children may be able to convert fractions mentally. These children should be encouraged to explain how they are able to do this, stating any useful methods they have found.

KEY LANGUAGE

In lesson: problem, total, difference, add, subtract, wholes, parts, fractions

Other language to be used by the teacher: calculation, operation, method, denominators, common, mixed number, improper fraction

STRUCTURES AND REPRESENTATIONS

Fraction strips, part-whole models, number lines

RESOURCES

Mandatory: paper for fraction strips

Optional: lengths of ribbon with fifths, tenths, quarters and twentieths marked on them

 In the eTextbook of this lesson, you will find interactive links to a selection of teaching tools.

Quick recap

Play fractions bingo to consolidate all the learning in this unit so far. Give children a series of fractions to choose from, and then present them with questions that cover both addition and subtraction of fractions and mixed numbers. If they have the answer, they cross it off. The first person to cross out all of their fractions wins.

Discover

WAYS OF WORKING Pair work

ASK

- Question ❶ a): *What does 'total' mean? What sort of number sentence will you need?*
- Question ❶ a): *How many wholes are there? Can you add the fractions as they are?*
- Question ❶ b): *What sort of calculation do you need to do? Do the fractions have the same denominators?*

IN FOCUS Question ❶ a) focuses on adding two mixed numbers. The numbers have denominators that are multiples of one another. Children will need to count the wholes and then find a common denominator in order to add the fractions. For question ❶ b), children are asked to find the difference between two mixed numbers. As in question ❶ a), the denominators of the fractions are multiples of one another. Children need to find a common denominator before finding the difference.

PRACTICAL TIPS You could use lengths of ribbon with fifths, tenths, quarters and twentieths marked on them. This will provide visual support for children when finding the equivalent fractions.

ANSWERS

Question ❶ a): Holly used $4\frac{3}{10}$ metres of ribbon in total.

Question ❶ b): Holly used $2\frac{2}{5}$ metres more of the dotty fabric.

Solve fraction problems

Discover

For these two dresses I used $1\frac{3}{5}$ metres of red ribbon and $2\frac{7}{10}$ metres of yellow ribbon.

Holly

❶ a) How much ribbon did Holly use in total to make the two dresses?

b) Holly used $4\frac{3}{4}$ metres of fabric for the dotty dress. She used $2\frac{7}{20}$ metres of fabric for the stripey dress. How much more dotty fabric did Holly use?

240

PUPIL TEXTBOOK 5A PAGE 240

Share

WAYS OF WORKING Whole class teacher led

ASK

- Question ❶ a): *What is the problem asking you to do with the lengths of ribbon?*
- Question ❶ a): *What can you do to the fractions so they can be added together?*
- Question ❶ b): *What method can you use to find the difference? Can you subtract the lengths as they are? Why not? What can you do to the fractions so they can be subtracted?*

IN FOCUS For question ❶ a), children could have their own fraction strips representing the mixed numbers. They could then group the wholes together and group the fractional parts together. Discuss and model why the fractions cannot be added as they are. As a class, find the common denominator and record an equivalent fraction, before adding the wholes and then adding the parts separately to find the total. Ask children whether the fraction is in its simplest form and encourage them to simplify it. For question ❶ b), children will have to work out what operation is needed to solve the problem and choose an appropriate method. Consider whether the fractions can be subtracted as they are and discuss why not. As with question ❶ a), create fraction strips and use them to find equivalent fractions. Encourage or model the use of fraction strips to count the new parts.

Share

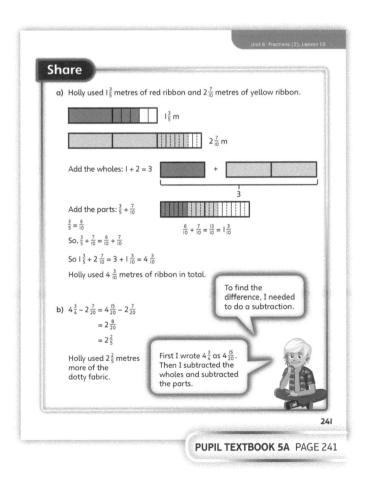

a) Holly used $1\frac{3}{5}$ metres of red ribbon and $2\frac{7}{10}$ metres of yellow ribbon.

$1\frac{3}{5}$ m

$2\frac{7}{10}$ m

Add the wholes: $1 + 2 = 3$

Add the parts: $\frac{3}{5} + \frac{7}{10}$

$\frac{3}{5} = \frac{6}{10}$

So, $\frac{3}{5} + \frac{7}{10} = \frac{6}{10} + \frac{7}{10}$

$\frac{6}{10} + \frac{7}{10} = \frac{13}{10} = 1\frac{3}{10}$

So $1\frac{3}{5} + 2\frac{7}{10} = 3 + 1\frac{3}{10} = 4\frac{3}{10}$

Holly used $4\frac{3}{10}$ metres of ribbon in total.

b) $4\frac{3}{4} - 2\frac{7}{20} = 4\frac{15}{20} - 2\frac{7}{20}$

$= 2\frac{8}{20}$

$= 2\frac{2}{5}$

Holly used $2\frac{2}{5}$ metres more of the dotty fabric.

To find the difference, I needed to do a subtraction.

First I wrote $4\frac{3}{4}$ as $4\frac{15}{20}$. Then I subtracted the wholes and subtracted the parts.

241

PUPIL TEXTBOOK 5A PAGE 241

Think together

WAYS OF WORKING Whole class teacher led (I do, We do, You do)

ASK

- Question **1**: *What does the problem ask you to do with the lengths? What method will you use to add the lengths?*
- Question **2**: *How much ribbon does Holly have? How many pieces has it been cut into? What operation can you use to find the length of the second piece? How can you make the fractions easier to subtract?*
- Question **3**: *What information do you have? What is the question asking you to do with the information?*
- Question **3**: *Can you add the amounts of fabric together as they are? How are you going to recombine the wholes and the fractions after adding?*

IN FOCUS Question **1** gives children another opportunity to find a total. Children will need to find a common denominator and change the fraction accordingly. Once the fractions have been added, children should remember to recombine with the wholes. For question **2**, children need to find the missing length by subtracting the given length from the total. They will need to convert the fractions so they have a common denominator. Children should realise that it is still difficult to subtract the fractions and that the easiest method is to convert the mixed numbers into improper fractions and then subtract. Question **3** requires the addition of four amounts. The amounts are made up of whole numbers and mixed numbers. The two fractions have different denominators but one is a multiple of the other.

STRENGTHEN The fraction strips in question **1** will help children visualise how many wholes there are and why the fractions cannot be added as they are. Some children may find it beneficial to make a fraction strip, folding or marking up the parts to identify the relationship between the fractions. Drawing or creating strips will also help children to see the relationship between denominators and convert fractions correctly.

DEEPEN Once children have answered question **3**, challenge them to find the totals needed for a range of orders (for example, two pairs of trousers and a shirt, four dresses). Alternatively (or as well), give children a total amount of fabric and ask them to investigate the different combinations of items Holly could make.

ASSESSMENT CHECKPOINT Do children know which operation to use and which methods to apply to find a solution? Can they recombine the wholes and fractions? Can they identify when it is not efficient to subtract certain fractions? Can children add more than two amounts?

ANSWERS

Question **1**: Holly needs $4\frac{1}{3}$ metres of fabric.

Question **2**: The other piece of ribbon is $1\frac{16}{25}$ metres long.

Question **3**: Holly uses $18\frac{2}{5}$ metres of fabric in total.

PUPIL TEXTBOOK 5A PAGE 242

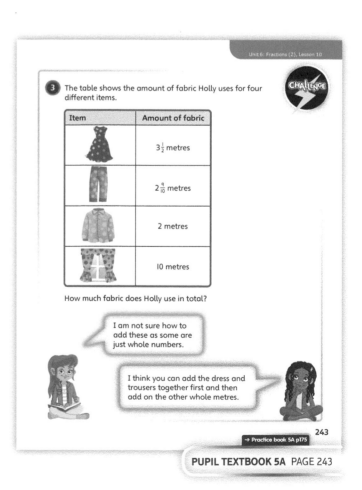

PUPIL TEXTBOOK 5A PAGE 243

Practice

WAYS OF WORKING Independent thinking

IN FOCUS Questions **1**, **2** a) and **3** provide opportunities to practise adding fractions and wholes to find totals. In all three questions, the fractions have denominators that are multiples of one another. Children must remember the final step of combining the wholes and the total of the fractional parts.

In question **2** b), children use their answer to part a) to find the difference between the full bag and the fraction already eaten. They need to understand that the full bag is one whole, or ten-tenths. Question **4** a) involves adding two mixed numbers. The fractions have different denominators but, as in earlier questions, these are multiples of one another. Question **4** b) asks children to find the difference between the two distances given. They may choose to subtract or count on to find the difference.

STRENGTHEN Children may create fraction strips to use as visual resources. For questions **3** and **4** a), children may benefit from putting the amounts into a part-whole model, so they can easily see what can be added as it is and what needs to be changed. For question **4** b), they may find it easier to use a fraction strip or number line.

When solving the Challenge question, children could convert all of the fractions into eighths initially to find answers. Some children may prefer to convert as and when necessary.

DEEPEN Once children have answers to questions **2** a) and b), they could use the information to reason further, for example, how many bags would be needed for a week, whether ☐ bags would last the rabbit ☐ days, or how much two rabbits would eat per day or week.

ASSESSMENT CHECKPOINT Do children understand vocabulary related to the four operations? Can they identify what operation is needed? Do children have secure methods for solving calculations? Can they partition mixed numbers into wholes and parts?

ANSWERS Answers for the **Practice** part of the lesson can be found in the *Power Maths* online subscription.

Reflect

WAYS OF WORKING Pair work

IN FOCUS This question will show children's understanding of why and when subtraction happens. Their ability to use the given numbers in the correct place will highlight their understanding. Children could give their problem to another pair to solve; this will give them another opportunity to practise subtracting mixed numbers.

ASSESSMENT CHECKPOINT Do children understand subtraction in real-life contexts? Can they choose an appropriate scenario to create a subtraction? Can children use the given mixed numbers in the correct places in their problem? Can they solve the calculation by finding a common denominator?

ANSWERS Answers for the **Reflect** part of the lesson can be found in the *Power Maths* online subscription.

After the lesson ⏸

- Can children choose the correct operation for a given problem?
- Can children add and subtract a range of fractions?
- Can children recombine whole numbers and fractions?

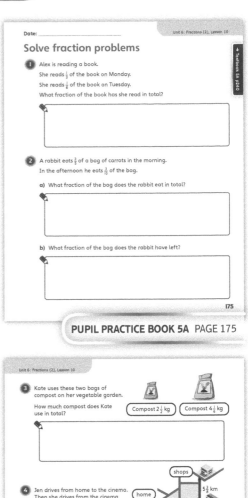

PUPIL PRACTICE BOOK 5A PAGE 175

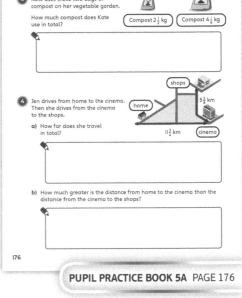

PUPIL PRACTICE BOOK 5A PAGE 176

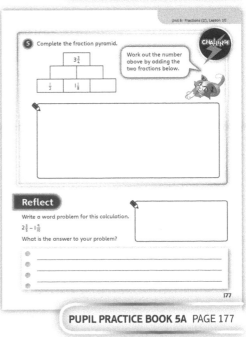

PUPIL PRACTICE BOOK 5A PAGE 177

Solve multi-step fraction problems

Learning focus

In this lesson, children will solve multi-step addition and subtraction word problems using fractions and mixed numbers. They will interpret what is being asked based on a real-life context, write the problem as a number sentence and achieve an answer.

Before you teach

- Do children recognise and understand mathematical vocabulary?
- Can children complete one-step word problems?
- Can children convert fractions to a common denominator?

NATIONAL CURRICULUM LINKS

Year 5 Number – fractions (including decimals and percentages)

Add and subtract fractions with the same denominator and denominators that are multiples of the same number.

ASSESSING MASTERY

Children can identify what a problem is asking and create appropriate calculations. They will break down multi-step problems into smaller, more manageable steps, and be secure with methods for adding and subtracting fractions to solve word problems.

COMMON MISCONCEPTIONS

Children may believe there is only one way to solve a problem. Allow children the freedom to explore the problem and find their own path to the solution. Supply them with the tools to explore; for example, make the problem practical, teach methods of modelling the problem and show them how to break a problem into manageable steps. These tools will help children to comprehend different word problems and to use appropriate strategies and operations to solve problems. Ask:

- *What information do you have? What are you trying to find? Can you draw a picture of the problem?*

STRENGTHENING UNDERSTANDING

The **Textbook** uses a variety of models to demonstrate problems, such as fraction strips, bar models and number lines. It is important for children to be able to visualise and then model each problem. Seeing the problem will strengthen their understanding of what information they have, what they are trying to find and the operations they need to perform. Allow ample time for modelling, exploration and discussion.

GOING DEEPER

Children can show their working out in a variety of ways – fraction strips, bar models, number lines, etc. Encourage them to verbalise their method of finding the solution, explaining their strategies, calculations and reasoning for each step.

KEY LANGUAGE

In lesson: problem, wholes, fractions, convert, common denominator, add, subtract, total

Other language to be used by the teacher: mixed number, improper fractions, difference, multi-step, small steps, manageable, operation, method

STRUCTURES AND REPRESENTATIONS

Bar models, fraction strips, number lines, part-whole models

RESOURCES

Mandatory: paper for fraction strips

Optional: blank part-whole models, blank number lines

 In the eTextbook of this lesson, you will find interactive links to a selection of teaching tools.

Quick recap 🔄

Draw a rectangle on the board with one of the lengths labelled 4 m and a square with one of the lengths labelled 3 m. To the side of the rectangle, write: perimeter = $12\frac{4}{5}$ m. Ask: *What do you know? What can you find out?*

Children will naturally start to answer multi-step problems but without the pressure of needing to find one specific answer.

Discover

WAYS OF WORKING Small group work

ASK

• Question ➊ a): *How can you write the problem as a number sentence? What is the most efficient way to subtract the amount poured into the three glasses? How can you convert the fraction?*

• Question ➊ b): *How much does the small carton hold? How can you find how much the large carton holds? How can you find the total?*

IN FOCUS In question ➊ a), children could use a variety of methods, such as converting the mixed number to an improper fraction and finding a common denominator before subtracting, or converting only the fractional part of the mixed number to an equivalent fraction and then subtracting. Encourage children to explore different strategies and discuss them as a class. In question ➊ b), children will need to convert the amount held in the small carton, so they can easily add the two amounts together to find the amount held in the large carton.

PRACTICAL TIPS Draw a model or flowchart of the problem, breaking it into manageable steps.

ANSWERS

Question ➊ a): There are $1\frac{1}{8}$ litres of milk left in the carton.

Question ➊ b): The large carton holds $4\frac{1}{4}$ litres. There are $5\frac{3}{4}$ litres of milk in total.

Solve multi-step fraction problems

Discover

➊ a) Jen pours 3 glasses of milk from the $1\frac{1}{2}$ litre carton. Each glass holds $\frac{1}{8}$ of a litre. How much milk is left in the carton?

b) The large carton holds $2\frac{3}{4}$ litres more than the small carton. How much milk do the two cartons hold in total?

244

PUPIL TEXTBOOK 5A PAGE 244

Share

WAYS OF WORKING Whole class teacher led

ASK

• Question ➊ a): *How much milk does the carton hold? Could you represent this problem using a model?*

• Question ➊ b): *What sort of calculation will this be? What will you work out first?*

IN FOCUS In question ➊ a), some children may calculate $\frac{1}{8} + \frac{1}{8} + \frac{1}{8}$ and subtract $\frac{3}{8}$ from $1\frac{4}{8}$. Others may subtract one-eighth at a time. For either method, children will need to convert the half into eighths.

Question ➊ b) is a 2-step problem. First, children will find the amount of milk in the large carton. Then they will find the total of the two cartons. The bar model gives a pictorial representation of the two parts. This illustrates that the large carton holds $2\frac{3}{4}$ litres *more*. Watch out for children who misinterpret the question and think the large carton holds $2\frac{3}{4}$ litres.

STRENGTHEN In question ➊ a), encourage children to explore different ways of subtracting. Is it more efficient to subtract each glass in turn, or to combine them and then subtract?

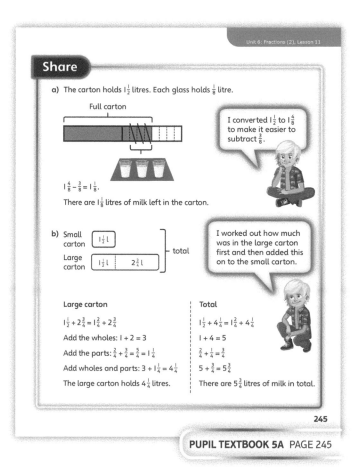

Share

a) The carton holds $1\frac{1}{2}$ litres. Each glass holds $\frac{1}{8}$ litre.

Full carton

$1\frac{4}{8} - \frac{3}{8} = 1\frac{1}{8}$.

There are $1\frac{1}{8}$ litres of milk left in the carton.

I converted $1\frac{1}{2}$ to $1\frac{4}{8}$ to make it easier to subtract $\frac{3}{8}$.

b) Small carton $1\frac{1}{2}$ l

Large carton $1\frac{1}{2}$ l $2\frac{3}{4}$ l — total

I worked out how much was in the large carton first and then added this on to the small carton.

Large carton

$1\frac{1}{2} + 2\frac{3}{4} = 1\frac{2}{4} + 2\frac{3}{4}$

Add the wholes: $1 + 2 = 3$

Add the parts: $\frac{2}{4} + \frac{3}{4} = \frac{5}{4} = 1\frac{1}{4}$

Add wholes and parts: $3 + 1\frac{1}{4} = 4\frac{1}{4}$

The large carton holds $4\frac{1}{4}$ litres.

Total

$1\frac{1}{2} + 4\frac{1}{4} = 1\frac{2}{4} + 4\frac{1}{4}$

$1 + 4 = 5$

$\frac{2}{4} + \frac{1}{4} = \frac{3}{4}$

$5 + \frac{3}{4} = 5\frac{3}{4}$

There are $5\frac{3}{4}$ litres of milk in total.

245

PUPIL TEXTBOOK 5A PAGE 245

Think together

WAYS OF WORKING Pair work

ASK

- Question ①: *What operations would you use to write the problem as a number sentence? Is it more efficient to subtract one at a time, or to add the amounts poured out and then subtract from the starting amount?*
- Question ②: *What does 'equally spaced' mean? How can you work out the amount between each marker? Now that you know the distance between the markers, how can you work out what C is pointing to?*
- Question ③: *What calculation is needed for perimeter? Do you need to subtract the sides one at a time or is there a more efficient way?*

IN FOCUS Question ① requires adding and/or subtracting to find the amount of juice left. Children may do this by converting each fraction to have the same denominator. In question ②, children will find the distance between A and B. They will then add this distance to B to determine the position of C. In question ③ a), children add three lengths to find the perimeter. They will need to find a common denominator and convert the fractions accordingly. For question ③ b), children are given the perimeter and the lengths of two sides. They can find the missing length by subtracting each length from the perimeter in turn or by adding the lengths of the two given sides and subtracting this total from the perimeter.

STRENGTHEN In question ②, children may find it useful to draw their own number line to help them count up from A to B. They may need prompting on how many eighths there are between 1 and 2. Children can count in eighths up to marker B and then work out how many eighths they 'jumped' in total. Similarly, when working out point C, children may find it easier to draw the jumps on their number line.

DEEPEN Extend children's understanding by presenting further problems after each question. For example, after question ①, you could ask: *Does Kate have enough juice left to fill another glass? How much more juice can be held in the jug than in the glass?* After question ②, you could ask: *If points D, E and F are also equally spaced on the same number line, where will they be placed?* (This question has many possible solutions – encourage children to discuss their reasoning.)

ASSESSMENT CHECKPOINT Do children recognise mathematical language within problems? Can they choose appropriate operations and methods for given problems? Can children make calculations more efficient? Do they know how to find a missing length?

ANSWERS

Question ①: There are $\frac{3}{20}$ of a litre of juice left in the bottle.

Question ②: The arrow at C is pointing to $3\frac{5}{8}$.

Question ③ a): The perimeter of the triangle is $7\frac{1}{12}$ cm.

Question ③ b): The missing length is $1\frac{1}{20}$ cm.

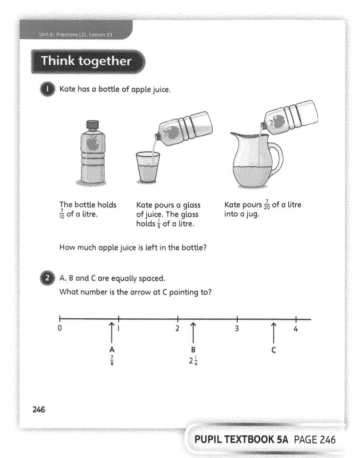

PUPIL TEXTBOOK 5A PAGE 246

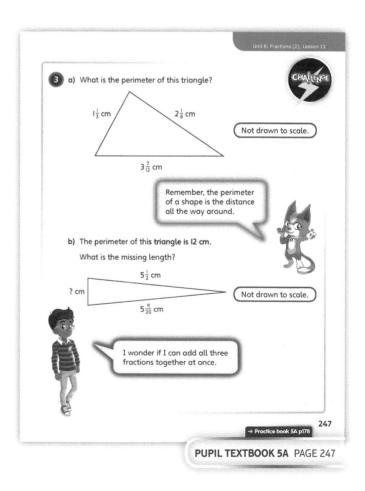

PUPIL TEXTBOOK 5A PAGE 247

Practice

WAYS OF WORKING Independent thinking

IN FOCUS Question **1** presents a word problem where two fractions are to be subtracted from 1. Children may find the total of the two given fractions and subtract this from 1, or they may subtract the two given fractions one at a time. To work out the answer, children must realise that the full amount is $\frac{9}{9}$ (equivalent to 1). Question **3** focuses on addition and subtraction within a word problem, although some children may just subtract. When subtracting from 3 kg, children may need to be prompted to show the equivalent of 3 kg as eighths. Question **6** is similar to question **2** of the **Think together** section. Once children have identified the fractions on the number line, they will need to find the difference by counting on or back once the fractions have been converted.

STRENGTHEN Continue to use concrete and visual models to support children to 'see' the problem. In question **1**, to work out how much pocket money Ebo has left, children could create a fraction strip or bar model split into ninths. They will then be able to shade what has been spent and see clearly how much is left. To subtract from 3 kg in question **3**, children could draw 3 as a fraction strip as a starting point. Some children may find it easier to convert 3 to $\frac{24}{8}$ and then subtract, or convert 1 to $\frac{8}{8}$ and subtract the parts and the wholes separately.

DEEPEN After children have solved the missing number problems in question **5**, present them with a similar question where both numerators are unknown. With two missing numbers, children can find a range of possible answers and prove their responses using bar models. After question **7**, extend children's understanding by asking similar questions but using different shapes.

ASSESSMENT CHECKPOINT Can children recognise mathematical language within problems? Can they correctly identify which operation needs to be used? Can they read fractions on a number line? Can children find missing numbers?

ANSWERS Answers for the **Practice** part of the lesson can be found in the *Power Maths* online subscription.

Reflect

WAYS OF WORKING Independent thinking

IN FOCUS These questions allow children to self-assess their learning journey within this lesson. Children can identify strengths and weaknesses and possibly set targets. Children's answers to this question will help you decide whether to provide extra support.

ASSESSMENT CHECKPOINT Can children reflect on the strategies and methods they have learnt and practised? Can they identify areas for development? Can they articulate what they found difficult and explain why?

ANSWERS Answers for the **Reflect** part of the lesson can be found in the *Power Maths* online subscription.

After the lesson ⏸

- Can children identify mathematical language within a problem?
- Can children identify the correct operation and a suitable, efficient method?
- Can children convert fractions in order to add and subtract?

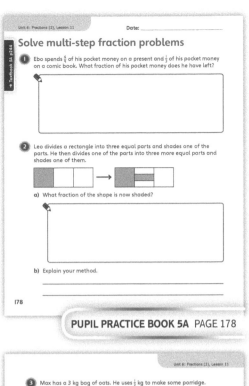

PUPIL PRACTICE BOOK 5A PAGE 178

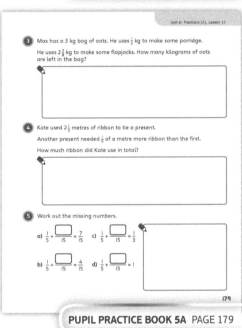

PUPIL PRACTICE BOOK 5A PAGE 179

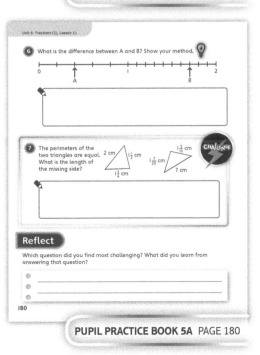

PUPIL PRACTICE BOOK 5A PAGE 180

End of unit check

Don't forget the unit assessment grid in your *Power Maths* online subscription.

WAYS OF WORKING Group work adult led and independent working

IN FOCUS

- Question **1** assesses children's ability to add two or more fractions with the same denominator where the answer is greater than 1. It also checks children's ability to convert an improper fraction to a mixed number.
- Questions **2** and **3** require children to add and subtract with fractions where one denominator is a multiple of the other. Children will demonstrate their understanding of subtracting from one whole.
- Questions **4** and **5** assess children's ability to identify the required calculation from a diagram, subtract mixed numbers and find equivalent fractions.
- Questions **6** and **7** are SATs-style questions. They encourage children to translate problems to identify the calculations needed to solve them.
- Encouraging children to use fraction strips to visualise what they need to do will be invaluable.

ANSWERS AND COMMENTARY Children who have mastered this unit are able to confidently add and subtract fractions and mixed numbers using written methods, solving a range of problems while explaining which method is most efficient.

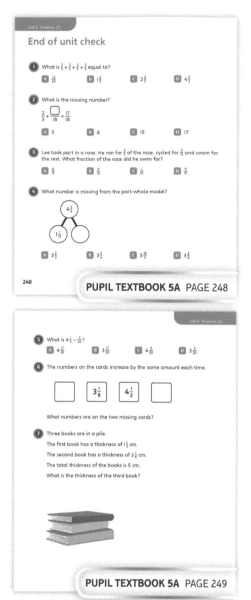

PUPIL TEXTBOOK 5A PAGE 248

PUPIL TEXTBOOK 5A PAGE 249

Q	A	WRONG ANSWERS AND MISCONCEPTIONS	STRENGTHENING UNDERSTANDING
1	C	D suggests counting how many $\frac{3}{5}$ there are.	Support children's understanding by using bar models and fraction strips to represent the questions.
2	A	B suggests children have worked out $3 \times 6 = 18$. C suggests children have not considered the denominators.	
3	C	A and D suggest children have added across the fraction parts. B suggests they have not subtracted from 1.	
4	A	C suggests they have added. D suggests they subtracted the numerators and denominators separately, each time subtracting the smaller number from the larger.	
5	B	C and D suggest they have found a common denominator but subtracted the smaller fraction from the larger, subtracting 1 in D.	
6	$1\frac{3}{4}$ and $5\frac{7}{8}$	Children may give an answer of $1\frac{6}{8}$ for the first number. Common mistakes include adding instead of subtracting.	Encourage children to identify what they can do first, i.e. find the increase by subtracting.
7	$1\frac{1}{3}$ cm	Children may give an unsimplified answer of $1\frac{2}{6}$. They may subtract both mixed numbers from 5.	Encourage children to use a fraction strip to help them decide if they need to add or subtract.

My journal

WAYS OF WORKING   Independent thinking

ANSWERS AND COMMENTARY   Question **1** assesses children's ability to choose and justify an efficient method. For example, adding $\frac{1}{6}$ and $\frac{1}{12}$ may be the most efficient way to work out the answer to question **1** b). It will be useful to observe whether children rely on fraction strips or are unsure of why one method is more efficient than another.

a): $8\frac{11}{12} + 7\frac{3}{4} = 16\frac{2}{3}$ b): $12\frac{1}{12} - 11\frac{5}{6} = \frac{3}{12} = \frac{1}{4}$

Question **2** assesses children's ability to identify the operations required to solve a problem. Be aware of children who make the common mistake of adding the two mixed numbers together. Encourage children to complete the problem in stages, i.e. find the amount of milk drunk this week, then find the total. Look for children clearly translating the problem into number sentences, fluently using common denominators and equivalent fractions, and confidently explaining the steps required to achieve the correct answer.

This week: $4\frac{1}{6} - 1\frac{2}{3} = 2\frac{3}{6}$ (= $2\frac{1}{2}$) litres of milk.

Total: $4\frac{1}{6} + 2\frac{3}{6} = 6\frac{4}{6} = 6\frac{2}{3}$ litres of milk.

Power check

WAYS OF WORKING   Independent thinking

ASK

- *How confident do you feel about finding equivalent fractions?*
- *Are you able to add fractions with different denominators?*

Power puzzle

WAYS OF WORKING   Pair work or small groups

IN FOCUS   Use this activity to assess whether children can add fractions and look for patterns, demonstrating understanding of the steps involved in adding related fractions using common denominators.

ANSWERS AND COMMENTARY   Children are given related fractions and asked to explore the patterns that occur when they are added together. Encourage them to discuss and share the patterns they discover.

Once they have worked out the first few calculations, children may be able to predict the sums of the next two lines before working them out.

$\frac{3}{4}, \frac{7}{8}, \frac{15}{16}, \frac{31}{32}$

$\frac{1}{2} + \frac{1}{4} + \frac{1}{8} + \frac{1}{16} + \frac{1}{32} + \frac{1}{64} = \frac{63}{64}$

$\frac{1}{2} + \frac{1}{4} + \frac{1}{8} + \frac{1}{16} + \frac{1}{32} + \frac{1}{64} + \frac{1}{128} = \frac{127}{128}$

After the unit ⏸

- Can children find equivalent fractions and convert between mixed numbers and improper fractions to add and subtract related fractions and mixed numbers?
- Are children confident answering multi-step addition and subtraction problems involving fractions?

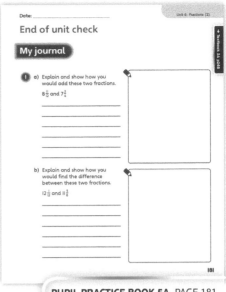

PUPIL PRACTICE BOOK 5A PAGE 181

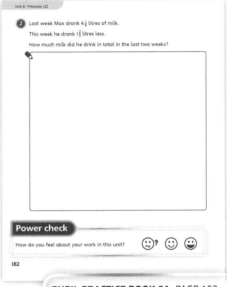

PUPIL PRACTICE BOOK 5A PAGE 182

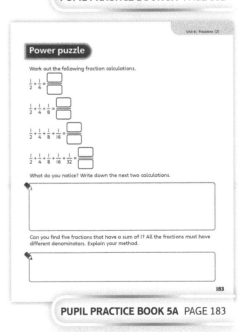

PUPIL PRACTICE BOOK 5A PAGE 183

Strengthen and **Deepen** activities for this unit can be found in the *Power Maths* online subscription.

Published by Pearson Education Limited, 80 Strand, London, WC2R 0RL.

www.pearsonschools.co.uk

Text © Pearson Education Limited 2018, 2022
Edited by Pearson and Florence Production Ltd
First edition edited by Pearson, Little Grey Cells Publishing Services and Haremi Ltd
Designed and typeset by Pearson and Florence Production Ltd
First edition designed and typeset by Kamae Design
Original illustrations © Pearson Education Limited 2018, 2022
Illustrated by Diego Diaz and Nadene Naude at Beehive Illustration, Emily Skinner at
Graham-Cameron Illustration, Kamae Design and Florence Production Ltd
Cover design by Pearson Education Ltd
Back cover illustration © Diego Diaz and Nadene Naude at Beehive Illustration
Series editor: Tony Staneff; Lead author: Josh Lury
Authors (first edition): Tony Staneff, Jian Liu, Josh Lury, Zhou Da, Zhang Dan, Zhu Dejiang,
Kate Henshall, Wei Huinv, Hou Huiying, Zhang Jing, Steph King, Stephanie Kirk, Huang Lihua,
Yin Lili, Liu Qimeng, Timothy Weal, Paul Wrangles and Zhu Yuhong
Consultants (first edition): Professor Jian Liu and Professor Zhang Dan

First published 2018
This edition first published 2022

26 25 24 23 22
10 9 8 7 6 5 4 3 2 1

British Library Cataloguing in Publication Data
A catalogue record for this book is available from the British Library

ISBN 978 1 292 45059 9

Printed in the UK by Ashford Press Ltd

For Power Maths online resources, go to:
www.activelearnprimary.co.uk

Note from the publisher
Pearson has robust editorial processes, including answer and fact checks, to ensure the
accuracy of the content in this publication, and every effort is made to ensure this publication
is free of errors. We are, however, only human, and occasionally errors do occur. Pearson is
not liable for any misunderstandings that arise as a result of errors in this publication, but it is
our priority to ensure that the content is accurate. If you spot an error, please do contact us at
resourcescorrections@pearson.com so we can make sure it is corrected.